Sustainable Environmental Waste Management Strategies

Series Editors

Alok Prasad Das, Department of Life Sciences, Rama Devi Women's University, Bhubaneswar, Odisha, India

Eric van Hullebusch, Équipe ACE, Institut de Physique du Globe de Paris, Paris, France

Sustainable waste management is an absolute imperative in a world that is already dealing with the effects of climate change. This book series seeks to publish an extensive collection of scientific books targeting researchers, scientists, professionals, academics, students, and anyone else involved in the wide, diverse field of sustainable waste management. The series accepts monographs, edited volumes, textbooks, and conference proceedings. Topics covered include but are not limited to environmental pollution and remediation, sources of environmental waste, environmental and human health, sustainable waste management, circular economy, nature-inspired environmental solutions, pollution prevention, and energy-efficient technologies for waste management. The series puts a particular emphasis on developing countries/ the majority world but accepts proposals globally as long as the topic is in scope.

Veer Singh · Ashish Kumar · Nidhi Singh ·
Vishal Mishra · Sachchida Nand Rai
Editors

Bioenergy and Value Added Byproducts from Biomass Waste

Springer

Editors
Veer Singh
Department of Biosciences
Manipal University Jaipur
Jaipur, Rajasthan, India

Nidhi Singh
Centre of Bioinformatics
University of Allahabad
Prayagraj, Uttar Pradesh, India

Sachchida Nand Rai
University Centre for Research
and Development (UCRD)
Chandigarh University
Mohali, Punjab, India

Ashish Kumar
Department of Biochemistry
Rajendra Memorial Research Institute
of Medical Sciences
Patna, Bihar, India

Vishal Mishra
School of Biochemical Engineering
Indian Institute of Technology (BHU)
Varanasi, Uttar Pradesh, India

ISSN 3005-1622 ISSN 3005-1630 (electronic)
Sustainable Environmental Waste Management Strategies
ISBN 978-3-032-22848-2 ISBN 978-3-032-22849-9 (eBook)
https://doi.org/10.1007/978-3-032-22849-9

This Springer imprint is published by the registered company Springer Nature Switzerland AG
The registered company address is: Gewerbestrasse 11, 6330 Cham, Switzerland

To my family: For their unwavering love and belief in my dreams.

To my mentors and colleagues: For inspiring and guiding me through this journey of discovery.

And to all those striving for a healthier, cleaner world: may this work contribute in some small way to your mission.

Preface

The transition toward sustainable and renewable energy sources is no longer a choice but an urgent global necessity. Accelerating climate change, depletion of fossil fuels, and increasing environmental degradation compel the scientific community, policymakers, and industries to seek cleaner alternatives that can support both economic development and ecological balance. Among these alternatives, bioenergy derived from biomass waste has emerged as a promising, scalable, and eco-friendly solution capable of reshaping the future energy landscape. This book, *Bioenergy and Value-Added Byproducts from Biomass Waste*, brings together recent advances, innovative technologies, and practical approaches for converting diverse waste resources into sustainable fuels and high-value products.

Biomass waste—ranging from agricultural residues, organic waste, food waste, lignocellulosic materials, seafood industry byproducts, and microalgal biomass—offers an abundant and renewable resource for biofuel production. However, the transformation of these raw materials into efficient and commercially viable fuels requires a deep understanding of their characteristics, suitable conversion technologies, catalytic processes, and sustainability assessments. Addressing this complexity, the book presents comprehensive insights through fourteen well-structured chapters authored by experts in energy, biotechnology, environmental science, and material engineering.

The chapters cover the full spectrum of bioenergy production. Beginning with advanced material characterization and novel approaches for generating value-added products (Chap. 1), the book transitions into biofuel production from agro-waste (Chap. 2), biomass selection and conversion pathways (Chap. 3), microbial bioenergy (Chap. 4), and food waste valorization (Chap. 5). The role of organic waste in sustainable fuel generation (Chap. 6) and the integration of green nanomaterials to enhance biofuel yield (Chap. 7) highlight technological innovations that improve efficiency and reduce environmental impacts.

Further, the book emphasizes the significance of life-cycle assessment (Chap. 8) and explores the environmental applications of byproducts generated during biofuel production (Chap. 9). Global policy frameworks, industrial participation, and bioenergy's alignment with Sustainable Development Goals are discussed through an international lens in (Chap. 10). Advances in catalysis for biomass conversion (Chap. 11), seafood waste valorization pathways (chap. 12), and microalgae-based biofuels (Chap. 13) showcase the dynamic and multidisciplinary nature of bioenergy research. The final chapter (Chap. 14) presents the environmental sustainability of second-generation biofuels and their critical role in achieving a circular economy.

This book aims to serve as a valuable resource for researchers, students, industry professionals, and policymakers working towards sustainable energy solutions. By integrating scientific advancements with real-world applications and sustainability principles, it provides a holistic understanding of how biomass waste can be efficiently converted into clean energy and value-added products—paving the way for a resilient, low-carbon future.

Jaipur, India Veer Singh
Patna, India Ashish Kumar
Prayagraj, India Nidhi Singh
Varanasi, India Vishal Mishra
Mohali, India Sachchida Nand Rai

Acknowledgments

The editors thank Manipal University Jaipur for providing the academic and research facilities essential for this work. Dr. Veer Singh would also like acknowledges Department of Health Research (DHR), Ministry of Health and Family Welfare, Government of India, for support through the Young Scientist award (File No. R.12014/37/2022-HR). Dr. Ashish Kumar would like to acknowledge to Indian Council of Medical Research, New Delhi for providing necessary support to the research. Dr. Nidhi Singh would like acknowledges University of Allahabad, Prayagraj. Dr. Vishal Mishra would like acknowledge to Indian Institute of Technology (BHU), Varanasi for necessary research support. Dr. Sachchida Nand Rai would like acknowledge to Banaras Hindu University, Varanasi—221005, India. It is also our great pleasure to acknowledge and express our enormous debt to all the contributors who have provided their quality material to prepare this book. We are grateful to our beloved family members, who joyfully supported and stood with us in many hours of our absence to finish this book project. We are also grateful to our friends and colleagues who offered their support and encouragement throughout the writing process. I would also like to acknowledge the publishing team Springer Nature for their professionalism, enthusiasm, and belief in this project. Their expertise in design, production, and marketing has been essential in bringing this book to life.

Veer Singh
Ashish Kumar
Nidhi Singh
Vishal Mishra
Sachchida Nand Rai

Contents

About the Editors

Dr. Veer Singh is currently working as Assistant Professor (Research) in the Department of Biosciences at Manipal University Jaipur. Prior to joining MUJ, he served as a Young Scientist at the ICMR–Rajendra Memorial Research Institute of Medical Sciences, Patna, India. He completed his Ph.D. in Biochemical Engineering from IIT (BHU), Varanasi. He also received funding/grant from the Department of Health Research (DHR), Government of India of ₹48.65 lakhs. His research is focused on wastewater treatment, development of eco-friendly adsorbents, material characterization, bioremediation, adsorption, environmental toxicity and human health and nanomaterial for wastewater treatment. He has published 36 research/review articles in SCI/Scopus-indexed journals, five patents, and contributed 17 book chapters in books. He serves as an editor for several reputed journals and has edited multiple books published by reputed international publishers.

Dr. Ashish Kumar is currently working as Scientist-D and Head of Department at the Department of Biochemistry, ICMR-Rajendra Memorial Research Institute of Medical Sciences, Patna, India. He has published more than 50 publications in Internation/national journals. He has received several awards and received several funding (sponsored projects) from many national and international agencies. Currently, five doctoral and one post-doctoral students are working under the supervision of Dr. Ashish Kumar. Dr. Kumar is working in the interdisciplinary area of life science. His work is mainly focused on human diseases, treatment, the role of environmental factors in disease development, synthesis of composite nanoparticles, bioenergy production and value-added products from its byproducts.

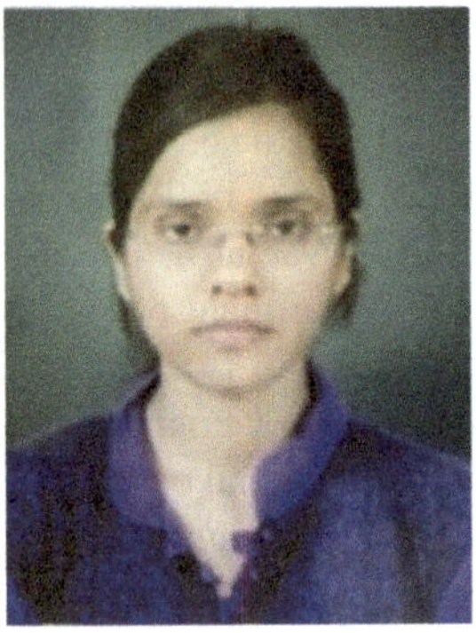

Dr. Nidhi Singh completed her Ph.D. in Bioinformatics (Metagenomics and its application environmental sustainability, Biomass conversion) from University of Allahabad, Prayagraj. Dr. Nidhi has published more than 20 publications in reputed publisher. She also filed patent under Intellectual Property India. She is also reviewer of several International Journals. She has received several awards during her academic career. The research area of Dr. Nidhi is focused on Metagenomics, Biomass characterization, Biofuel production, Waste management, Waste to value-added products and microbial biofuel production and wastewater treatment. Dr. Nidhi is a bioinformatician with extensive knowledge of data analysis used in biomass conversion and biofuel research. Her contributions will significantly enhance the quality of our proposed book.

Dr. Vishal Mishra is currently working as Associate Professor in the School of Biochemical Engineering, Indian Institute of Technology (BHU), Varanasi, India. He got his Ph.D. degree from IIT Roorkee, India. He has more than 80 potential publications and his area of work involved Algal fuel and Biodiesel; Life cycle assessment, agro-waste biomass, and Carbon Materials/green synthesized nanomaterials for enhanced capacitance.

Dr. Sachchida Nand Rai Ph.D. is Assistant Professor at the University Centre for Research and Development (UCRD), Chandigarh University, Mohali, Punjab, India. He earned his Ph.D. in Neuroscience from Banaras Hindu University (BHU) in 2019, where he also served as an ICMR Senior Research Associate from 2023 to 2026, working on epigenetic and neuroinflammatory mechanisms in Parkinson's disease. He completed a UGC–Dr. D. S. Kothari Postdoctoral Fellowship from 2020 to 2023 at the University of Allahabad focusing on autophagy-mediated neuroprotection. His research interests include neurodegenerative diseases, toxicology, epigenetics, and nutraceuticals. He serves on several international journal editorial boards, has published 100+ articles, reviewed 2000+ manuscripts, and edits multiple books and series with Elsevier, Springer, Bentham and CRC Press.

Chapter 1
Novel Approaches for Characterization and Its Conversion into Value-Added Products

Ananda Chettupalli, Jai Vardhan, Shikha Yadav, Abdul Althaf Shaik, and Sarad Pawar Naik Bukke

Abstract Material characterization and subsequently enormous advancements owe their claim to such capability and unparalleled comprehension of the myriad chemical, structural, and functional properties of materials. With this, the capacity to understand, modify, and optimize these materials for a vast spectrum of applications has received an unprecedented boost. The chapter discusses advanced characterization techniques while emphasizing that converting raw materials into useful products is of utmost importance. Spectroscopic and imaging-computational advances allow for extremely high precision in the hands of materials scientists in analysing materials, thereby fuelling innovations in various fields. For the material characterization part alone, other inter-disciplined technologies including nanotechnology, artificial intelligence (AI), and machine learning are excelling with their respective newfound opportunities in the domain of material functionality and design. Harmonized interdisciplined research is looking at custom-designing materials with specific properties, consequently enabling applications in pharmaceuticals, agriculture, energy storage, environmental sustainability, and more. High-throughput

A. Chettupalli (✉)
Department of Pharmaceutical Technology, School of Health and Medical Sciences, Adamas University, Kolkata, West Bengal, India
e-mail: anandphd88@gmail.com

J. Vardhan
Faculty of Pharmacy, Vidya University, Meerut, Uttar Pradesh, India

S. Yadav
Department of Pharmacy, School of Medical and Allied Sciences, Galgotias University, Greater Noida, Uttar Pradesh, India

A. A. Shaik
Department of Cell Biology and Biochemistry, Texas Tech University Health Science Center, Lubbock, TX, USA
e-mail: abdul.shaik@ttuhsc.edu

S. P. N. Bukke
Department of Pharmaceutics and Pharmaceutical Technology, Kampala International University, Ishaka, Bushenyi, Uganda

V. Singh et al. (eds.), *Bioenergy and Value Added Byproducts from Biomass Waste*, Sustainable Environmental Waste Management Strategies,
https://doi.org/10.1007/978-3-032-22849-9_1

screening methods, together with nanoscale imaging and computational modelling, would further solidify the possibilities for developing other value-added products, such as biodegradable polymers, renewable energy device applications, and targeted drug delivery systems. In this chapter, the key importance of employing sustainable practices in understanding the properties of materials and adding value to them is discussed. Green chemistry and circular economy principles are increasingly being incorporated to ensure that these technologies conform to and are compatible with international sustainability standards. By means of simple examples and a critical analysis of emerging technologies, this chapter connects scientific research and industrial development and gives an exhaustive guide on how to land in sustainable and value-efficient additions.

Graphical Abstract

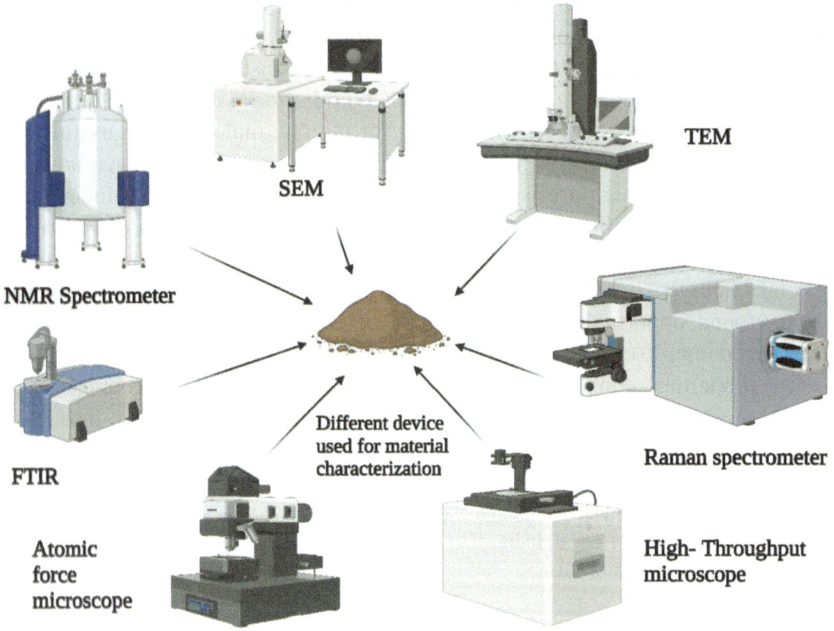

Overview of material characterization methods.

Keywords Advanced material characterization · Nanotechnology · Artificial intelligence · Value-added products · Sustainable innovation

1.1 Introduction

In materials science and engineering, material characterization is of utmost importance. By studying it, we can understand materials' physical, chemical, and structural properties. Material characterization has evolved over the last few decades, which has drastically changed how we perceive, investigate, and utilize materials in various industries: It was, until then, a study where simple observations and rudimentary tests were performed. Today material characterization utilizes an array of advanced tools and techniques that probe materials at the molecular and atomic level (Callister & Rethwisch, 2020). Material sciences have advanced to open new frontiers of innovation, specifically in the area of transforming raw materials into products with added value. The relevance of this intervention vis-a-vis global challenges across most fields such as health care, energy, food technology, and sustainability cannot be overstated. On the one hand, the ever-increasing-level advancements concerning the development of material characterization from very simple techniques to those with high technologies are symbolic of that very pace across scientific knowledge and technological ability. On the other hand, primitive microscopy and simple analytical tools provided a very minimal understanding of material properties. Entirely contrasting advanced spectroscopic techniques such as Raman Spectroscopy or FTIR Spectroscopy and high-resolution imaging modalities like Scanning Electron Microscopy and Atomic Force Microscopy allow telling the materials under experimental conditions very accurately (Binnemans et al., 2015). Furthermore, high-throughput screening and computational modelling have transformed the material development process, making it faster, more accurate, and more scalable.

Value addition is seen as a very critical dimension of modern materials science, whereby raw materials can be subjected to functional, economic, or environmental demands. Value addition is seen as a very critical process in various industries, such as pharmaceuticals, where careful characterization and optimization of materials along with formulating targeted drug delivery systems go hand in hand. Another shining example would be the use of characterization to enhance the manufacturing process of biodegradable polymers from renewable plant feedstocks, an environmentally friendly high-value product. With regards to value addition, besides optimizing performance, this process also contributes towards sustainable goals through waste minimization, resource conservation, and a lesser environmental footprint. Besides, new technologies, including nanotechnology, artificial intelligence (AI), and machine learning, further provide an impetus for the evolution of the field of material characterization (Hasan et al., 2022). Nanotechnology allows for observing the behaviour of materials on a very small level, which helps in the easier material design with peculiar properties. Meanwhile, AI and Machine Learning have created much havoc in interpreting complex data, helping researchers to predict how materials behave, fine-tune their recipes, and fast-track the path to market for new products. These tools not only help in identifying materials but also enable conceptualizing new ideas in renewable energy, food safety, and better healthcare options (Chang, 1996). One of the goals of this chapter is to show how material science is being transformed by the interplay of various professions through modern spectroscopy, imaging, and computation techniques. Sustainable development is the other focus, with a case made for green chemistry and circular economy principles in material processing. Finally, the chapter covers challenges and opportunities in this fast-changing field while giving views on future advances and possible breakthroughs. This chapter aims to explain to the reader how newer methods can help accommodate sustainable development and economic growth through setting material research and use side by side (Bhushan, 2017).

1.2 State-of-the-Art Characterization Techniques

Since the advent of increasingly refined and versatile analyzing tools, the field of material characterization science has experienced an unprecedented growth spurt in the last few decades. Such advanced methods enable scientists to analyze materials from atomic dimensions up to macroscopic dimensions, generating vital information concerning a material's structure, constitution, and properties (Guddati et al., 2019). This section explores the latest spectroscopic, imaging, and computational approaches that are revolutionizing material science. Table 1.1 shows a detailed comparison of characterization technique. Figure 1.1 illustrates the characterisation procedures.

1.2.1 Spectroscopic Advancements

Indeed, the spectroscopic methods are indispensable and bear the ability for non-invasive, high-resolution analyses on the chemical makeup, molecular structure, and electronic properties of materials. Applications abound, in nanotechnology, pharmaceuticals, catalysis, and biomaterials, for bonding, phase transitions, and material behaviour information. Over the past few years, the major improvements in resolution and sensitivity have provided the possibility for in situ real-time monitoring. Single-molecule detection has been afforded by methods like SERS and TERS, while FTIR and XPS enable better chemical analysis of polymers and thin films. Synchrotron-based X-ray methods like XAS, XRF provide in situ analysis that is critical to energy storage applications like lithium-ion batteries and fuel cells.

The collaboration of Artificial Intelligence with Machine Learning has upgraded the spectral analysis to its full power of pattern recognition and molecular structure prediction in drug discovery and material design. Coming advancements will miniaturize the tools, enable real-time analysis, and develop portable spectrometers for environmental monitoring, food safety, and forensic applications. Besides, as the field of spectroscopy upgrades itself, it will directly foster advancements of sustainable materials, energy improvements, and biomedical developments (Bhaskar, 2023).

Table 1.1 Comparison of characterization techniques

Technique	Principle	Resolution	Advantages	Applications	References
Raman Spectroscopy	Inelastic light scattering to study molecular vibrations	~ 1 μm	Non-destructive, high specificity	Drug analysis, nanomaterials, biological samples	Chen et al. (2024)
FTIR Spectroscopy	Absorption of IR radiation to determine functional groups	~ 5–10 μm	Rapid, qualitative and quantitative analysis	Polymers, proteins, catalysts	Errico et al. (2023)
X-ray Diffraction (XRD)	Diffraction of X-rays by crystalline structures	Sub-nm	High accuracy in phase identification	Crystalline materials, alloys, minerals	Ameh (2019)
Scanning Electron Microscopy (SEM)	Electron beam scanning for surface morphology	~ 1–10 nm	High-resolution imaging, elemental analysis	Nanomaterials, coatings, biomaterials	Agrawal et al. (2024)
Transmission Electron Microscopy (TEM)	Electron transmission through ultrathin samples	~ 0.1 nm	Atomic resolution, internal structure imaging	Catalysts, semiconductors, biological cells	Musso et al. (2020)
Atomic Force Microscopy (AFM)	Mechanical probe scans surface topology	~ 0.1 nm	Measures mechanical and electrical properties	Thin films, biomolecules, composites	Joshi et al. (2022)

1.2.1.1 Raman Spectroscopy and Its Applications

Raman spectroscopy has established itself as a powerful method for studying molecular vibrations and crystal structures. Through the study of inelastic scattering of light, detailed information about chemical bonds, phase transitions, and stresses in the material is obtained. Recent innovations, includes surface-enhanced Raman spectroscopy (SERS) and tip-enhanced Raman spectroscopy (TERS), have pushed detection to the single-molecule level sensitivity (Chen et al., 2024). These advancements have enabled applications in diverse fields, including pharmaceuticals (e.g., drug polymorph identification), nanotechnology (e.g., carbon nanotube characterization), and environmental science (e.g., pollutant detection).

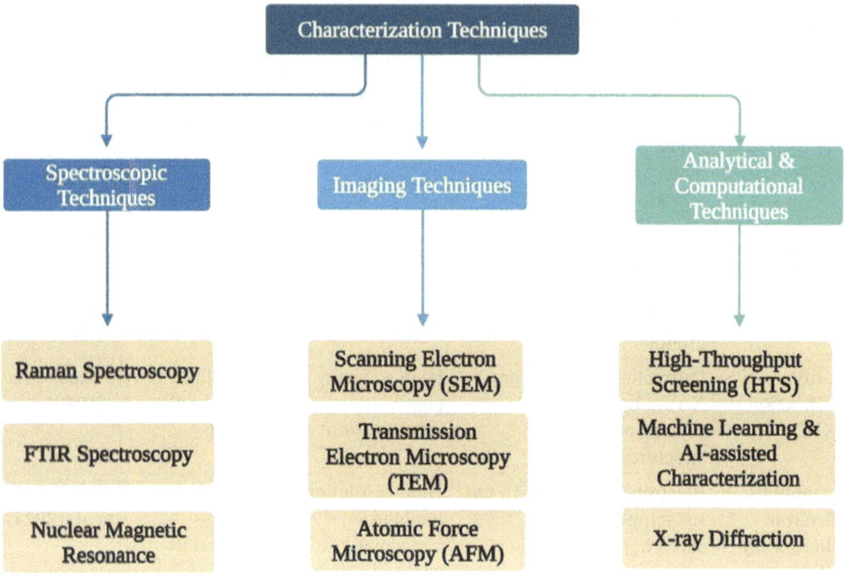

Fig. 1.1 The characterization techniques

1.2.1.2 FTIR and X-Ray Diffraction (XRD)

FTIR spectra are critical in identifying functional groups and chemical bonds in organic and inorganic substances. Given the fast generation of high-resolution spectral data, FTIR finds its application in domains like as polymer science, catalysis, and biomedicine. FTIR has been paired with advanced data analytical tools, enabling real-time monitoring of chemical reactions and degradation processes (Errico et al., 2023).

Grayscale X-ray diffraction (XRD) is still the fossil reference standard for crystallographic examination. Modern XRD apparatus with elaborate high-intensity X-ray sources and detectors can slice through the most complicated crystal structures with atomic resolution. Recent improvements in synchrotron radiation X-ray diffraction and small-angle X-ray scattering (SAXS) have allowed their application to amorphous materials, thin films, and biological macromolecules (Ameh, 2019).

1.2.1.3 Nuclear Magnetic Resonance (NMR)

NMR spectroscopy shows enormous potential for the purpose of investigation the atomic and molecular behaviour of compounds. Due to the magnetic properties present in atomic nuclei, NMR provides comprehensive insights into molecular structure, chemical environment, and intermolecular interactions. Solid-state NMR, in particular, serves as a key technique for the study of materials that are poorly crystalline or have complex morphologies, such as polymers, glasses, and porous materials (Huang et al., 2022). Recent advancements in high-field NMR and dynamic nuclear polarization (DNP) have significantly improved its sensitivity and resolution, enabling the study of previously inaccessible systems.

1.2.2 Imaging and Visualization Tools

Now we provide you with an opportunity to think about how we can argue that imaging distinguishes the material characterization realm by showing direct visual evidence on microstructural characterize—structure–property correlations leading to synthesizing newer materials and enhancing their performances for applications ranging from nanotechnology to electronics, biomaterials, and energy storage. The imaging techniques have provided resolution and vast length scales, with pertinent information about grain boundaries, defects, phase distributions, and surface changes, all of which are critical for optimization.

SEM and TEM are widely utilized because of their ability to generate high-resolution visuals of surfaces as well as internal structures. SEM offers high-resolution topography information through secondary electron imaging, while TEM facilitates atomic-scale inspection of internal structures, which is necessary for the investigation of nanomaterials, catalysts, and biomolecules. Improvement in aberration-corrected TEM has recently pushed resolution boundaries to sub-angstrom levels, thus making it feasible to determine atomic arrangements and defects.

Atomic Force Microscopy (AFM) allows three-dimensional profiling of surfaces with nanometer resolution, making it especially suitable for probing mechanical, electrical, and magnetic properties on the nanoscale. Compared to this, 3D imaging methods, including X-ray computed tomography (X-ray CT) and Focused Ion Beam (FIB)-SEM, allow for volumetric reconstruction of intricate material architectures, vital for the exploration of porous materials, composite material structures, and battery electrodes. The coupling of AI and machine learning with imaging methods is also propelling computerized image analysis, defect analysis, and structure quantification. As imaging science continues to mature, it will continue to stay at the epicentre of powering materials research, facilitating innovation in material design, failure analysis, and industrial use (Álvarez López et al., 2018).

1.2.2.1 Scanning Electron Microscopy (SEM) and Transmission Electron Microscopy (TEM)

Scanning Electron Microscopy (SEM) and Transmission Electron Microscopy (TEM) are among the foremost common imaging techniques used in materials science, which give outstanding and distinct images of the structural, morphological, and compositional characteristics of various materials (Agrawal et al., 2024). Because of high-resolution surface imaging capabilities from scanning the sample across with a finely focused beam of electrons, the SEM unique features allow to obtain, by registering secondary and backscattered electrons, information about topography and elemental composition. The SEM is, therefore, ideally suited for analyses related to surface roughness, particle size, phase distribution, and microstructural defects in materials, from metals and ceramics to biological tissues and polymer (Tüysüz et al., 2008). SEM combined with Energy-Dispersive X-ray Spectroscopy (EDS) increases the range of capabilities that this technology offers for qualitative and quantitative elemental analysis at micro- and nanoscale resolutions.

In contrast, the technique of TEM uses the passage of a electron beam through thin specimens to generate high-resolution pictures of inner structures on an atomic level. It is more suitable for examining lattice structures, grain boundaries, dislocations, and faults of crystalline materials (Musso et al., 2020). Just recently, advancements in aberration-corrected transmission electron microscopy and cryo-transmission electron microscopy have pushed the limits to sub-angstrom resolution, allowing for direct imaging of single atoms, interfaces between nanoparticles, and defects in catalyst and semiconductor materials. Cryo-TEM has altered the fate of soft matter and biological sample studies by preserving their native states via vitrification, thus preventing electronic beam-induced structural artifacts (Vincent et al., 2020). In addition, the gradual improvements of in situ TEM techniques enabled researchers to observe dynamic phenomena like phase transformation, nanoparticular growth, and chemical reactions in real-time under controlled environmental conditions (Wang et al., 2025).

The ability of SEM and TEM to provide detailed imaging at micro- and nanoscale levels has made them indispensable tools in nanotechnology, materials science, catalysis, and biomedical research (Samrot et al., 2019). These techniques enable the characterization of nanomaterials such as carbon nanotubes, quantum dots, graphene, and metal–organic frameworks and facilitate the innovation of next-generation electronic, optical, and energy storage devices. Additionally, the continuous enhancement of detector sensitivity, automation, and artificial intelligence-assisted processing of images is improving electron microscopy's accuracy and efficiency, leading to vast advances in materials discovery and design (Kumar et al., 2019).

1.2.2.2 Atomic Force Microscopy (AFM) and 3D Imaging Techniques

Presently, atomic force microscopy (AFM) is a great workhorse for characterization at the nanoscale with the highest resolution of surface topography and mechanical properties. In this case, the sample is raster's-canned with a sharp probe, imaging to a resolution below one nanometer and force detection to a piconewton (Joshi et al., 2022). Recent innovations, such as multifrequency AFM and conductive AFM, have expanded its capabilities to include electrical, magnetic, and thermal properties.

For the reconstruction of 3D structures of materials, X-ray computed tomography (CT) or focused ion beam (FIB)-SEM methods of 3D imaging have come into the limelight. These will particularly help unravel complex architectures such as porous materials, composites, or bio tissues, in areas where 2D imaging methods may not be sufficient (Rzhevskii, 2019).

1.2.3 High-Throughput Screening and Analytical Tools

The growing demand for rapid material discovery and optimization has driven the development of high-throughput screening (HTS) techniques, which have revolutionized the way researchers identify and evaluate novel materials (Gómez-Bombarelli et al., 2016). High-throughput screening (HTS) encompasses a variety of techniques capable of screening thousands of samples at a time, thus accelerating research and development within a broad spectrum of scientific and industrial applications. Materials discovery via trial-and-error methods is oftentimes slow, tedious, and expensive, and therefore it is in this arena that HTS plays an important role in 21st-century materials science. By screening in an automated fashion, HTS enables fast screening of huge material libraries, quick ascertainment of important properties, and selection of best candidates for further in-depth study (Manna et al., 2022).

HTS equipment essentially involves robotics, miniaturized assays, and sophisticated data analysis to best optimize the experimentation within the given constraints. Automated liquid handlers and microplate assays allow the reproducible simultaneous study of many samples under controlled conditions, reducing variability and providing for greater reproducibility. Consequently, with the interfacing of high-throughput analytical techniques, including spectroscopic analysis, X-ray diffraction (XRD), and mass spectrometry, HTS can rapidly relay the structural, optical, and chemical properties of materials (Kaspersetz et al., 2022). Additionally, the analysis of the huge datasets created from HTS experimentation is considerably aided by machine learning algorithms and AI-driven analytics for pattern recognition, predictive modelling, and optimization of synthesis parameters.

HTS finds an application in several fields of science, thereby demonstrating its versatility. In drug discovery, HTS allows a pharmaceutical company to screen large compound libraries in search of bioactive molecules that might have therapeutic effects (Stefano et al., 2022). In catalysis science, HTS is employed to develop the best catalyst compositions for commercial reactions, enhancing efficiency and selectivity. In energy materials science, HTS is employed to discover next-generation battery materials, photovoltaic materials, and fuel cell materials, facilitating the shift to green energy technologies. In addition, HTS is being applied increasingly in polymer science, nanomaterials, and biomaterials research to enable the design of high-functionality materials with customized properties (Craig, 2021).

With advancing technology, it will become integrated with artificial intelligence, robotic automation, and digital twin modelling to improve its performance. Utilization of in situ and operando techniques in HTS platforms will enable real-time monitoring of material performance under various environmental conditions with enhanced insight into structure–property relationships (Apurwa et al., 2024). With these advancements, HTS is poised to remain a cornerstone of modern material discovery, paving the way for breakthroughs in healthcare, renewable energy, and sustainable manufacturing.

1.2.4 Computational Approaches and AI in Material Characterization

Integration of computational modelling with artificial intelligence (AI) has significantly impacted the material characterization process, and researchers can now predict the material properties, optimize experimental parameters, and interpret complex data sets with unprecedented accuracy (Fanni et al., 2020). Experimental methods historically entail vast amounts of trial and error, and material discovery thus becomes time- and resource-intensive. Computational methods, such as density functional theory (DFT) and molecular dynamics (MD) simulations, have revolutionized the study of atomic and molecular interactions by providing a thorough insight into electronic configurations, mechanisms of bonding, phase transitions, and thermodynamic stability (Au Yeung et al., 2024). These techniques help researchers understand material behaviour at the atomic level, facilitating the design of novel materials with tailored properties for applications in energy storage, catalysis, and nanomedicine.

The convergence of artificial intelligence (AI) and machine learning (ML) technologies has greatly enhanced computational materials science via the automation of data analysis procedures and the unearthing of hidden patterns within large datasets. On one hand, the supervised techniques, which include support vector machines and decision trees, are capable of predicting and classifying the material property, while the unsupervised machine learning techniques facilitate clustering and analysis of high-dimensional data without any bias. In addition, advanced AI algorithms, represented by deep learning and generative adversarial networks (GANs), are increasingly being employed for novel material synthesis through the generation of molecular structures and the prediction of their physical and chemical properties (Xiao & Bordas, 2023). The following AI models possess the ability to screen millions of prospective material candidates much quicker than conventional experimental screening techniques, thereby accelerating the identification of high-performance materials for next-generation technologies. Alongside predictive modelling, AI-aided automation is also revolutionizing high-throughput experimentation and robotic synthesis platforms to facilitate in-situ material synthesis and real-time optimization of processing conditions. Reinforcement learning-based algorithms are able to dynamically optimize reaction parameters for optimal yield, efficiency, and waste minimization during material production. Further, advances in AI-boosted image recognition methodologies are enabling enhanced electron microscopy and spectroscopy data analysis to achieve accurate defect identification, phase detection, and microstructural characterization (Monteiro da Silva et al., 2024).

The combination of artificial intelligence and computational modelling is not only reducing the cost and time of material development, but it is also opening up new avenues for innovation. These approaches are critical to the creation of sustainable materials, including biodegradable polymers, energy-efficient semiconductors, and high-performance battery materials. With the evolution of computational power and AI technologies, the combination of machine learning, quantum computing, and material informatics is poised to continue to revolutionize the field, enabling rapid discovery and optimization of materials for new applications in medicine, electronics, and renewable energy (Felipe Hernández-Pérez et al., 2019).

1.3 Integration of Characterization with Nanotechnology

Nanotechnology has revolutionized material science by enabling matter to be controlled at the atomic and molecular level. This has enabled new materials with tailored properties and functions to be created. But nanoscale materials' very unique challenges—like their small size, high surface-to-volume ratio, and quantum effects—require equally advanced characterization techniques (Advances in Materials Science and Engineering, 2023). The integration of nanotechnology with state-of-the-art characterization tools has not only enhanced our understanding of nanomaterials but also facilitated the development of multifunctional materials with applications in diverse fields.

1.3.1 Nanoscale Imaging and Analysis

The capability to visually inspect and investigate materials at the nanoscale is necessary in order to have an understanding of their structure, properties, and behaviour. Nanomaterials have distinct mechanical, electrical, and optical properties that are measurably different from the bulk counterparts of the materials, making characterization essential to application in electronics, energy storage, biomedicine, and manufacturing. Characterization methods with standard capabilities, however, tend to be not applicable when employed on nanomaterials owing to resolution limits, sample preparation problems, and inability to observe dynamic phenomena at atomic or molecular scales. Such limitations have precipitated extensive technological innovation in imaging and analysis instruments to allow researchers to study nanomaterials with unprecedented sensitivity.

TEM and Scanning Transmission Electron Microscopy (STEM) now attain sub-angstrom resolution, which allows the direct observation of atomic structures, crystal defects, and phase boundaries. Aberration-corrected TEM has significantly improved spatial resolution, thereby allowing the imaging of single atoms in nanomaterials, which gives valuable information regarding lattice distortion and structural defects. AFM offers high-resolution topographic maps and nanoscale measurements of mechanical, electrical, and thermal properties, making it an indispensable tool for the characterization of nanocomposites and soft materials.

Moreover, in situ and operando imaging methods like environmental TEM (ETEM) and in situ AFM enable real-time monitoring of structural and chemical changes in operational circumstances, essential for catalysis, battery materials, and biomaterials studies. AI and machine learning integration are also transforming nanoscale characterization by automating defect detection and image analysis, enhancing efficiency and accuracy. Nanoscale imaging will be a major enabler in next-generation material design with these advances, facilitating new innovations in many industries (Almeida et al., 2022).

1.3.1.1 High-Resolution Imaging Techniques

- **Transmission Electron Microscopy (TEM)**: TEM has become a necessary nanoscale imaging tool, providing atomic-scale information on the nanomaterial's structure and composition. The improvements of aberration-corrected TEM and in-situ TEM enable researchers to visualize dynamic phenomena, e.g., nanoparticle synthesis and chemical reactions, in real-time (Eswara et al., 2019).
- **Scanning Electron Microscopy (SEM)**: SEM can yield high-resolution images of nanomaterial surfaces and is therefore best suited for morphology and topography studies. ESEM and FIB-SEM have only added to its list, allowing analysis of fragile materials and 3D reconstructions to be built (Nazarov et al., 2017).
- **Atomic Force Microscopy (AFM)**: AFM is a general-purpose tool for nanoscale imaging and measurement of mechanical properties. It can image surface topography at sub-nanometer resolution and measure properties like elasticity, adhesion, and conductivity. Extended modes like Kelvin probe force microscopy (KPFM) and conductive AFM offer more information on electrical and magnetic properties (Joshi et al., 2022).

1.3.1.2 Spectroscopic Techniques for Nanoscale Analysis

- **Raman Spectroscopy**: Raman spectroscopy is extensively used to probe the vibrational modes of nanomaterials and acquire information regarding their crystalline structure and chemical makeup. Techniques like surface-enhanced Raman spectroscopy (SERS) and tip-enhanced Raman spectroscopy (TERS) have pushed the sensitivity boundary to the point of detecting a single molecule (Logan et al., 2022).
- **X-Ray Photoelectron Spectroscopy (XPS)**: XPS is a versatile method for nanomaterial surface chemistry study. It gives precise information on chemical states, elemental composition, and bonding environments and is thus a necessity for catalyst study, thin films, and nanostructured surfaces (Radnik et al., 2021).

1.3.1.3 In-Situ and Operando Characterization

Operando and in-situ techniques enable nanomaterials to be studied in real environments, such as high pressures, temperatures, electromagnetic fields, or chemically reactive conditions. Unlike conventional ex-situ characterization, where sample removal from native operating conditions is required, operando and in-situ techniques enable real-time observation of structural, chemical, and electronic behaviour, and thus a more complete understanding of material properties. Operando and in-situ techniques are vitally important to study nanomaterials for catalysis, energy storage, semiconductors, and medical therapies, all with performance significantly dependent on environmental conditions.

Different techniques that are well recognized as in situ methods include in situ Transmission Electron Microscopy or real-time observation of atomic-scale nano structural evolution. In cables made of nanowires, for example, in situ TEM can explore nucleation and growth phenomena, witness phase transitions of battery electrodes during charge/discharge cycles, and assess catalytic structural degradation during operation. Operando X-ray absorption spectroscopy (XAS) is another one which gives crucial information into the electronic environment of the active sites of a catalyst and its role in improving the efficiency and stability of reactions.

Supplementary procedures such as in-situ Atomic Force Microscopy (AFM) and operando Raman Spectroscopy allow the observation of surface morphology, mechanical stress, and nanoscale evolution of chemical bonding. Moreover, machine learning and artificial intelligence-driven data analysis are enhancing real-time in-situ data processing and interpretation, thereby enabling better and faster characterization. With advancing methodologies, they will become instrumental in the development of next-generation materials for energy, electronics, and nanomedicine applications, thereby completing theoretical modelling and experimental demonstration (Li et al., 2009).

1.3.2 Role in Designing Multifunctional Materials

The marriage of nanotechnology with sophisticated characterization methods has been pivotal in the design and development of multifunctional materials with a variety of properties, including mechanical toughness, electrical conductivity, thermal stability, and optical properties. These materials are transforming industries such as electronics, healthcare, energy storage, and aerospace, where high-performance, light-weight, and durable materials are critical. By designing nanostructures at either the molecular or atomic level, researchers can create materials that accomplish several functions at once, including self-healing coatings, energy-harvesting surfaces, and smart nanocomposites.

Sophisticated characterization methods are of paramount importance to explain the structure–property relationship in multifunctional materials to ensure accurate synthesis and control of functionality. Transmission Electron Microscopy (TEM) and

Atomic Force Microscopy (AFM) enable nanoscale probing of material morphology, defects, and interfacial properties. X-ray Photoelectron Spectroscopy (XPS) and Raman Spectroscopy provide essential chemical and electronic information important for the design of more conductive and reactive materials. X-ray Diffraction (XRD) and Small-Angle X-ray Scattering (SAXS) enable probing of crystalline structures and phase compositions, particularly in polymer nanocomposites and hybrid materials.

Furthermore, operando and in-situ approaches make it possible to observe in real time structural and functional changes upon external stimuli, such as mechanical stress or temperature change. Artificial intelligence and machine learning incorporation is further driving material discovery by predicting optimal compositions and processing parameters. With continued advances in nanotechnology and characterization methods, multifunctional materials are set to drive next-generation applications, ranging from bendable electronics, wearable sensors, energy-efficient devices, and biocompatible implants, to ensure enhanced performance and sustainability in industries (Sarma & Rao, 2024).

1.3.2.1 Nanocomposites

Nanocomposites, composed of a matrix phase reinforced with nanoscale fillers, are one of the most widespread examples of multifunctional materials. TEM, SEM, and AFM characterization methods are employed to investigate the dispersion, orientation, and interfacial interactions of nanofillers within the matrix. This information is essential to tailor the mechanical, thermal, and electrical properties of nanocomposites to meet aerospace, automotive, and electronics applications (Micard et al., 2020).

1.3.2.2 Nanostructured Catalysts

Nanostructured catalysts are made to have the maximum surface area and active sites in order to increase their efficiency and selectivity. Sophisticated characterization methods like XPS, XRD, and in-situ spectroscopy are employed to examine the surface chemistry, crystal structure, and reaction mechanism of the catalysts (Du et al., 2020). This knowledge enables the rational design of catalysts for applications in energy conversion, environmental remediation, and chemical synthesis.

1.3.2.3 Smart and Responsive Materials

Smart materials are nano-structured up to control the interaction with stimuli, such as temperature, light, or electric fields. The dynamic behaviour of these materials is then evaluated using characterization methods, including AFM, Raman, in-situ

TEM, and others, to maximize responsiveness (Lin et al., 2022). Examples include shape-memory polymers, self-healing materials, and photochromic coatings.

1.3.2.4 Biomedical Applications

In recent times, nanotechnology has facilitated the development of multifunctional materials for various applications in the realm of biomedicine, that is, in systems for drug delivery, biosensing, and scaffolding for tissue engineering. Various characterization techniques are employed to study the biocompatibility, drug release kinetics, and mechanical properties of such materials (Hoque Apu et al., 2022). For example, TEM and AFM are used to analyse the structure and stability of nanoparticles, while spectroscopic techniques provide insights into their interactions with biological systems.

1.4 Conversion into Value-Added Products

This chapter deliberates on the contribution of material characterization in driving innovation in pharmaceuticals, food and agriculture, renewable energy, and biodegradable materials, where the end goal of material characterization is to convert raw materials into value-added products that can solve real problems in modern industries. The characterization techniques applied to advance material properties, functionality, and sustainability applications are taught to scientists and industries (Kute & Singh, 2023).

1.4.1 Applications in Pharmaceuticals

Characterization of materials during the development processes for safe, effective, and targeted drug delivery systems are crucially dependent on the pharmaceutical sector. Characterization methods provide invaluable information concerning the physical, chemical, and biological properties of drug products that promote their optimization of performance and stability. Table 1.2 provides information of applications of Value-Added Materials. Figure 1.2 shows the application of characterization techniques.

1.4.1.1 Drug Delivery Systems

- **Nanoparticles and Liposomes**: As drug carriers, nanoparticles and liposomes can encapsulate and deliver therapeutic drugs to specific sites of action. Characteristics such as size, shape, and surface properties critical to optimal drug loading

Table 1.2 Applications of value-added materials

Industry	Value-added product	Characterization techniques	Benefits	References
Pharmaceuticals	Targeted drug delivery systems	TEM, DLS, NMR, XPS	Enhanced drug efficacy, reduced side effects	Hussein and Abdullah (2022)
Food packaging	Biodegradable polymer films	FTIR, SEM, DSC, XRD	Sustainable, reduces plastic waste	Farhoodi (2016)
Renewable energy	Perovskite solar cells	XRD, SEM, UV–Vis Spectroscopy	High efficiency, low-cost production	Li et al. (2009)
Energy storage	Lithium-ion battery materials	TEM, XPS, Electrochemical Analysis	Improved charge retention, longer lifespan	Kute and Singh (2023)
Nanotechnology	Carbon nanotubes, graphene	Raman, SEM, TEM, AFM	High conductivity, lightweight materials	Samrot et al. (2019)

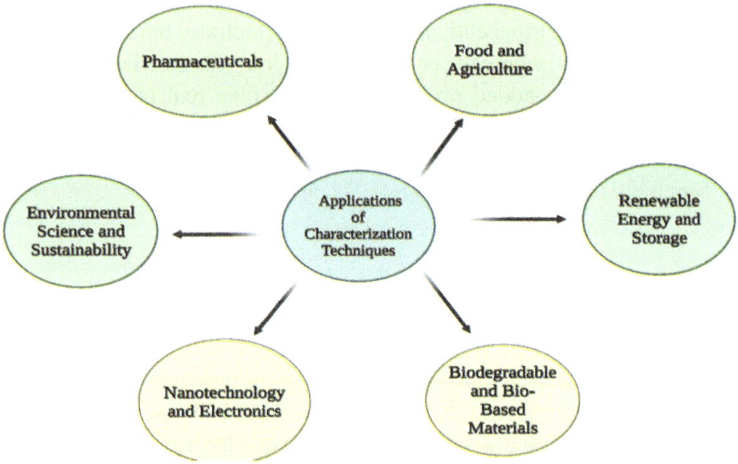

Fig. 1.2 The application of characterization techniques

and release are examined by techniques such as dynamic light scattering (DLS), transmission electron microscopy (TEM), and atomic force microscopy (AFM) (Hussein & Abdullah, 2022).

- **Polymer-Based Drug Delivery**: Biodegradable polymers such as polylactic acid (PLA) and poly(lactic-co-glycolic acid) (PLGA) are used in the manufacture of controlled release drug delivery systems. FTIR, XRD, and SEM are techniques used for the study of polymer structure, crystallinity, and degradation behaviour with a view to ensuring reproducible drug release profiles (Liu et al., 2022).

1.4.1.2 Characterization of Drug Polymorphs

Polymorphism is a phenomenon whereby a drug can exist in a different crystalline form that may have different solubility and consequently bioavailability and stability due to transformation. Among the techniques suitable for polymorph characterization are x-ray diffraction (XRD), Raman spectroscopy, and differential scanning calorimetry (DSC) to ensure the choice of the most stable form, which will guarantee efficacy.

1.4.1.3 Targeted Therapeutics

The advanced characterization tools like NMR and XPS are used to study the surface chemistry of drug carriers and their interactions with biological systems. This knowledge provides the basis for designing targeted therapeutics with minimal side effects and maximal efficiency.

1.4.2 Food and Agricultural Innovations

Material characterization is essentially concerned with food safety, quality, and sustainability, and enhancing agricultural productivity. The understanding of the structure and properties of food and agricultural materials gives an opportunity for researchers to find various creative and innovative solutions to face with the contemporary worldwide challenges (Farhoodi, 2016).

1.4.2.1 Food Packaging

- **Active and Intelligent Packaging**: Advanced packaging materials can have active components, such as oxygen scavengers or antimicrobial agents, to prolong shelf life and ensure food safety. The study of the interaction between packaging materials and food products involves characterization methods like FTIR and SEM (Hemmati et al., 2021).
- **Biodegradable Packaging**: Biodegradable polymers such as starch-based films and cellulose derivatives are being developed as eco-friendly alternatives to normal plastics. A battery of tests, including DSC and AFM, are deployed to optimize their mechanical-barricade properties (Zoumaki et al., 2019).

1.4.2.2 Agricultural Inputs

- **Nano pesticides and Nano fertilizers**: Thanks to nanotechnology, agricultural inputs are being made more efficient and eco-friendlier. Their size, stability, and release behaviour are studiously analysed with characterization tools, like TEM and DLS to ensure that these products are effective (Usman et al., 2019).
- **Soil and Crop Analysis**: This study used advanced methods for the analysis of soil composition and crop health, thus making way for precise farming and sustainable agricultural practices: amongst others, NIR spectroscopy applied to the spectral range covering near-infrared up to visual wavelengths (Farber et al., 2020).

1.4.3 Renewable Energy and Storage Solutions

The world energy challenges cannot be solved without transitioning to renewable energy sources and developing efficient energy storage systems. Optimizing the performance and durability of energy materials relies on material characterization.

1.4.3.1 Solar Energy

- **Photovoltaic Materials**: Characterization techniques like XRD, SEM, and UV–Visible spectroscopy help elucidate the structure, morphology, and optical properties of the studied photovoltaic materials such as perovskite and silicon-based solar cells. This knowledge aids in designing solar panels with higher efficiency and stability (Hoye et al., 2017).
- **Thin-Film Solar Cells**: The efficiency performance and deposition of thin-film solar cells are optimized using advanced imaging and spectroscopic tools to allow for high efficiency and low production costs (Karl et al., 2017).

1.4.3.2 Energy Storage

- **Batteries**: Characterization tools, like TEM, XPS, and electrochemical impedance spectroscopy (EIS), are aimed at studying the structure, composition, and electrochemical behaviour of battery materials such as lithium-ion and solid-state batteries. This information is vital in enhancing energy density, cycle life, and safety (Abdelrahim et al., 2023).
- **Supercapacitors**: Supercapacitor materials area characterization with respect to surface area, porosity, and conductivity is performed by AFM and Raman techniques that collectively assist in the growth of high-performance energy storage devices (Yesuraj et al., 2020).

1.4.4 Biodegradable and Bio-Based Materials

Sustainable development for biodegradable and bio-based materials is critical in minimizing environmental pollution. Characterization is crucial in elaborating these materials and optimizing properties and performances.

1.4.4.1 Biodegradable Polymers

- **Polylactic Acid (PLA)**: Considering PLA is a well-known biodegradable polymer derived from renewable sources, its characterization is analysed using FTIR, DSC, and SEM for thermal, mechanical, and degradation properties to predict applications (Singhvi & Gokhale, 2013).
- **Polyhydroxyalkanoates (PHAs)**: PHAs are a family of biodegradable polymers assembled by microorganisms. Characterization tools like NMR and XRD are engaged to help in the optimization of production and properties for application in packaging, agriculture, and biomedicine (Johnston et al., 2017).

1.4.4.2 Bio-based Composites

- **Natural Fiber Composites**: The natural fiber composites are being developed as sustainable alternatives to synthetic composites. The interfacial bonding and mechanical properties of these composites are evaluated mostly through the use of SEM and AFM (Faiyaz Ahmed et al., 2017).
- **Chitosan and Alginate**: These bio-based polymers find applications in the fields of wound healing, water purification, and food packaging. These polymers are characterized with the help of various facilities, such as FTIR and TEM, to enable tuning of their properties and functionalities (Leceta et al., 2020).

1.5 Sustainability in Material Conversion

With increasing global concerns over environmental degradation, resource depletion, and climate change, sustainable practices in material conversion are of utmost importance. Sustainable material conversion emphasizes environmental friendliness, waste reduction, and optimal resource utilization during the life cycle of the materials. This section discusses two major pillars of sustainability in material conversion: green chemistry approaches and model circular economy (Titirici et al., 2022).

1.5.1 Green Chemistry Approaches

Green chemistry, described as sustainable chemistry, is a scientific philosophy with the mandate to design chemical products and processes to either reduce or eliminate the use and generation of hazardous substances. Green chemistry is applied within material conversion for eco-friendly materials and optimization processes with minimal environmental impact (Fitria et al., 2022).

1.5.1.1 Principles of Green Chemistry in Material Conversion

1. **Atom Economy**: Designing processes that maximize the incorporation of all raw materials into the final product, minimizing waste.
2. **Use of Renewable Feedstocks**: Prioritizing the use of biomass, agricultural waste, and other renewable resources over fossil fuels.
3. **Energy Efficiency**: Developing low-energy processes, such as room-temperature reactions or microwave-assisted synthesis.
4. **Reduction of Hazardous Substances**: Replacing toxic solvents and reagents with safer alternatives, such as water or ionic liquids.
5. **Design for Degradation**: Creating materials that can break down into non-toxic byproducts at the end of their lifecycle (Santos et al., 2015).

1.5.1.2 Applications of Green Chemistry

- **Biodegradable Polymers**: Green chemistry principles are used to design polymers, such as polylactic acid (PLA) and polyhydroxyalkanoates (PHAs), that are derived from renewable resources and can degrade naturally.
- **Eco-Friendly Catalysts**: The development of non-toxic, reusable catalysts, such as enzymes or metal–organic frameworks (MOFs), reduces the environmental impact of chemical reactions.
- **Solvent-Free Processes**: Innovations in solvent-free synthesis, such as mechanochemistry and solid-state reactions, eliminate the need for harmful solvents.
- **Waste Valorisation**: Green chemistry enables the conversion of industrial waste, such as carbon dioxide or lignin, into valuable products like fuels, chemicals, and construction materials (Hong & Chen, 2017).

1.5.1.3 Role of Characterization in Green Chemistry

The advanced characterization techniques, FTIR, NMR, and XRD, are accepted as required for sustaining processes of green chemistry. They could serve to verify the renewable nature of feedstocks, measure the efficiency of reactions, and assess the

Table 1.3 Sustainability and circular economy approaches in material science

Sustainable approach	Description	Examples	Environmental benefits	References
Green chemistry synthesis	Use of eco-friendly solvents and catalysts	Biodegradable plastics, bio-based coatings	Reduces toxic waste, safer production	Bhushan (2017)
Recycling of electronic waste	Recovery of metals and rare-earth elements	Lithium recovery from batteries, circuit board recycling	Conserves resources, reduces landfill waste	Radnik et al. (2021)
Biodegradable materials	Materials that decompose naturally	PLA, PHA, starch-based packaging	Reduces plastic pollution	Li et al. (2009)
Circular economy in textiles	Recycling and upcycling fabric waste	Recycled polyester, organic cotton	Lowers carbon footprint, less water use	Hoque Apu et al. (2022)

biodegradability of various materials (Annu et al., 2017). Table 1.3 gives detailed information sustainability and Circular Economy Approaches in Material Science.

1.5.2 Circular Economy and Resource Optimization

The circular economy represents an overarching approach to economic development that aims to eradicate waste and keep resources in use. The circular economy opposes the linear model of economy (take-make-dispose) and focuses instead on the reuse, recycling, and regeneration of materials, thereby forming a closed-loop system. The intent is to reduce maximum environmental impact.

1.5.2.1 Key Principles of the Circular Economy

1. **Design for Longevity**: Creating durable, repairable, and upgradable products to extend their lifespan.
2. **Recycling and Upcycling**: Transforming waste materials into new products of equal or higher value.
3. **Resource Recovery**: Extracting valuable materials from waste streams, such as metals from electronic waste or nutrients from agricultural residues.
4. **Industrial Symbiosis**: Promoting collaboration between industries to use waste from one process as a resource for another (Suárez-Eiroa et al., 2019).

1.5.2.2 Applications in Material Conversion

- **Recycled Plastics**: Advanced sorting and processing technologies, such as near-infrared (NIR) spectroscopy and automated separation systems, enable the efficient recycling of plastics into high-quality materials for packaging, textiles, and construction (Ding & Zhu, 2023).
- **Metals Recovery**: Techniques like hydrometallurgy and pyrometallurgy are used to recover precious and base metals from electronic waste, reducing the need for mining and conserving natural resources.
- **Bio-Based Materials**: The circular economy promotes the use of bio-based materials, such as cellulose and lignin, which can be sustainably sourced and recycled.
- **Construction Materials**: Waste materials, such as fly ash, slag, and recycled concrete, are being used to produce eco-friendly construction materials, reducing the demand for virgin resources (Adetunji et al., 2023).

1.5.2.3 Role of Characterization in the Circular Economy

Characterization tools play a critical role in enabling the circular economy by:

- **Identifying Material Composition**: Techniques like XRF (X-ray fluorescence) and ICP-MS (inductively coupled plasma mass spectrometry) are used to analyze the composition of waste streams and identify recoverable materials.
- **Optimizing Recycling Processes**: Tools such as SEM, TEM, and DSC are used to study the microstructure and properties of recycled materials, ensuring their quality and performance (Table 1.3).
- **Monitoring Environmental Impact**: Life cycle assessment (LCA) and environmental monitoring tools help evaluate the sustainability of circular economy practices (Hoang et al., 2022).

1.6 Challenges and Future Perspectives

While material characterization has advanced significantly, challenges persist in achieving high-resolution, real-time, and multi-scale analysis. Traditional techniques often require complex sample preparation, expensive equipment, and specialized expertise, limiting accessibility and scalability. Key challenges include improving nanoscale imaging techniques like TEM and AFM, enhancing trace analysis for detecting low-concentration elements, and integrating AI-driven computational modelling to optimize material discovery.

Interdisciplinary collaboration, combining nanotechnology, AI, and green chemistry, is crucial for developing sustainable, efficient characterization methods. Future advancements, including quantum imaging, 4D microscopy, and AI-powered automated laboratories, will push the boundaries of precision and material innovation. By embracing these emerging technologies, material characterization will continue to drive progress in industries such as healthcare, energy, and electronics, making analysis faster, more accurate, and more sustainable (Briers, 2015).

1.6.1 Technical and Analytical Challenges

Despite the advancements in characterization techniques, several technical and analytical challenges persist, limiting their applicability and effectiveness in certain scenarios.

1.6.1.1 Resolution and Sensitivity Limitations

- **Nanoscale Characterization**: While techniques like TEM and AFM offer high resolution, they often require complex sample preparation and are limited to small areas. Improving resolution while maintaining ease of use remains a challenge.
- **Trace Analysis**: Detecting trace elements or low-concentration components in complex matrices, such as environmental samples or biological tissues, requires highly sensitive instruments and advanced data analysis methods (Santos et al., 2015).

1.6.1.2 Dynamic and In-Situ Characterization

- **Real-Time Monitoring**: Capturing dynamic processes, such as chemical reactions, phase transitions, or mechanical deformation, in real time is challenging due to the limitations of current in-situ techniques.
- **Environmental Conditions**: Many characterizations tools struggle to operate under extreme conditions, such as high temperatures, pressures, or corrosive environments, which are often relevant for industrial applications (Yang et al., 2021).

1.6.1.3 Data Complexity and Interpretation

- **Big Data Challenges**: The increasing use of high-throughput and automated techniques generates vast amounts of data, requiring advanced computational tools and algorithms for analysis and interpretation.
- **Multimodal Integration**: Combining data from multiple characterization techniques to obtain a comprehensive understanding of materials is often complex and time-consuming (Shockley, 2015).

1.6.1.4 Cost and Accessibility

- **High Costs**: Advanced characterization tools, such as synchrotron-based techniques or cryo-TEM, are expensive to acquire and maintain, limiting their accessibility for many researchers and institutions.
- **Standardization**: The lack of standardized protocols and reference materials can lead to inconsistencies in data interpretation and hinder reproducibility (Zdora et al., 2020).

1.6.2 Opportunities for Interdisciplinary Research

Interdisciplinary approaches remain fundamental in solving problems pertaining to material characterization, therefore opening prospects for innovations in materials science. By combining the skill sets of experts from different fields such as physics, chemistry, biology, engineering, and computer science, new solutions, methodologies, and applications are built to stretch the frontiers of materials science. The complex nature of modern materials, from nanomaterials, biomaterials, and smart polymers to quantum dots, necessitates collaboration so that knowledge from one discipline adds to other insights for better understanding and real-world application.

The convergence of nanotechnology and biomedical science has led to breakthroughs in drug delivery systems, biosensors, and regenerative medicine. Researchers in physics and chemistry collaborate with biologists to create nanocarriers that enhance bioavailability and minimize the toxicity of pharmaceuticals. In the same way, the meld of materials science and artificial intelligence (AI) has fundamentally altered high-throughput material discovery, wherein AI-driven predictive modeling effectively accelerates the identification of new materials possessing desired properties. Machine learning algorithms evaluate vast volumes of experimental and computational data, picking out patterns that would be impossible to identify manually. These techniques would greatly accelerate and drive down the cost of materials synthesis and optimization.

Sustainable materials development and environmental science is yet another important area profiting from interdisciplinary research. There have been joint efforts among polymer chemists, environmental scientists, and mechanical engineers in the development of biodegradable polymers, green energy storage systems, and recyclable composites. Advancements in circular economy models, whereby materials are engineered for reuse and recyclability from the very start, act as a transformative force through the cut back of waste generation and optimization of resource utilization. In addition, quantum computing developments and semiconductor physics have gone hand in hand in laying the foundation for next-generation electronic materials to disrupt computing power and energy efficiency significantly.

In addition, interdisciplinary approaches are required to address challenges in real-time and in-situ material characterization. Innovations such as correlative microscopy, which combine techniques including scanning electron microscopy,

Table 1.4 Advancements in nanotechnology and AI for material characterization

Innovation	Technology used	Key advantages	Application areas	References
AI-driven microscopy	Deep learning models for image analysis	Automated defect detection, real-time analysis	Material defect detection, medical diagnostics	Monteiro da Silva et al. (2024)
Nanorobotics for characterization	Nanoscale robotic systems	Precise material manipulation, high accuracy	Atomic-scale imaging, drug delivery	Joshi et al. (2022)
Cryo-TEM for soft materials	Cryogenic electron microscopy	Preserves native structure of materials	Biological materials, polymers	Vincent et al. (2020)
High-throughput AI screening	Machine learning-based material prediction	Rapid material discovery, cost reduction	Battery materials, catalysis	Apurwa et al. (2024)
Quantum sensing	Quantum-based imaging techniques	Ultra-high resolution imaging	Semiconductor research, bioimaging	Almeida et al. (2022)

transmission electron microscopy, and Raman spectroscopy, need expertise from different scientific disciplines to enable correct interpretation of data generated and useful conclusions derived. Moreover, research on bioinspired materials—where nature is considered as a prototype for high-performance material design—calls for biomimicry that lays a bridge between biology, chemistry, and engineering. Table 1.4 elaborate advancements in Nanotechnology and AI for Material Characterization.

Collaboration among these disciplines will allow researchers to overcome traditional barriers in material characterization and thus allow for truly transformative advances. Future material science will increasingly depend on boring interdisciplinary synergies to develop smarter, more efficient, and greener materials that serve the demands of industries such as healthcare, energy, electronics, and aerospace.

1.6.2.1 Integration with Artificial Intelligence (AI)

- **Machine Learning for Data Analysis**: AI-driven algorithms can analyse complex datasets, identify patterns, and predict material properties, enabling faster and more accurate characterization.
- **Automated Experimentation**: AI can optimize experimental parameters, control instruments, and even design new materials, reducing the time and cost of research (Rodrigues et al., 2021).

1.6.2.2 Collaboration with Biology and Medicine

- **Biomimetic Materials**: Insights from biological systems can inspire the design of new materials with unique properties, such as self-healing or adaptive behaviour.
- **Medical Applications**: Collaboration with the medical field can lead to the development of advanced diagnostic tools, drug delivery systems, and biocompatible materials (Kan et al., 2021).

1.6.2.3 Synergy with Environmental Science

- **Sustainable Materials**: Interdisciplinary research can drive the development of eco-friendly materials and processes, addressing global challenges such as climate change and resource depletion.
- **Environmental Monitoring**: Advanced characterization tools can be used to detect pollutants, monitor ecosystems, and assess the impact of human activities on the environment (Tripathi et al., 2022).

1.6.2.4 Cross-Disciplinary Innovation

- **Nanotechnology and Quantum Computing**: The integration of nanotechnology with quantum computing can lead to breakthroughs in material characterization, such as ultra-sensitive sensors and quantum imaging techniques (Table 1.4).
- **Advanced Manufacturing**: Collaboration with the manufacturing industry can enable the development of new materials and processes for additive manufacturing, 3D printing, and smart factories (Nadeem et al., 2024).

1.6.3 Future Trends in Characterization

The future of material characterization is shaped by emerging technologies, innovative approaches, and evolving research priorities. Several trends are expected to drive the field forward in the coming years.

1.6.3.1 Advanced Imaging Techniques

- **4D Imaging**: The development of 4D imaging techniques, which capture spatial and temporal changes in materials, will provide new insights into dynamic processes and structural evolution.
- **Quantum Imaging**: Quantum-based imaging techniques, such as quantum sensors and entangled photon microscopy, promise unprecedented resolution and sensitivity (Nughays et al., 2024).

1.6.3.2 AI-Driven Characterization

- **Autonomous Laboratories**: The integration of AI with robotic systems will enable autonomous laboratories that can design, conduct, and analyse experiments with minimal human intervention.
- **Predictive Modelling**: AI-driven predictive models will accelerate material discovery by simulating properties and behaviour, reducing the need for extensive experimental testing (Ohahiduzzaman et al., 2024).

1.6.3.3 Sustainable and Green Characterization

- **Eco-Friendly Techniques**: The development of characterization techniques that minimize energy consumption, waste generation, and environmental impact will align with global sustainability goals.
- **Circular Economy Integration**: Characterization tools will play a key role in enabling the circular economy by optimizing recycling processes and resource recovery (Katamesh et al., 2024).

1.6.3.4 Personalized and On-Demand Characterization

- **Portable Devices**: The development of portable and low-cost characterization devices will enable on-site analysis in remote or resource-limited settings, such as field research or industrial inspections.
- **Personalized Medicine**: Advanced characterization tools will support the development of personalized medicine by enabling the precise analysis of biological samples and drug formulations (Yenilmez et al., 2016).

1.7 Conclusion

Advances in material characterization have transformed material science, allowing for accurate analysis of micro- and nanoscale structural, chemical, and functional properties. These advances have made it possible to create value-added products in fields such as pharmaceuticals, renewable energy, and environmental sustainability. Coupling spectroscopy, imaging, and computational methods, scientists can optimize material performance more efficiently. Addition of nanotechnology, AI, and machine learning has opened the field to an even broader scope, making material discovery more affordable, efficient, and predictive. AI-based methods and high-throughput screening have sped up the discovery of novel materials, and imaging methods such as SEM and TEM offer atomic-scale insights. Green chemistry and the circular economy have also encouraged sustainable material development, focusing on resource efficiency and waste minimization.

Still, challenges such as the demand for greater resolution characterization, real-time analysis, and affordable techniques persist. Future breakthroughs in quantum computing, 4D imaging, and AI-based automation will continue to enrich material characterization and industrial processes. Interdisciplinary research and technology integration will drive further sustainable innovations, offering greater efficiency, accuracy, and environmental responsibility in material science.

Acknowledgements Authors are thankful to the Galgotias University, Greater Noida for providing necessary research support.

References

Abdelrahim, A. M., Abd El-Moghny, M. G., El-Shakre, M. E., & El-Deab, M. S. (2023). Promoted glucose electrooxidation at Ni(OH) $_2$ /graphene layers exfoliated facilely from carbon waste material. *RSC Advances, 13*(3), 1811–1822.

Adetunji, A. I., Oberholster, P. J., & Erasmus, M. (2023). Bioleaching of metals from E-waste using microorganisms: A review. *Minerals, 13*(6), 828.

Agrawal, M., Prasad, V., Nijhawan, G., Jalal, S. S., Rajalakshmi, B., & Dwivedi, S. P. (2024). A comprehensive review of electron microscopy in materials science: technological advances and applications. In *E3S Web of Conferences* (vol. 505, p. 01029).

Almeida, T. P., Palomino, A., Lequeux, S., Boureau, V., Fruchart, O., Prejbeanu, I. L., et al. (2022). Off-axis electron holography for the direct visualization of perpendicular shape anisotropy in nanoscale 3D magnetic random-access-memory devices. *APL Materials, 10*(6).

Álvarez López, Y., García Fernández, M., Grau, R., & Las-Heras, F. (2018). A synthetic aperture radar (SAR)-based technique for microwave imaging and material characterization. *Electronics (Basel)., 7*(12), 373.

Ameh, E. S. (2019). A review of basic crystallography and x-ray diffraction applications. *The International Journal of Advanced Manufacturing Technology, 105*(7–8), 3289–3302.

Annu, Sahu, M., Singh, S., Prajapati, S., Verma, D. K., & Shin, D. K. (2024). From green chemistry to biomedicine: The sustainable symphony of cobalt oxide nanoparticles. *RSC Advance, 14*(45), 32733–32758.

Apurwa, S., Shaikh, M., Devhad, S., Patel, S., Chougule, R., Shejwalkar, P., et al. (2024). Enhancing cancer detection through AI-driven high content analysis of circulating tumor cells. *Journal of Clinical Oncology, 42*(16_suppl), 3054–3054.

Au Yeung, V. P. W., Obrezanova, O., Zhou, J., Yang, H., Bowen, T. J., Ivanov, D., et al. (2024). Computational approaches identify a transcriptomic fingerprint of drug-induced structural cardiotoxicity. *Cell Biology and Toxicology, 40*(1), 50.

Bhaskar, S. (2023). Biosensing technologies: A focus review on recent advancements in surface Plasmon coupled emission. *Micromachines (Basel), 14*(3), 574.

Bhushan, B. (2017). *Nanotribology and nanomechanics: An introduction*. Springer.

Binnemans, K., Jones, P. T., Blanpain, B., Van Gerven, T., & Pontikes, Y. (2015). Towards zero-waste valorisation of rare-earth-containing industrial process residues: A critical review. *Journal of Cleaner Production, 99*, 17–38.

Briers, W. J. (2015). *Overcoming challenges in material characterization of polymers at intermediate strain rates*, pp. 153–164.

Callister, Jr. W. D., & Rethwisch, D. G. (2020). *Materials science and engineering: An introduction*. Wiley.

Chang, T. T. (1996). *Novel approaches for characterization and its conversion into value.*

Chen, C., Qi, J., Li, Y., Li, D., Wu, L., Li, R., et al. (2024). Applications of Raman spectroscopy in the diagnosis and monitoring of neurodegenerative diseases. *Frontiers in Neuroscience, 18* (2024).

Craig, M. (2021). High-throughput screening and rational design to drive discovery in molecular water oxidation catalysis. *ioChem-BD Computational Chemistry Datasets.*

De Stefano, P., Bianchi, E., & Dubini, G. (2022). The impact of microfluidics in high-throughput drug-screening applications. *Biomicrofluidics, 16*(3).

Ding, Q., & Zhu, H. (2023). The key to solving plastic packaging wastes: Design for recycling and recycling technology. *Polymers (Basel), 15*(6), 1485.

dos Santos, T. R., Nilges, P., Sauter, W., Harnisch, F., & Schröder, U. (2015). Electrochemistry for the generation of renewable chemicals: Electrochemical conversion of levulinic acid. *RSC Advances, 5*(34), 26634–26643.

dos Santos, A. C. V. D., Tranchida, D., Lendl, B., & Ramer, G. (2022). Nanoscale chemical characterization of a post-consumer recycled polyolefin blend using tapping mode AFM-IR. *Analyst, 147*(16), 3741–3747.

Du, J., Wang, L., Li, J., Cao, L., Dang, S., & Qin, X. (2020). Highly active and durable nanostructured Ni-CNTs-HG composite electrocatalyst for hydrogen production. *Current Nanoscience, 16*(2), 259–267.

Errico, S., Moggio, M., Diano, N., Portaccio, M., & Lepore, M. (2023). Different experimental approaches for Fourier-transform infrared spectroscopy applications in biology and biotechnology: A selected choice of representative results. *Biotechnology and Applied Biochemistry, 70*(3), 937–961.

Eswara, S., Pshenova, A., Yedra, L., Hoang, Q. H., Lovric, J., Philipp, P., et al. (2019). Correlative microscopy combining transmission electron microscopy and secondary ion mass spectrometry: a general review on the state-of-the-art, recent developments, and prospects. *Applied Physics Reviews, 6*(2).

Fanni, B. M., Sauvage, E., Celi, S., Norman, W., Vignali, E., Landini, L., et al. (2020). A proof of concept of a non-invasive image-based material characterization method for enhanced patient-specific computational modeling. *Cardiovascular Engineering and Technology, 11*(5), 532–543.

Farber, C., Mahnke, M., Sanchez, L., & Kurouski, D. (2019). Advanced spectroscopic techniques for plant disease diagnostics. A review. *TrAC Trends in Analytical Chemistry, 118*, 43–49.

Farhoodi, M. (2016). Nanocomposite materials for food packaging applications: Characterization and safety evaluation. *Food Engineering Reviews, 8*(1), 35–51.

Felipe Hernández-Pérez, A., de Arruda, P. V., Sene, L., da Silva, S. S., Kumar Chandel, A., & de Almeida Felipe, M. G. (2019). Xylitol bioproduction: State-of-the-art, industrial paradigm shift, and opportunities for integrated biorefineries. *Critical Reviews in Biotechnology, 39*(7), 924–43.

Fitria, A., Trisunaryanti, W., & Triyono, S. I. (2024). Synthesis, characterization and performance of Ni/mesoporous silica—NH_2/mesoporous silica and Ni-NH_2/mesoporous silica as bifunctional catalyst in one step conversion of waste palm oil to biodiesel. *Communications in Science and Technology, 9*(2), 430–441.

Gómez-Bombarelli, R., Aguilera-Iparraguirre, J., Hirzel, T. D., Duvenaud, D., Maclaurin, D., Blood-Forsythe, M. A., et al. (2016). Design of efficient molecular organic light-emitting diodes by a high-throughput virtual screening and experimental approach. *Nature Materials, 15*(10), 1120–1127.

Guddati, S., Kiran, A. S. K., Leavy, M., & Ramakrishna, S. (2019). Recent advancements in additive manufacturing technologies for porous material applications. *The International Journal of Advanced Manufacturing Technology, 105*(1–4), 193–215.

Hasan, M. H., Sagor, J. A., & Agarwala, I. (2022). A systematic trend analysis of 3D printing techniques used in specific soft robotic elements. *Materials Today: Proceedings, 50*, 1088–1099.

Hemmati, F., Bahrami, A., Esfanjani, A. F., Hosseini, H., McClements, D. J., & Williams, L. (2021). Electrospun antimicrobial materials: Advanced packaging materials for food applications. *Trends in Food Science & Technology, 111*, 520–533.

Hoang, A. T., Varbanov, P. S., Nižetić, S., Sirohi, R., Pandey, A., Luque, R., et al. (2022). Perspective review on municipal solid waste-to-energy route: Characteristics, management strategy, and role in circular economy. *Journal of Cleaner Production, 359*, 131897.

Hong, M., & Chen, E. Y. X. (2017). Chemically recyclable polymers: A circular economy approach to sustainability. *Green Chemistry, 19*(16), 3692–3706.

Hoque Apu, E., Nafiujjaman, M., Sandeep, S., Makela, A. V., Khaleghi, A., Vainio, S., et al. (2022). Biomedical applications of multifunctional magnetoelectric nanoparticles. *Materials Chemistry Frontiers, 6*(11), 1368–1390.

Hoye, R. L. Z., Schulz, P., Schelhas, L. T., Holder, A. M., Stone, K. H., Perkins, J. D., et al. (2017). Perovskite-inspired photovoltaic materials: Toward best practices in materials characterization and calculations. *Chemistry of Materials, 29*(5), 1964–1988.

Huang, C., Peng, Y., Lin, E., Ni, Z., Lin, X., Zhan, H., et al. (2022). Adaptable singlet-filtered nuclear magnetic resonance spectroscopy for chemical and biological applications. *Analytical Chemistry, 94*(10), 4201–4208.

Hussein, H. A., & Abdullah, M. A. (2022). Novel drug delivery systems based on silver nanoparticles, hyaluronic acid, lipid nanoparticles and liposomes for cancer treatment. *Applied Nanoscience, 12*(11), 3071–3096.

Johnston, B., Jiang, G., Hill, D., Adamus, G., Kwiecień, I., Zięba, M., et al. (2017). The molecular level characterization of biodegradable polymers originated from polyethylene using non-oxygenated polyethylene wax as a carbon source for polyhydroxyalkanoate production. *Bioengineering, 4*(3), 73.

Joshi, J., Homburg, S. V., & Ehrmann, A. (2022). Atomic Force Microscopy (AFM) on biopolymers and hydrogels for biotechnological applications—Possibilities and limits. *Polymers (Basel), 14*(6), 1267.

Kan, X., Wu, C., Wen, L., & Jiang, L. (2022). Biomimetic nanochannels: From fabrication principles to theoretical insights. *Small Methods, 6*(4).

Karl, A., Osvet, A., Li, N., & Brabec, C. J. (2020). A general guideline for vertically resolved imaging of manufacturing defects in organic tandem solar cells. *Advanced Materials Interfaces, 7*(13).

Kaspersetz, L., Waldburger, S., Schermeyer, M. T., Riedel, S. L., Groß, S., Neubauer, P., et al. (2022). Automated bioprocess feedback operation in a high-throughput facility via the integration of a Mobile Robotic Lab Assistant. *Frontiers in Chemical Engineering, 4*.

Katamesh, N. S., Abbas, A. E. F., Halim, M. K., Abdel-Lateef, M. A., & Mahmoud, S. A. (2024). Green micellar UPLC and complementary eco-friendly spectroscopic techniques for simultaneous analysis of anti-COVID drugs: A comprehensive evaluation of greenness, blueness, and whiteness. *BMC Chemistry, 18*(1), 149.

Kumar, P. S., Pavithra, K. G., & Naushad, Mu. (2019). Characterization techniques for nanomaterials. In *Nanomaterials for solar cell applications* (pp. 97–124). Elsevier.

Kute, V. G., & Singh, M. (2023). Development and assessment of immediate release nitrofurantoin tablet and sustained release trimethoprim tablet: Formulation, characterization and comparative evaluation. *International Journal of Membrane Science and Technology, 10*(5), 867–877.

Leceta, I., Alcalde, C., Urdanpilleta, M., Guerrero, P., Caba, K., & de la Burusco, A. (2020). *L-fuzzy concept analysis as a decision-making tool in the development of bio-based materials.*

Li, C., Gao, M., Ding, C., Zhang, X., Zhang, L., Chen, Q., et al. (2009). *In situ* comprehensive characterization of optoelectronic nanomaterials for device purposes. *Nanotechnology, 20*(17), 175703.

Lin, Z., Gan, Y., & Li, L. (2022). Preparation and characterization of smart (solid–gel) shape-stable phase change materials with photoluminescent and thermochromic properties. *Journal of Thermal Analysis and Calorimetry, 147*(13), 7171–7181.

Liu, Y., Li, X., & Liang, A. (2022). Current research progress of local drug delivery systems based on biodegradable polymers in treating chronic osteomyelitis. *Frontiers in Bioengineering and Biotechnology, 10*.

Logan, N., Cao, C., Freitag, S., Haughey, S. A., Krska, R., & Elliott, C. T. (2024). Advancing mycotoxin detection in food and feed: novel insights from surface-enhanced Raman Spectroscopy (SERS). *Advanced Materials, 36*(15).

Manna, S., Roy, D., Das, S., & Pathak, B. (2022). Capacity prediction of K-ion batteries: A machine learning based approach for high throughput screening of electrode materials. *Materials Advances, 3*(21), 7833–7845.

Micard, Q., Pellegrino, A. L., Lo Nigro, R., Bartasyte, A., Condorelli, G. G., & Malandrino, G. (2020). Piezoelectric Ba and Ti co-doped $BiFeO_3$ textured films: Selective growth of solid solutions or nanocomposites. *Journal of Materials Chemistry C, 8*(45), 16168–16179.

Monteiro da Silva, G., Cui, J. Y., Dalgarno, D. C., & Lisi, G. P. (2024). Rubenstein BM. High-throughput prediction of protein conformational distributions with subsampled AlphaFold2. *Nature Communications, 15*(1), 2464.

Musso, F., Volpe, M. A., Faraoni, M. B., & Yañez, M. J. (2020). Investigation on $Pd/SeO_2/SiO_2$ for α-Pinene Oxidation. Characterization by Transmission Electron Microscopy (TEM). *Microscopy and Microanalysis, 26*(S1), 83–4.

. Faiyaz Ahmed, A., Latif, K., Mudassir, S., & Salim Al-Amri, K. R. (2024). Fabrication of green composite materials using the weight sum method (WSM) method. *International Journal of Innovative Science and Research Technology (IJISRT)* 200–207.

Nadeem, T., Rafique, T., Khan, W., Khandakar, S., Alkhayyat, A., Mahesh Lohith, K. S., et al. (2024). Nanotechnology meets quantum computing: advancing quantum bits (qubits) and devices for next generation computing. *Nanotechnol Percept*, 248–68.

Nazarov, D., Zemtsova, E., Solokhin, A., Valiev, R., & Smirnov, V. (2017). Modification of the surface topography and composition of ultrafine and coarse grained titanium by chemical etching. *Nanomaterials, 7*(1), 15.

Nughays, R. O., Yang, C., Nematulloev, S., Yin, J., Harrison G. T., Zhao, J., et al. (2023). Visualization of surface charge carrier diffusion lengths in different perovskite crystal orientations using 4D electron imaging. *Advanced Optical Materials, 11*(10).

Ohahiduzzaman, M., et al. (2024). IoT based live streaming and object detecting AI Robotic Car. *Journal of Networking and Communication Systems (JNACS), 7*(3), 20–34.

Radnik, J., Kersting, R., Hagenhoff, B., Bennet, F., Ciornii, D., Nymark, P., et al. (2021). Reliable surface analysis data of nanomaterials in support of risk assessment based on minimum information requirements. *Nanomaterials, 11*(3), 639.

Retracted. (2023). The impact of the development of modern biotechnology and nanotechnology on the swimming sports industry. *Advances in Materials Science and Engineering* 1–1.

Rodrigues, J. F., Florea, L., de Oliveira, M. C. F., Diamond, D., & Oliveira, O. N. (2021). Big data and machine learning for materials science. *Discov Materials, 1*(1), 12.

Rzhevskii, A. (2019). The recent advances in Raman microscopy and imaging techniques for biosensors. *Biosensors (Basel), 9*(1), 25.

Samrot, A. V., Shobana, N., Suresh Kumar, S., & Narendrakumar, G. (2019). Production, optimization and characterisation of chitosanase of Bacillus sp. and its applications in nanotechnology. *Journal of Cluster Science, 30*(3), 607–20.

Sarma, S., & Rao, V. R. (2024). Emerging synthesis and characterization techniques for hybrid polymer nanocomposites. *Nanotechnology, 35*(1), 012002.

Shockley, K. R. (2015). Quantitative high-throughput screening data analysis: Challenges and recent advances. *Drug Discovery Today, 20*(3), 296–300.

Singhvi, M., & Gokhale, D. (2013). Biomass to biodegradable polymer (PLA). *RSC Advances, 3*(33), 13558.

Suárez-Eiroa, B., Fernández, E., Méndez-Martínez, G., & Soto-Oñate, D. (2019). Operational principles of circular economy for sustainable development: Linking theory and practice. *Journal of Cleaner Production, 214*, 952–961.

Titirici, M., Baird, S. G., Sparks, T. D., Yang, S. M., Brandt-Talbot, A., Hosseinaei, O., et al. (2022). The sustainable materials roadmap. *Journal of Physics: Materials., 5*(3), 032001.

Tripathi, D., Sharma, S., Mishra, G., Gupta, T., Prabhu, B. H. M., & Sahoo, J. (2024). Unleashing the power of green chemistry in phytoconstituent extraction by hydrotropic solubilization. *Current Materials Science, 17*.

Tüysüz, H., Lehmann, C. W., Bongard, H., Tesche, B., Schmidt, R., & Schüth, F. (2008). Direct imaging of surface topology and pore system of ordered mesoporous silica (MCM-41, SBA-15, and KIT-6) and nanocast metal oxides by high resolution scanning electron microscopy. *Journal of the American Chemical Society, 130*(34), 11510–11517.

Usman, M., Farooq, M., Wakeel, A., Nawaz, A., Cheema, S. A., Rehman, H., et al. (2020). Nanotechnology in agriculture: Current status, challenges and future opportunities. *Science of The Total Environment, 721*, 137778.

Vincent, J., Mohan, S., Fernandez-Granda, C., & Crozier, P. (2020). Denoising large in situ TEM image datasets: A convolutional neural network-based approach. *Microscopy and Microanalysis, 26*(S2), 1720–1721.

Wang, B. J., Wu, W. L., Wei, X. L., & Chen, Q. (2025). Mechanical and electromechanical properties of 2D materials studied *via in situ* microscopy techniques. *Nanoscale, 17*(4), 1722–1763.

Xiao, S., Bordas, S. P. A., & Kim, T. Y. (2023). Editorial: deep learning in computational materials science. *Front Mater, 10*.

Yang, K., Shen, Y., He, K., Zhang, T., Xu, R., Zhao, S., et al. (2021). An optical fiber temperature sensor based on fluorescence intensity ratio used for real-time monitoring of chemical reactions. *Ceramics International, 47*(23), 33537–33543.

Yenilmez, B., Knowlton, S., & Tasoglu, S. (2016). Self-contained handheld magnetic platform for point of care cytometry in biological samples. *Advanced Materials Technologies, 1*(9).

Yesuraj, J., Padmaraj, O., & Suthanthiraraj, S. A. (2020). Synthesis, characterization, and improvement of supercapacitor properties of $NiMoO_4$ nanocrystals with polyaniline. *Journal of Inorganic and Organometallic Polymers and Materials, 30*(2), 310–321.

Zdora, M. C., Zanette, I., Walker, T., Phillips, N. W., Smith, R., Deyhle, H., et al. (2020). X-ray phase imaging with the unified modulated pattern analysis of near-field speckles at a laboratory source. *Applied Optics, 59*(8), 2270.

Zoumaki, M., Tzetzis, D., & Mansour, G. (2019). Development and characterization of starch-based nanocomposite materials. *IOP Conference Series: Materials Science and Engineering, 564*(1), 012037.

Chapter 2
Biofuel Production from Agro-waste Materials

Chandrabhan Prajapati, Supriya, Sachchida Nand Rai, and Santosh Kumar Singh

Abstract The pursuit of sustainable energy solutions has led to a surge in research on biofuels derived from agricultural waste. This chapter investigates the feasibility of converting agricultural residues into biofuels, highlighting the potential of wheat straw, rice husks, and sugarcane bagasse. The biofuel industry has shifted towards non-food biomass sources to address food security concerns. Advanced technologies, including hydrothermal carbonization and enzymatic hydrolysis, have enhanced biofuel production. Microbial fermentation, facilitated by genetically engineered microorganisms, has improved bioethanol yields. However, challenges persist, including high production costs, feedstock availability, and process inefficiencies. To overcome these hurdles, innovative strategies, such as improving enzymatic efficiency and integrating hybrid conversion technologies, are essential. This study emphasizes the importance of integrating agricultural waste into biofuel production, promoting a circular economy, enhancing energy security, and mitigating environmental impact. By addressing these challenges, we can unlock the full potential of biofuels and create a more sustainable energy future.

Keywords Agro-waste · Biofuel · Bioethanol · Biogas · Pyrolysis · Sustainability

C. Prajapati · Supriya · S. K. Singh (✉)
Centre of Experimental Medicine and Surgery, Institute of Medical Sciences, Banaras Hindu University, Varanasi, India
e-mail: santosh.singh01@bhu.ac.in

S. N. Rai
University Centre for Research and Development (UCRD), Chandigarh University, Mohali 140413, Punjab, India

V. Singh et al. (eds.), *Bioenergy and Value Added Byproducts from Biomass Waste*, Sustainable Environmental Waste Management Strategies,
https://doi.org/10.1007/978-3-032-22849-9_2

2.1 Introduction

The growing importance of biofuels can be attributed to their emergence as a renewable and environmentally friendly substitute for fossil fuels. Biofuels are produced from diverse organic sources, including residual plant matter, algae, and microorganisms. A key advantage of biofuels is their potential to minimize greenhouse gas emissions, thereby contributing to global efforts to combat climate change (Rodionova et al., 2017). Biofuels can be broadly classified into distinct generations, with the first generation being primarily sourced from conventional food crops such as corn as well as sugarcane (Rodionova et al., 2017) (Table 2.1). The second generation of biofuels represents a significant advancement, as it leverages non-food biomass sources, including lignocellulosic materials along with agricultural by-products such as wheat straw, rice husks, as well as sugarcane bagasse (Panpatte & Jhala, 2019). The third generation of biofuels marks a further innovation, with production cantered on microscopic organisms, notably cyanobacteria and various strains of algae, which offer a promising source of renewable energy (Rodionova et al., 2017). The first generation of biofuels has encountered criticism due to the contentious issue of competing with the global food supply, prompting researchers to shift their focus towards second and third-generation biofuels. Specifically, there is a growing interest in exploring the potential of biofuels derived from agricultural waste, which is perceived as a more feasible and environmentally sustainable solution (Panpatte & Jhala, 2019) (Fig. 2.1). Over their five generations, biofuels can improve efficiency along with sustainability. Fifth-generation combines engineered microbes via carbon capture to increase yields. By improving biomass conversion, genetic engineering raises the production of biofuel in addition to reduces costs. Although cellulosic ethanol is still expensive, developments in microbial and plant manipulation can guarantee a sustainable and affordable substitute for fossil fuels (Mushtaq et al., 2023). Various agricultural waste materials, such as potato peel waste, fruit peels, banana residues, and municipal solid waste, have shown remarkable potential for producing substantial amounts of biofuel. This is achieved through advanced processing techniques, including enzymatic hydrolysis, anaerobic digestion, in combination with hydrothermal processing, which efficiently convert these waste materials into valuable biofuels (Srivastava et al., 2021). A notable example of the bioenergy potential of potato peel waste (PPW) represent a study that successfully converted PPW into bioethanol, achieving an impressive conversion efficiency of 71.37%. This process yielded a substantial amount of energy, equivalent to 7.15×10^3 megajoules (MJ) per ton of dry PPW (Barampouti et al., 2023). In another significant development, activated carbons derived from bio-waste have proven to be effective catalysts in the production of biofuels. By utilizing these bio-waste-derived activated carbons, researchers have been able to improve conversion rates and optimize the quality of the resulting fuels (Ledesma & Beltramone, 2021). The world's escalating energy requirements, forecasted to surge by more than 50% by 2050, underscore the imperative of transitioning towards sustainable energy alternatives, including biofuels. This shift is critical to mitigating our increasing reliance on finite fossil fuels and ensuring a more environmentally conscious energy future (Panpatte & Jhala, 2019). The pervasive dependence on fuels derived from petroleum has resulted in a trifecta of pressing issues: escalating carbon emissions, dwindling resources, and also heightened concerns regarding energy security. In response, researchers have been driven to explore and develop innovative, alternative energy solutions that can mitigate these challenges and provide a more sustainable future (Panpatte & Jhala, 2019). India has made notable strides in incorporating biofuels into its energy mix, driven by the National Policy on Biofuels of 2018, which aims for 20% ethanol blending with petrol by 2025 (Panpatte & Jhala, 2019). Government initiatives such as second-generation bioethanol facilities and the "RUCO" program, which promotes biodiesel production from used cooking oil, are helping shape a sustainable biofuel ecosystem. Additionally, biogas plants in rural areas enhance energy security and promote sustainable development (Kothari et al., 2020). Globally, countries like the United States, Brazil, the European Union, and China are leading biofuel production. The U.S. focuses on corn-based ethanol, Brazil on sugarcane ethanol, the EU on rapeseed biodiesel, and China on bioethanol and biogas from crop residues (Ji, 2015). Ongoing advancements in synthetic biology, metabolic engineering, and microbial optimization are improving biofuel production efficiency, making them increasingly attractive for industrial-scale adoption. By using agricultural waste as feedstock, biofuels not only reduce waste but also lessen reliance on polluting fossil fuels, contributing to a circular economy and sustainability (Stephen & Periyasamy, 2018)s. A comprehensive case study conducted in China revealed that harnessing 930.8 million tons of crop residues for the production of bioethanol and bio-oil could have a profound impact on the country's energy landscape. Specifically, the study estimated that this approach could effectively displace 26.9 million tons of gasoline and 58.2 million tons of diesel fuel on an annual basis (Ji, 2015). A notable example from Mexico illustrates the economic viability of repurposing food industry waste as a bioethanol feedstock. According to projections, the annual conversion of 3.27 million tons of such waste could result in the production of approximately 72.7 million liters of bioethanol, generating an estimated annual revenue of $43.7 million USD (Murillo-Alvarado & Flores Russell, 2022). Hydrothermal processing of biomass waste, particularly from high-moisture content sources like oil palm fronds and municipal solid waste, has proven effective in upgrading fuel quality. This process increases the calorific value of the fuel while reducing nitrogen and chlorine levels, resulting in a more efficient and refined energy product (Yuliansyah et al., 2010). Compared to fossil fuels, biofuels offer a more sustainable and environmentally friendly energy solution. They emit fewer greenhouse gases, can be replenished rapidly, and support sustainability, making them a promising alternative for climate change mitigation (Huzir et al., 2018). Research suggests that biofuels derived from agricultural waste can reduce greenhouse gas emissions by 20–70%, highlighting their potential to combat climate change (Srivastava et al., 2021). Biofuels, unlike fossil fuels, are cleaner, with less carbon dioxide, nitrogen oxides, and sulfur dioxide released into the atmosphere, thus improving air quality (Swain, 2017). Economic analyses support the commercialization of biofuels, particularly when underutilized waste materials are used as feedstocks. This repurposing of waste reduces production costs, making biofuels more economically viable (Stephen & Periyasamy, 2018). However, the widespread adoption of biofuels is impeded by challenges such as high processing costs, technological limitations, and feedstock availability. To address these, researchers are focusing on advanced technologies like genetic engineering, microbial fermentation, enzymatic hydrolysis, and catalytic processing to optimize biofuel production and enhance cost-efficiency (Ambaye et al., 2021). Recent innovations, including the integration of solid-state fermentation (SSF) and hydrothermal treatment, have shown promise in boosting biofuel yields, offering a path toward more sustainable energy solutions (Sadh et al., 2018). Biofuels derived from food waste, crop residues, and municipal waste offer both environmental and economic benefits by reducing waste disposal issues and generating economic value (Zhao et al., 2014). Technological innovations in hydrothermal processing, enzymatic hydrolysis, and catalytic conversion have led to higher biofuel yields, particularly in biodiesel production, with catalytic processes achieving near 100% conversion efficiencies (Basumatary et al., 2018). These advancements reinforce the potential of biofuels to contribute to global energy security and climate change mitigation, supporting international climate action efforts and promoting a circular economy (Procentese et al., 2019). Cutting-edge developments including genetic manipulation, fermentation by microbes, enzymatic hydrolysis, along with catalytic processes represent the main focus of researchers' ongoing investigation into innovative ways to get overcome aforementioned obstacles. As we address further throughout this chapter, several new technologies have enormous potential to optimize the generation of biofuel, opening the door towards more effective, economical, along with sustainability large-scale application.

2.2 Categories of Agro-waste Employed for Biofuel Production

Material from agriculture-related waste, which are sometimes seen as leftovers as well as by-products regarding farming operations, have a lot of promise towards sustainable uses (Raisa et al., 2024). Wastage from agriculture resources, which could be categorized according to their origin, distinct content, possible uses, along with degradability (Kuthiala et al., 2022). Crop leftovers, process leftovers, compost from animals, as well as the aquaculture industry waste are groups of agro-waste products according to the source they come from (Gupta et al., 2023). The residual components of cultivated crops, including stubble, straw, leaves, husks, bagasse, seed pods, cobs, ears, peels shells as well as stalks, are referred to simply as residues from the crop (Awogbemi & Von Kallon, 2023). A few examples of these are straw from rice, wheat, maize cobs, as well as shells from coconuts and others (Arvanitoyannis & Tserkezou, 2008). The manufacturing sector of food crops produces waste products including processed food material leftover, oily cakes, as well as bagasse made from

Table 2.1 Various generational sources of biofuels with their advantages and challenges

Biofuel generation	Primary sources	Key strengths	Significant hurdles	References
First	Traditional food crops (e.g., corn, sugarcane, vegetable oils)	Benefits from existing infrastructure and widespread availability	Raises concerns about food security and land allocation	de Souza et al. (2014), Kauffman et al. (2011) and Vásquez et al. (2017)
Second	Agricultural residues (such as straw, husks, and sugarcane bagasse	Offers a sustainable solution by utilizing waste biomass, reducing environmental impact	Poses technical challenges due to intricate processing and pretreatment requirements	Nunta et al. (2023), Waheed and Williams (2013) and Yadav et al. (2024)
Third	Microscopic algae and cyanobacteria	Boasts impressive yields and thrives on marginal land, reducing competition for food crops	Faces significant economic and logistical hurdles, including steep production costs and scalability concerns	de Farias Silva and Bertucco (2016) and Parmar et al. (2011)
Fourth	Bioengineered microorganisms and crops	Unlocks improved efficiency and harnesses CO_2, offering a promising pathway	Raises important questions about ethics, safety, and compliance, necessitating careful regulatory oversight	Aro (2016) and Lü et al. (2011)

sugarcane, which are known as processed residues (Nair et al., 2022; Singhania et al., 2008). The rich in nutrients waste product from cattle, chickens, as well as pigs can sometimes be used like alternate form of natural fertilizer (Upadhyay et al., 2024). Similar to this, trash from aquaculture—which comprises crustacean shells as well as the processing of fish waste—can be recycled either feed, composting purposes or pharmacological extraction (Khiari, 2022; Rathore & Pandey, 2020). Depending on their composition, agro-waste substances can be divided into subcategories such as fatty trash, protein-rich waste products, starch-rich as well as lignocellulosic wastage (Álvarez-Castillo et al., 2021; Aruwajoye et al., 2020; Mushtaq et al., 2024). The straw from wheat as well as sawdust are instances of lignocellulosic residue, which is comprised of cellulose, hemicellulose, together with lignin while is capable of being used either biofuels or recyclable packaging (Taherzadeh & Karimi, 2008). Nitrogen-rich protein-containing wastage, like oilseed cakes along with animal dung, could be utilized both animal feed as well as enriching the soil (Chhetri et al., 2020; Sarkar et al., 2021). Peels as well as processing leftovers are examples of starch-rich trash that comes primarily from crops including maize and potatoes in addition to used in fermenting procedures to produce bioethanol (Hamdi et al., 2024). The considerable level of oil of fatty waste products, which originates from the processing of oilseeds as

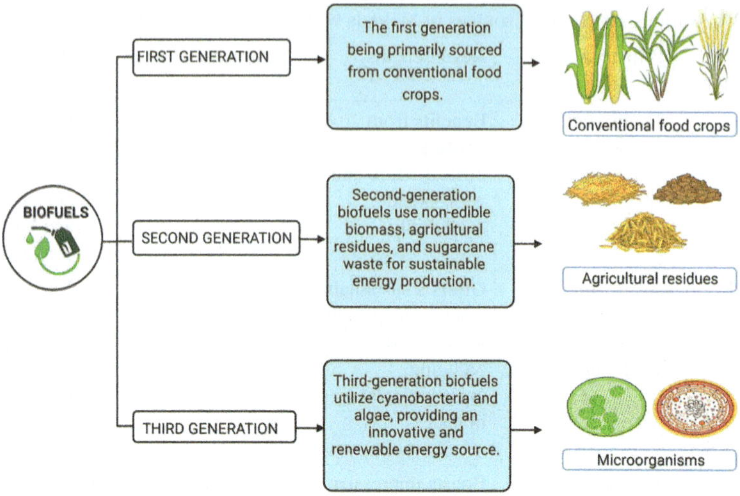

Fig. 2.1 Biofuel categorization using generational advancements. The diagram differentiates biofuels under three generations depending on feedstock sources along with sustainability. In order to produce energy, first-generation biofuels use sugars, starches, as well as vegetable oils coming from traditional food crops like maize and sugarcane. Their dependence on food crops, however, poses questions over competitiveness with land use along with the food supply. By utilizing non-edible biomass, such as lignocellulosic waste as well as agricultural wastes, second-generation biofuels solve these problems while improving sustainability and lowering competition against food production. Third-generation biofuels can be generated from microorganisms such as algae and cyanobacteria, while they have a high yield, without the need for agricultural land, and represent an environmentally friendly renewable energy source. This grouping emphasizes the significance of ecologically friendly energy solutions by highlighting the transition compared to food-based biofuels with greater sustainable alternatives

well as animal byproducts, makes it useful for the creation of bio-lubricants together with biodiesel (Hájek et al., 2021; Perera et al., 2022). Regarding their prospects for energy generation, making compost, commercial uses, along with feed for livestock, agro-waste products also have a lot of promise. Rice husks as well as sugarcane bagasse constitute two examples of agro-wastes that are capable of being directly burned or transformed into natural gas as well as bioethanol, making them great sources regarding bioenergy (Chieng & Kuan, 2022; Konde et al., 2021). Composting organic wastage, such as agricultural residues as well as waste from animals, can increase soil fertility while aiding agricultural sustainability (Noor et al., 2024). To create bio-based goods like paper, charcoal, as well as bioplastics, the manufacturing industry massively uses agro-wastage (Naitam et al., 2022). The usage of banana fiber regarding textiles continues to increase, for instance, whereas cornstarch offers an essential ingredient manufacturing biodegradable polymers (Ortega et al., 2016; Tabasum et al., 2019). Furthermore, processing certain agro-wastages, such as oilseed cakes as well as grain stover, producing nutrient-rich feeding materials can lower feed costs as well as promote livestock production (Popp et al., 2016). Furthermore, materials from agricultural trash could be divided into three sub-categories according to how easily, moderately, as well as difficultly they decompose. For instances, vegetable leftovers as well as fruit peels are examples of trash that disintegrate quickly because of their simplistic organic constituents along with their substantial level of moisture (Râpă et al., 2024; Sharma et al., 2022). The breakdown of moderately decomposing materials, such as lignin-based waste (rice husks as well as corn stover), necessitates a certain degree of processing as well (Banerjee et al., 2009; Wojcieszak et al., 2022). Unless undergo specific processes including pyrolysis as well as microbial breakdown, hard-to-degrade agricultural wastes—such coconut along with nut shells—decompose slowly due to their elevated content of lignin together with additional resistant components (Gupta et al., 2019; Singh et al., 2024). Comprehending these categories facilitates the development of efficient waste management plans as well as the investigation of creative ways to reduce the adverse effects on the natural world.

2.3 Processes Along with Technologies for Biofuel Production

Various conversion procedures, primarily categorized as biochemical as well as thermochemical methods, have been employed for producing biofuel. Each approach for creating sustainable energy involves distinctive functioning, benefits, along with limitations (Shahzad et al., 2024). These methods produce biofuels involving syngas, biogas, additionally bioethanol, along with bio-oil by using a variety of raw materials, especially organic matter, agricultural wastes, together with crops cultivated specifically for energy (Durán-Valle & López-Coca, 2024; Mohan et al., 2006). In order to convert biomass, produce biofuels, biochemical transformation depends on microbes as well as enzymes (Pattanaik et al., 2019). One popular method for producing biogas called anaerobic digestion, which breaks down agricultural materials without the presence of oxygen (Achinas et al., 2019; Bardiya et al., 1996). Hydrolysis is one of the many discrete phases of AD, in which complex compounds containing carbohydrates or/and proteins decompose leaving simplest sugars alongside amino acids, correspondingly (Odedina et al., 2017). Acidogenesis, another process applied to agro-waste processing, transforms organic matter producing volatile fatty acids that act as intermediates facilitating the synthesis of biogas, accelerating waste valorisation as well (Sun et al., 2021). Furthermore, methanogenesis, which creates biogas with significant methane through the action of methanogenic microbes (Rogeri et al., 2024). Another important biochemical approach for the processing of agro-waste called fermentation, in which microorganism such as *Saccharomyces cerevisiae* turn residual sugars produce bioethanol, therefore facilitating the creation of sustainable energy (Mondal et al., 2024). However, in order for breaking down its hard structure, lignocellulosic agricultural residue needs to be pretreated prior to fermentation and facilitate the efficient transformation of sugar toward bioethanol (Samantaray et al., 2024). Bioethanol output is increased by processes like steam explosion as well as acid pretreatment, which increase the accessibility of cellulosic waste for digestion regarding enzymes (de Brito et al., 2024). On the other hand, a technique known as thermochemical conversion uses elevated temperatures to turn agricultural waste produce biofuels and other useful byproducts. Crop leftovers, husks, straws, along with wood waste are examples of lignocellulosic biomass that can be broken into more accessible, energy-rich molecules using this process (Rahimi et al., 2022). Pyrolysis represents one of the most thoroughly explored thermochemical conversion processes. The process of pyrolysis entails heating biomass among 300–700 °C with little to no oxygen present. Three main products are produced from this consequence is that syngas, a form of gaseous fuel; biochar, a solid substance rich in carbon; along with bio-oil, a liquid form of fuel (Kan et al., 2016; Salehi et al., 2011). The kind of pyrolysis that is employed determines the yield along with composition of the resulting products. The generation of bio-oil can be maximized by fast pyrolysis, and this may be used for liquid fuel systems since it functions at about 500 °C and has a brief residence period (Ali et al., 2016). In contrast, slow pyrolysis produces greater biochar, which may be utilized as an additive to the soil or as carbon sequestration because it runs at temperatures that are lower over an extended period of time (Fryda & Visser, 2015). Gasification represents another crucial thermochemical process that partly oxidizes agro-residue between temperatures 800–1200 °C resulting in syngas, a blend of carbon dioxide (CO_2), hydrogen (H_2), Nitrogen (N_2) and methane (CH_4), as well as carbon monoxide (CO) (Kumar & Shukla, 2016). Fischer-Tropsch formation serves as a technique that could be employed to further transform syngas producing liquid fuels as well as utilize them directly to generate energy (Alsudani et al., 2023). A variety of gasification reactor patterns, including fluidized-bed and even fixed-bed gasifiers, have been utilized to maximize fuel production along with energy efficiency (da Silva et al., 2024; Manu et al., 2023). This process is very helpful for creating highly adaptable and sustainable energy sources since syngas could potentially be transformed to artificial fuels including methanol as well as biodiesel or optimized to applications such as internal combustion engines as well as gas turbines (Kim et al., 2011; Paykani et al., 2022; Rauch et al., 2016). Regarding thermochemical transformation, combustion remains an immensely easy and simple process. It involves burning biomass completely under abundance of oxygen with temperatures mostly higher than 1000 °C. The process of combustion produces thermal energy, that can often be used by manufacturing heating systems along with power plants. While combustion works well for producing energy on huge amounts, it is not very effective like gasification as well as pyrolysis for producing biofuel. It also produces emissions of carbon dioxide (CO_2) along with additional impurities including nitrogen oxides (NO_x) together with particulate matter, that necessitate the use of emission control technology in order to operate cleaner (Akhtar et al., 2018; Atimtay, 2010; Díez et al., 2024). Hydrothermal carbonization (HTC) acts as a relatively recent and effective thermochemical method that uses higher-pressure, moisturized circumstances to function at low temperatures between 180–250 °C (Shen, 2020). This method speeds up the normal coal manufacturing process by a few hours. HTC is a good alternative to fossil coal since it creates hydrochar, a form of fuel that resembles coal and has an excellent calorific value (Fakudze & Chen, 2023). Furthermore, because HTC does not require the expensive drying procedures needed for remaining other thermochemical procedures, it is especially helpful to handle wet biomass, for instance wasted food material, algal biomass, along with sewage sludge (Czerwińska et al., 2022). Thermochemical conversion provides numerous advantages, although it also has drawbacks, such as increased initial expenses, complicated biological oils refining procedures, emission control specifications, along with feedstock quality deviations (Al-Qadri et al., 2025; Paudel et al., 2024). Work is being done for better catalysts, as well reactor that builds, and integration via technology for carbon capture in order to render these processes more efficient and sustainable, as well as the creation of hybrid approaches that combine thermochemical as well biochemical methods could further increase bio fuel yield along with process efficiency (Nimmas et al., 2024; Poornima et al., 2024; Xie et al., 2024). Overall comparing sustainability along with efficiency, thermochemical processes provide a wider range of feedstocks together with energy products, whereas biochemical methods are more efficient and ecologically friendly for managing waste. Environmental factors, energy demand, along with feedstock accessibility all influence one's decision for biofuel production.

2.4 The Role of Microbes in the Production of Biofuels

Microorganisms are the unsung heroes of biofuel production, harnessing their metabolic and biochemical prowess to transform organic matter into energy-rich fuels. In this context, heterotrophic microorganisms like yeasts and bacteria have emerged as key players in the production of bioethanol and biogas. Meanwhile, photosynthetic microorganisms, comprising microalgae and cyanobacteria, are leveraging their ability to fix inorganic carbon, thereby generating precursors that can be converted into biofuels. By elucidating the intricate mechanisms underlying these microbial processes, researchers can unlock novel strategies for optimizing biofuel production, ultimately paving the way for a more sustainable energy landscape (Majidian et al., 2018). Notwithstanding substantial breakthroughs in the field, the commercialization of microbial biofuel production processes remains an elusive goal, with only a handful of exceptions. A major stumbling block in this pursuit is the need to fine-tune microbial metabolic pathways, thereby enhancing the microorganisms' ability to tolerate fuels, boost productivity, and reduce costs. This optimization challenge is a critical hurdle that must be overcome in order to render microbial biofuel production a viable and economically sustainable proposition (Singh et al., 2022a, 2022b). The dwindling reserves of fossil fuels, exacerbated by escalating costs and mounting environmental concerns stemming from greenhouse gas emissions, have galvanized research into microbial biofuels. Certain microorganisms, notably Escherichia coli, Clostridia species, Saccharomyces cerevisiae, and *Yarrowia lipolytica*, have emerged as promising candidates for biofuel production. This is attributable to their versatility in metabolizing a wide range of feedstocks, encompassing pentose sugars, lignocellulosic biomass, and even atmospheric carbon dioxide. By harnessing the metabolic capabilities of these microorganisms, researchers aim to develop innovative, sustainable solutions for biofuel synthesis (Singh et al., 2022a, 2022b). Although yeast-based systems continue to hold sway in the realm of bioethanol production, the emergence of genetically engineered bacterial strains has ushered in a new era of possibilities for augmenting fuel synthesis. By harnessing the power of genetic engineering, researchers have been able to create novel bacterial strains that exhibit enhanced capabilities for fuel production, thereby expanding the repertoire of available biofuel production strategies (Clomburg & Gonzalez, 2010).

2.4.1 Biofuel Synthesis: The Role of Key Microorganisms

A diverse array of microbial species has been recognized as pivotal contributors to biofuel synthesis, boasting distinct biochemical signatures characterized by specialized metabolic pathways and substrate utilization profiles.

2.4.1.1 Escherichia Coli

Escherichia coli (*E. coli*) has emerged as a preferred host organism in biofuel research, owing to its exceptional amenability to genetic modification and remarkable metabolic adaptability. By leveraging metabolic engineering techniques, *E. coli* can be optimized to efficiently convert a broad spectrum of feedstocks—encompassing simple and complex carbohydrates, as well as non-carbohydrate carbon sources—into an array of biofuels, including ethanol, butanol, and biodiesel, thereby underscoring its versatility and potential in sustainable biofuel production (Clomburg & Gonzalez, 2010). Furthermore, breakthroughs in synthetic biology have enabled researchers to fine-tune both fermentative and non-fermentative metabolic pathways, thereby unlocking the potential for producing advanced biofuels with higher carbon content. This optimization has been achieved by harnessing the biochemical capabilities of fatty acid and isoprenoid pathways, which has paved the way for the development of novel, high-carbon biofuels with enhanced energy density (Peralta-Yahya et al., 2012).

2.4.1.2 *Clostridia* Species

Microorganisms belonging to the *Clostridia* genus have emerged as exemplary producers of biohydrogen and biobutanol, primarily attributable to their anaerobic metabolic mode. By leveraging dark fermentation, a process that occurs in the absence of light, these microbes can efficiently synthesize fuels while concurrently mitigating the accumulation of organic waste residues, thereby rendering this process a promising strategy for sustainable biofuel production (Singh et al., 2022a, 2022b).

2.4.1.3 Yeast-Based Systems

Traditional yeast strains, prominently exemplified by *Saccharomyces cerevisiae*, have long been the cornerstone of bioethanol production. Nevertheless, unconventional yeast species, such as *Yarrowia lipolytica*, have garnered attention for their superior ethanol tolerance, thermal resilience, and genetic diversity. These attributes render them attractive alternatives for augmenting biofuel yields, thereby expanding the repertoire of yeast-based bioethanol production strategies (Singh et al., 2022a, 2022b).

2.4.1.4 Microalgae and Cyanobacteria

Photosynthetic microorganisms have emerged as viable contenders for the production of lipid-based biofuels, owing to their capacity to sequester atmospheric carbon dioxide and synthesize hydrocarbons that closely resemble fossil fuel-derived counterparts. This inherent ability positions them as promising sources for sustainable biofuel generation. Nevertheless, significant hurdles persist, notably the scalability and economic viability of these microorganisms, which must be surmounted to facilitate their widespread commercial adoption (Majidian et al., 2018).

2.4.2 Second-Generation Microbial Hosts

Researchers have been investigating the potential of genetically engineered microorganisms to convert lignocellulosic biomass into advanced biofuels, such as n-butanol, terpenoids, and higher lipids. However, a significant obstacle remains: the development of robust microbial catalysts that can efficiently produce these biofuels while withstanding the stresses inherent to the production process, including toxicity, fermentation inhibitors, and other detrimental factors that can impact yield, productivity, and overall process efficiency (Fischer et al., 2008).

2.5 Genetic Enhancements for Boosting Biofuel Yields: Recent Advances and Future Directions

2.5.1 CRISPR/Cas9-Based Genome Editing

The advent of CRISPR/Cas9 technology has transformative implications for microbial genome engineering, facilitating precise and targeted modifications that significantly enhance microbial robustness. Specifically, this technology enables the introduction of desirable traits such as elevated fuel tolerance, improved resistance to inhibitory compounds, and enhanced thermotolerance, thereby optimizing microbial performance for biofuel production (Javed et al., 2019). Furthermore, this methodology has substantially enhanced the catalytic prowess of pivotal enzymes, including cellulases and hemicellulases, which are indispensable for the deconstruction of lignocellulosic biomass, a critical precursor to biofuel production (Javed et al., 2019).

2.5.2 Metabolic Engineering for Fuel Optimization

Substantial breakthroughs in metabolic engineering have markedly elevated the efficiency of biofuel production. Through strategic redirection of carbon flux towards biofuel synthesis pathways, researchers have successfully amplified the yields of various biofuel types, including alcohol-based biofuels, fuels derived from fatty acids, and isoprenoid-based biofuels, thereby advancing the frontiers of sustainable biofuel production (Clomburg & Gonzalez, 2010). A notable illustration of this progress is the genetic engineering of *Zymomonas mobilis*, which has been successfully modified to ferment pentose sugars, a pivotal breakthrough in the production of bioethanol from lignocellulosic biomass, particularly agricultural waste, thereby expanding the potential feedstock options for sustainable biofuel production (Majidian et al., 2018).

2.5.3 Enhancing Microbial Fuel Tolerance and Productivity

A significant bottleneck in the microbial production of biofuels is the buildup of toxic compounds that impede cellular growth and metabolic function. To overcome this limitation, researchers have implemented genetic modifications that target specific cellular processes, including membrane transport mechanisms and stress response pathways, thereby enhancing microbial resilience and tolerance to biofuel-related stresses, ultimately leading to improved biofuel production efficiency (Peralta-Yahya et al., 2012).

2.5.4 Multi-feedstock Utilization

The capacity to harness a diverse array of carbon-rich substrates, encompassing sugars, lignocellulosic biomass, and syngas, is a critical determinant of economically viable biofuel production. Recent advances in microbial engineering have yielded strains with augmented substrate promiscuity, thereby enabling the production of biofuels in a more sustainable and economically favorable manner (Fischer et al., 2008).

2.6 Role of Synthetic Biology in Biofuel Optimization

2.6.1 *Engineering Synthetic Metabolic Pathways*

The advent of synthetic biology has paved the way for the design and construction of innovative biosynthetic pathways tailored to biofuel production. Through strategic manipulation of fatty acid and isoprenoid metabolic pathways, scientists have successfully engineered novel microbial strains with the capacity to produce alkanes and other hydrocarbon-based fuels, thereby expanding the repertoire of microbial biofuel production (d'Espaux et al., 2015).

2.6.2 *Synthetic Microbial Consortia for Enhanced Biofuel Production*

In a departure from traditional approaches that rely on single microorganisms, researchers are now designing synthetic microbial consortia, where multiple microbial species synergistically interact to optimize biofuel synthesis. This innovative strategy not only enhances the efficiency of substrate utilization but also mitigates the metabolic burden imposed on individual microbial strains, thereby fostering a more robust and efficient biofuel production process (Jang et al., 2012).

2.6.3 *Systems Metabolic Engineering for Process Optimization*

A multidisciplinary approach, coined systems metabolic engineering, has emerged as a transformative strategy for establishing large-scale biofuel production platforms. This integrated framework combines the principles of metabolic engineering, systems biology, and synthetic biology, and exploits the synergies between computational modeling, high-throughput screening, and rational design to create high-performing microbial strains. These optimized strains exhibit enhanced productivity and yield, thereby paving the way for the development of efficient and scalable biofuel production processes (Jang et al., 2012).

2.6.4 *Addressing Pathway Bottlenecks and Toxicity Issues*

A major hurdle in the microbial production of biofuels is the existence of metabolic constraints that impede efficient fuel synthesis. To address this challenge, researchers have leveraged the tools of synthetic biology to pinpoint and alleviate these bottlenecks, thereby achieving substantial improvements in biofuel yields while concurrently minimizing toxicity, ultimately leading to more efficient and sustainable biofuel production processes (Kung et al., 2012). The production of biofuels through microbial means offers a viable, eco-friendly solution for harnessing energy from agricultural waste. Although ethanol currently dominates the biofuel landscape, next-generation alternatives, including butanol and lipid-based biofuels, are gaining traction due to their enhanced fuel characteristics. Recent advancements in the intersection of metabolic engineering, synthetic biology, and CRISPR-mediated genome editing have substantially elevated microbial biofuel production yields. Nevertheless, to render microbial biofuels economically competitive with fossil fuels, further investigations are necessary to optimize production protocols, augment microbial resilience, and improve cost-effectiveness (Kung et al., 2012). As the biofuel industry transitions from laboratory-scale research to large-scale industrial applications, innovations driven by synthetic biology will play a crucial role in shaping the future of renewable energy, ultimately paving the way for a more sustainable energy paradigm.

2.7 Factors Affecting Biofuel Yield

Various key factors affecting biofuel yield from agro-waste, including substrate composition, pretreatment, microbial efficiency, process optimization, scalability challenges, cost effectiveness, and recent advancements for industrial-scale production are discussed further in detail.

2.7.1 Substrate Composition and Pretreatment Methods

The effectiveness of biofuel production hinges largely on two critical factors: the inherent characteristics of the biomass feedstock and the specific pretreatment techniques utilized. Biomass sources such as lignocellulosic and microalgal materials, as well as residual agricultural matter, necessitate a meticulous preprocessing step to dismantle intricate polymer structures into readily fermentable sugars (Shanmugam et al., 2021). Pretreatment techniques can be broadly categorized into three distinct approaches: physical, chemical, and biological methods. Notably, research has demonstrated that chemical pretreatment utilizing alkaline solutions, such as sodium hydroxide, substantially augments methane production, yielding remarkable improvements of up to 338 milliliters per gram of volatile solids during the digestion of rice straw (Mirmohamadsadeghi et al., 2021). In addition, thermochemical pretreatment has emerged as a particularly potent strategy for microalgal biomass, significantly bolstering methane production efficiency by effectively dismantling the robust cell walls that typically hinder digestion (de Oliveira et al., 2022). An eco-friendlier approach to pretreatment involves harnessing the power of fungi and enzymatic hydrolysis, offering a sustainable alternative that aligns with environmental concerns. However, the practical implementation of this method is hindered by the notably longer processing times required, which can be a significant drawback (Zabed et al., 2019). By integrating biological and chemical pretreatment approaches, researchers have found that the breakdown of fibrous materials and the release of sugars can be significantly enhanced, ultimately leading to optimized conversion efficiency and a more effective utilization of biomass resources (Meenakshisundaram et al., 2021). Furthermore, the diversity of microbial populations is a vital factor in the production of biogas and biofuels, as alterations to the substrate can significantly impact the composition and structure of microbial communities, ultimately influencing the overall yield of these renewable energy sources (Mirmohamadsadeghi et al., 2021). Beyond modifying substrate properties, pretreatment also has a profound impact on the metabolic pathways employed by microorganisms, a phenomenon that has been extensively observed in research focused on hydrogenotrophic methane production, where subtle changes in microbial metabolism can significantly influence the overall efficiency of the process (Mirmohamadsadeghi et al., 2021).

2.7.2 *Process Conditions (pH, Temperature, Retention Time)*

To achieve optimal biofuel production, it is essential to fine-tune process conditions, including pH levels, temperature, and retention time. Interestingly, different microbial populations exhibit distinct preferences for pH conditions, with certain species thriving in specific ranges. For instance, oleaginous yeasts, which are known for their lipid-producing capabilities, tend to accumulate the highest lipid content within a narrow pH range of 5–6, coupled with a temperature of approximately 30 °C, highlighting the importance of tailored process optimization for maximizing biofuel yields (Robles-Iglesias et al., 2023). In stark contrast, the anaerobic digestion process employed for biogas production necessitates a narrowly defined neutral pH range of 6.5–7.5, which is crucial for maintaining the stability and balance of the microbial community, thereby ensuring optimal biogas yields (Klassen et al., 2016). The duration of microbial activity, dictated by retention time, is a critical parameter that significantly impacts process efficiency. Notably, batch processing, a common approach in research settings, often falls short in achieving optimal efficiencies compared to continuous processes, primarily due to the inherent instability of microbial populations and limitations in substrate availability, which can compromise the overall performance of the system (Klassen et al., 2016). Research findings suggest that transitioning from traditional batch-based digestion to continuous digestion processes, facilitated by cutting-edge monitoring technologies, can substantially augment methane production, leading to improved process efficiency and enhanced yields (Klassen et al., 2016). Temperature is a critical factor that significantly influences enzyme activity and microbial metabolic processes. Notably, thermophilic conditions, characterized by temperatures ranging from 50 to 60 °C, have been shown to dramatically accelerate hydrolysis rates during anaerobic digestion, resulting in enhanced biofuel production and improved overall yields (Tsegaye et al., 2019). A noteworthy investigation into the conversion of microalgal biomass revealed that pre-treating the biomass with microwave energy at a specific input of 5 MJ/kg yielded the most impressive results, with significantly enhanced production of bioethanol, higher alcohols, and biodiesel. This finding underscores the critical importance of employing energy-efficient pretreatment techniques to optimize biofuel production (Ha et al., 2020).

2.7.3 Microbial Efficiency and Enzyme Availability

The efficacy of microbial catalysts and the availability of enzymes play a pivotal role in determining the viability of biofuel production. Specifically, the microbial strains employed in fermentation processes must possess a remarkable resilience to withstand the inhibitory compounds generated during the pretreatment of biomass. Recent studies emphasize that the development of robust, inhibitor-tolerant microbial strains is essential for augmenting biofuel yields and ensuring the stability of the production process, thereby underscoring the critical importance of microbial robustness in biofuel production (Wang et al., 2018). The process of enzymatic hydrolysis plays a vital role in decomposing lignocellulosic biomass into sugars that can be fermented, thereby serving as a crucial step in the production of biofuels. Nevertheless, the commercial production of enzymes required for this process remains a costly endeavor, which poses a significant barrier to the widespread adoption and large-scale implementation of this technology (Singh et al., 2022a, 2022b). The rapid progress in genetic engineering has significantly enhanced the stability and efficiency of enzymes, leading to substantial improvements in biofuel production. Notably, genetic modifications have been shown to increase lipid accumulation efficiencies by a remarkable 400%, thereby revolutionizing the biofuel production process and opening up new avenues for sustainable energy development (Robles-Iglesias et al., 2023). The use of immobilized microbial cells has emerged as a promising strategy in fermentation processes, with notable successes achieved in studies employing *Clostridium acetobutylicum*. Specifically, these immobilized cells have demonstrated remarkable resilience, retaining viability across multiple fermentation cycles, which in turn has led to significant enhancements in biofuel production yields, highlighting the potential of this approach to improve process efficiency and productivity (Efremenko et al., 2012). In addition, microbial delignification has emerged as a promising eco-friendly alternative for pretreatment, leveraging the capabilities of lignin-degrading bacteria and fungi to break down complex lignocellulosic structures. By harnessing the power of these microorganisms to degrade lignin and hydrolyze cellulose, researchers have demonstrated improved substrate digestibility, paving the way for more efficient biofuel production. However, despite its potential, large-scale implementation of microbial delignification remains in its infancy, hindered by cost constraints and process inefficiencies that must be addressed through further research and development (Tsegaye et al., 2019). Tailoring microbial strains to excel in lignin degradation and hydrolysis holds considerable promise for elevating biological pretreatment to a more industrially viable and scalable solution for biofuel production. By fine-tuning the genetic makeup of these microorganisms, researchers can unlock their full potential, paving the way for more efficient and cost-effective biological pretreatment processes that can meet the demands of large-scale industrial biofuel production (Tsegaye et al., 2019).

2.7.4 Challenges in Scaling up Production

Although substantial advancements have been made in harnessing agro-waste for biofuel production, the path to widespread industrial adoption remains fraught with significant obstacles. The prohibitively high capital and operational expenditures associated with critical process components, including pretreatment, enzyme production, and microbial strain optimization, pose a major hurdle to the economic viability of large-scale biofuel production, thereby underscoring the need for innovative solutions to drive down costs and enhance process efficiency (Cheah et al., 2020). Furthermore, the scalability of biofuel production processes is constrained by several key factors, including the availability of suitable substrates, inconsistencies in the composition of feedstocks, and the substantial infrastructure investments required to support large-scale processing operations. These challenges underscore the need for strategic planning, innovative solutions, and collaborative efforts to overcome the hurdles impeding the widespread adoption of biofuel production technologies (Cheah et al., 2020).

Recent breakthroughs in technology have led to the development of innovative integrated biorefinery models, which synergistically combine multiple biofuel production pathways to create a more robust and resilient process. By harnessing the benefits of this integrated approach, researchers propose that it is possible to significantly enhance the economic viability of biofuel production, thereby addressing a major hurdle to widespread adoption and paving the way for a more sustainable energy future (Cheah et al., 2020). To achieve economically viable biofuel production, it is imperative to adopt a multi-faceted approach that incorporates the recycling of waste and wastewater, implementation of energy recovery strategies, and development of cost-effective pretreatment methodologies. By integrating these strategies, biofuel producers can significantly reduce overall production expenditures, thereby enhancing the competitiveness of biofuels in the global energy market and paving the way for a more sustainable energy future (Cheah et al., 2020). Research indicates that shifting towards more energy-efficient pretreatment processes, such as those leveraging enzymatic or thermochemical approaches, can be a pivotal step in surmounting the economic hurdles that have long impeded the widespread adoption of biofuel production technologies. By adopting these more efficient pretreatment strategies, biofuel producers can potentially reduce energy expenditures, lower costs, and enhance overall process viability, thereby paving the way for a more sustainable and economically resilient biofuel industry (Meenakshisundaram et al.,

2021). In addition to technological advancements, the biofuel industry's continued growth and progress are contingent upon the convergence of two critical factors: supportive policy frameworks and collaborative interdisciplinary research endeavors. By fostering a conducive policy environment and facilitating cross-disciplinary cooperation among experts from diverse fields, we can unlock the full potential of the biofuel industry, drive innovation, and accelerate the transition towards a more sustainable energy future (Rame et al., 2023). Moreover, to propel the biofuel industry forward, it is essential to tackle the complex challenges associated with large-scale biofuel production, including the high costs, variability in substrate quality, and the need for microbial systems to tolerate inhibitory compounds. This can be achieved through collaborative, interdisciplinary research endeavors and the development of innovative technologies. The long-term viability of producing biofuels from agro-waste materials hinges on the creation of integrated biorefinery models that synergistically combine multiple processes, policy interventions, and advancements in microbial and enzymatic systems, ultimately paving the way for a more sustainable and circular bioeconomy (Rame et al., 2023). Effectively mitigating these factors will be a pivotal step in transforming biofuel into a commercially viable and environmentally sustainable energy solution, ultimately enabling it to emerge as a credible alternative to traditional fossil fuels and contribute meaningfully to a low-carbon future. The cumulative insights gleaned from these studies underscore the imperative of optimizing pretreatment protocols, microbial efficacy, and process parameters to unlock the full potential of biofuel production. While substrate modification is indeed a critical factor, the dynamics of microbial communities, enzyme efficiency, and process stability also play equally vital roles in determining the overall efficiency of biofuel production. Ultimately, the research underscores that the pursuit of cost-effective and sustainable biofuel production necessitates a holistic approach, one that seamlessly integrates cutting-edge pretreatment strategies, microbial engineering, and process optimization to drive innovation and advancement in the field.

2.8 Advances, Sustainability, and Future Outlook

The world's escalating energy needs, coupled with the alarming environmental repercussions of relying on fossil fuels, have ignited a quest for cleaner, alternative energy solutions. Within this landscape, biofuels sourced from agricultural waste have emerged as a beacon of hope. Their allure lies in three pivotal benefits: fostering sustainability, tackling waste management challenges, and curbing greenhouse gas emissions (Stephen & Periyasamy, 2018) (Fig. 2.2).

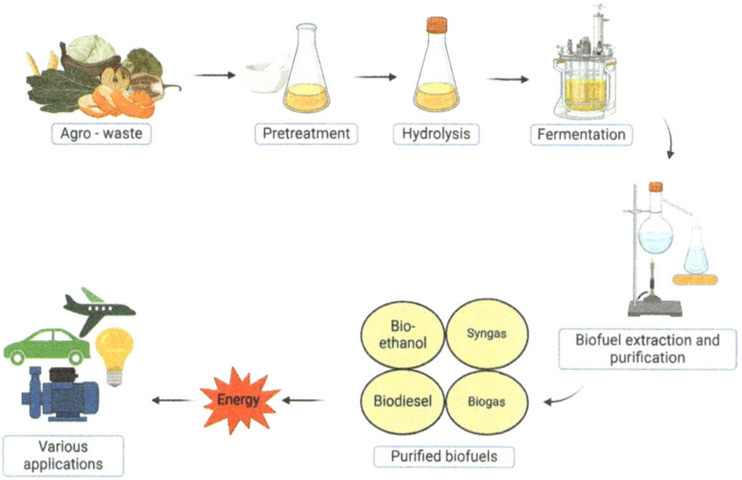

Fig. 2.2 The process of transforming agricultural waste into biofuels along with their application. The illustration highlights essential manufacturing steps and uses as it demonstrates the bioconversion of agricultural waste into biofuels. Fruit peels, vegetable scraps, along with agricultural wastes are examples of agro-waste that is pretreated in order to break down its structural components as well as increase accessibility for complex biomolecules. Enzymes hydrolyse carbohydrates to generate fermentable sugars, allowing microbes then ferment to create syngas, bioethanol, as well as biogas. In order to eliminate contaminants and improve fuel quality, the fermented product is extracted and purified into biofuel. The ultimate refined biofuels—bioethanol, biodiesel, syngas, and biogas—are sources of renewable energy for industry, transportation, and power generation. This procedure lessens dependency on fossil fuels, supporting sustainable environmental practices along with renewable energy

2.8.1 Advances in Biofuel Production from Agro-waste

Early investigations into biofuel production explored the potential of organic waste materials, such as crop residues and agricultural by-products, as viable feedstocks. Initial results revealed that liquid biofuels, particularly biodiesel, held great promise as a fossil fuel alternative, offering improved efficiency and reduced environmental impact. However, several hurdles hindered the commercial viability of this approach, including prohibitively high production costs, inadequate feedstock management strategies, and energy-intensive processing methods (Stephen & Periyasamy, 2018).

2.8.2 Thermo-biochemical and Enzymatic Innovations

Recent breakthroughs in waste-to-energy technologies have enabled the harnessing of agricultural waste through advanced thermal and biochemical processes. Notably, pyrolysis, gasification, and enzymatic hydrolysis have emerged as effective methods for converting agro-waste into sustainable biofuels (Malode et al., 2021). The incorporation of bio-based energy alternatives, including biodiesel, bio-compressed natural gas (CNG), and biogas, into the transportation sector and power generation infrastructure has proven to be a promising strategy for mitigating greenhouse gas (GHG) emissions (Malode et al., 2021).

2.8.3 The Rise of Next-Generation Biofuels

As the constraints of first-generation biofuels became apparent, the scientific community pivoted its attention to second- and third-generation alternatives. Next-generation biofuels, encompassing cellulosic ethanol, biomethane, and biohydrogen, have shown considerable promise in achieving enhanced efficiency and minimizing environmental footprint (Zinoviev et al., 2010). The integration of lignocellulosic biomass as a feedstock, combined with advances in microbial fermentation and synthetic biology, has significantly optimized bioethanol production. This multi-faceted approach effectively mitigates pressing concerns surrounding land utilization and food security (Raud et al., 2019).

2.8.4 Microbial Engineering and Synthetic Biology

Metabolic engineering and synthetic biology have emerged as pivotal tools in boosting biofuel production, as underscored by recent research. By cleverly reprogramming microbial pathways, scientists have made significant strides in harnessing engineered microorganisms to manufacture an array of biofuels, including alcohols, biodiesel, and even jet fuel (Cheon et al., 2016). Synthetic biology has revolutionized the field of biofuel production by allowing researchers to reengineer key metabolic pathways, notably the fatty acid synthesis (FAS) pathway and polyketide synthesis pathway. By tailoring these pathways, scientists can now optimize biofuel yields, paving the way for more efficient and sustainable biofuel production methods (Cheon et al., 2016).

2.8.5 The Role of Nanotechnology in Biofuel Production

The advent of nanotechnology has been a game-changer in the biofuel landscape, yielding substantial improvements in enzymatic efficiency, accelerating reaction kinetics, and bolstering the performance of biofuel cells (BFCs) (Zhao et al., 2017). The strategic application of nanostructured materials, encompassing nanoparticles, nanotubes, and nanocomposites, has significantly elevated enzyme immobilization, ultimately culminating in substantially enhanced biofuel conversion efficiencies (Verma et al., 2016). Furthermore, the integration of nanomaterials into microbial fuel cells (MFCs) has opened up new avenues for the sustainable generation of bioelectricity, offering a promising solution for environmentally friendly energy production (Zhao et al., 2017).

2.9 Sustainability in Biofuel Production from Agro-waste

2.9.1 Environmental Benefits and Greenhouse Gas Reduction

Converting agricultural waste into biofuels has emerged as a vital strategy for substantially decreasing the carbon footprint, thereby offering a more environmentally friendly alternative to traditional fossil fuels (Ambaye et al., 2021). The transition from fossil fuels to cleaner-burning biofuels, including methane, ethanol, and algal biodiesel, has yielded a dual environmental benefit: a notable reduction in CO_2 emissions and a significant improvement in air quality (Raheem et al., 2018). Moreover, biofuels derived from microalgae have shown remarkable potential in dual functionality: not only do they provide a sustainable energy source, but they also offer a viable means of capturing and sequestering CO_2 emissions from industrial sources (Raheem et al., 2018).

2.9.2 Waste Valorization and Circular Economy

Effective waste management continues to play a pivotal role in advancing biofuel research, with a particular focus on mitigating the environmental consequences of open-field burning of crop residues. Research underscores the critical importance of harnessing agro-residues as a valuable feedstock for biofuel production, thereby reducing waste and promoting sustainability (Divyabharathi et al., 2024). Beyond traditional biofuel conversion methods, innovative approaches are emerging that leverage agro-waste in diverse applications, including water purification systems, the production of bio-based chemicals, and the development of sustainable green construction materials, thereby significantly broadening the potential uses of agro-residues (Divyabharathi et al., 2024).

2.9.3 Socio-economic Challenges and Ethical Considerations

While biofuels offer numerous benefits, their widespread adoption is hindered by several obstacles. A pressing concern is the rivalry between food crops and biofuel feedstocks, which raises complex ethical dilemmas and underscores the need for carefully crafted policies that strike a balance between ensuring energy security and maintaining global food availability (Sikarwar et al., 2017). Despite their potential, the commercial viability of advanced biofuels is impeded by a range of economic hurdles, notably the high production costs and inadequate infrastructure, which collectively hinder their widespread adoption and large-scale commercialization (Raud et al., 2019). Fortunately, strategic government interventions, including targeted incentives and research funding, have proven instrumental in bridging these gaps and facilitating the advancement of advanced biofuels (Sikarwar et al., 2017).

2.10 Future Outlook in Biofuel Production from Agro-waste

2.10.1 Process Optimization and Cost Reduction Strategies

Recent studies have underscored the importance of optimizing biofuel production processes to improve their economic viability. One promising approach involves leveraging advanced techniques, such as the utilization of coalescer reactors, to synthesize biodiesel from waste cooking oil, thereby enhancing the overall efficiency and cost-effectiveness of the process (Stephen & Periyasamy, 2018). Comprehensive lifecycle assessments and techno-economic evaluations have revealed that adopting a multifaceted approach, wherein multiple conversion pathways are integrated, holds significant potential for substantial cost reductions and notable enhancements in energy efficiency, thereby optimizing the overall sustainability and viability of the process (Sikarwar et al., 2017).

2.10.2 Advanced Biofuels and Hydrogen Economy

The emergence of innovative technologies, including Fischer-Tropsch fuels, biohydrogen production, and syngas purification methods, signifies a paradigm shift towards the development of high-energy-density biofuels, which promise to revolutionize the renewable energy landscape (Zinoviev et al., 2010). Despite these advances, significant obstacles persist, particularly with regards to the safe and efficient storage and distribution of hydrogen, which continue to pose substantial technical and logistical challenges (Sikarwar et al., 2017). Looking ahead, future research endeavors are anticipated to concentrate on seamlessly integrating the hydrogen economy, with anticipated innovations in fuel cell technology and sustainable hydrogen production methods poised to drive transformative breakthroughs (Sikarwar et al., 2017).

2.10.3 Policy, Market Implications, and Global Energy Transition

The European Union's proactive stance on renewable energy has had a profound impact on the trajectory of biofuel research and development, driving significant advancements in the commercialization of these sustainable energy sources (Raud et al., 2019). Furthermore, developing nations are increasingly embracing biofuel strategies, taking into account the intricate interplay of socio-economic factors and market forces, as they strive to create a sustainable and equitable energy future (Zinoviev et al., 2010). To facilitate widespread adoption of biofuels, future policy frameworks must adopt a multifaceted approach, incorporating financial incentives to stimulate investment, ensuring the reliable availability of feedstocks, and establishing stringent sustainability regulations to guarantee environmentally responsible large-scale production (Raud et al., 2019).

2.11 Conclusion

The conversion of agro-waste into biofuels offers a compelling strategy for achieving sustainable energy solutions. Recent breakthroughs in microbial engineering, nanotechnology, and process integration have substantially enhanced biofuel productivity and efficiency. Nevertheless, to facilitate widespread adoption, it is imperative to address the economic, environmental, and policy hurdles that currently impede large-scale implementation. Future research endeavors should prioritize optimizing feedstock utilization, further enhancing process efficiency, and establishing supportive regulatory frameworks, ultimately paving the way for a viable and sustainable biofuel-driven energy transition. In conclusion, thermochemical conversion provides an effective way to transform agricultural waste into biofuels, reducing dependence on fossil fuels while addressing environmental concerns. With advancements in technology and supportive waste-to-energy policies, these processes have the potential to play a key role in sustainable energy production and global carbon management in the future. Comparing efficiency and sustainability, biochemical methods are environmentally friendly and effective for waste management, while thermochemical processes offer versatility in feedstocks and energy products. The choice of biofuel production depends on feedstock availability, energy demand, and environmental considerations.

Acknowledgments SKS lab would like to acknowledge funding from the Institution of Eminence Incentive Grant Scheme, Trans-Disciplinary Research Scheme, Department of Science and Technology, SEED grant, New Delhi and funding from Institute of Medical Sciences intramural grant, Banaras Hindu University, Varanasi, India. CP & SNR would like to acknowledge CSIR (09/0013(12277)/2021-EMR-I) and ICMR (3/1/3/196/Neuro/2021-NCD-I) for providing fellowship assistance respectively. The author would also like to acknowledge that Figs. 2.1 and 2.2 in this manuscript were created using BioRender software.

Disclosure Statement The author declares no conflict of interest.

References

Achinas, S., Li, Y., Achinas, V., & Euverink, G. J. W. (2019). Biogas potential from the anaerobic digestion of potato peels: Process performance and kinetics evaluation. *Energies, 12*(12), 2311.

Akhtar, A., Krepl, V., & Ivanova, T. (2018). A combined overview of combustion, pyrolysis, and gasification of biomass. *Energy & Fuels, 32*(7), 7294–7318.

Ali, N., Saleem, M., Shahzad, K., Hussain, S., & Chughtai, A. (2016). Effect of operating parameters on production of bio-oil from fast pyrolysis of maize stalk in bubbling fluidized bed reactor. *Polish Journal of Chemical Technology, 18*(3), 88–96.

Al-Qadri, A. A., Ahmed, U., Razzak, S. A., Shahbaz, M., Hussain, M., & Zein, S. H. (2025). Technical and economic assessment of hydrogen production from heavy fuel oil via thermochemical conversion routes. *Fuel, 385*, 134054.

Alsudani, F. T., Saeed, A. N., Ali, N. S., Majdi, H. S., Salih, H. G., Albayati, T. M., et al. (2023). Fisher–Tropsch synthesis for conversion of methane into liquid hydrocarbons through gas-to-liquids (GTL) process: a review. *Methane, 2*(1), 24–43.

Álvarez-Castillo, E., Felix, M., Bengoechea, C., & Guerrero, A. (2021). Proteins from Agri-food industrial biowastes or co-products and their applications as green materials. *Food, 10*(5), 981.

Ambaye, T. G., Vaccari, M., Bonilla-Petriciolet, A., Prasad, S., van Hullebusch, E. D., & Rtimi, S. (2021). Emerging technologies for biofuel production: A critical review on recent progress, challenges and perspectives. *Journal of Environmental Management, 290*, 112627.

Aro, E.-M. (2016). From first generation biofuels to advanced solar biofuels. *Ambio, 45*(Suppl 1), 24–31.

Aruwajoye, G. S., Kassim, A., Saha, A. K., & Gueguim Kana, E. B. (2020). Prospects for the improvement of bioethanol and biohydrogen production from mixed starch-based agricultural wastes. *Energies, 13*(24), 6609.

Arvanitoyannis, I. S., & Tserkezou, P. (2008). Corn and rice waste: a comparative and critical presentation of methods and current and potential uses of treated waste. *International Journal of Food Science and Technology, 43*(6), 958–988.

Atimtay, A. T. (2010). Combustion of agro-waste with coal in a fluidized bed. *Clean Technologies and Environmental Policy, 12*, 43–52.

Awogbemi, O., & Von Kallon, D. V. (2023). Application of biochar derived from crops residues for biofuel production. *Fuel Communications, 15*, 100088.

Banerjee, S., Sen, R., Pandey, R. A., Chakrabarti, T., Satpute, D., Giri, B. S., & Mudliar, S. (2009). Evaluation of wet air oxidation as a pretreatment strategy for bioethanol production from rice husk and process optimization. *Biomass and Bioenergy, 33*(12), 1680–1686.

Barampouti, E. M., Christofi, A., Malamis, D., & Mai, S. (2023). A sustainable approach to valorize potato peel waste towards biofuel production. *Biomass Conversion and Biorefinery, 13*(9), 8197–8208.

Bardiya, N., Somayaji, D., & Khanna, S. (1996). Biomethanation of banana peel and pineapple waste. *Bioresource Technology, 58*(1), 73–76.

Basumatary, S., Nath, B., & Kalita, P. (2018). Application of agro-waste derived materials as heterogeneous base catalysts for biodiesel synthesis. *Journal of Renewable and Sustainable Energy, 10*(4).

Cheah, W. Y., Sankaran, R., Show, P. L., Ibrahim, T., Baizura, T. N., Chew, K. W., et al. (2020). Pretreatment methods for lignocellulosic biofuels production: Current advances, challenges and future prospects. *Biofuel Research Journal, 7*(1), 1115–1127.

Cheon, S., Kim, H. M., Gustavsson, M., & Lee, S. Y. (2016). Recent trends in metabolic engineering of microorganisms for the production of advanced biofuels. *Current Opinion in Chemical Biology, 35*, 10–21.

Chhetri, R. K., Aryal, N., Kharel, S., Poudel, R. C., & Pant, D. (2020). Agro-based industrial wastes as potent sources of alternative energy and organic fertilizers. In *Current developments in biotechnology and bioengineering* (pp. 121–136). Elsevier.

Chieng, S., & Kuan, S. H. (2022). Harnessing bioenergy and high value–added products from rice residues: A review. *Biomass Conversion and Biorefinery, 12*(8), 3547–3571.

Clomburg, J. M., & Gonzalez, R. (2010). Biofuel production in Escherichia coli: The role of metabolic engineering and synthetic biology. *Applied Microbiology and Biotechnology, 86*, 419–434.

Czerwińska, K., Śliz, M., & Wilk, M. (2022). Hydrothermal carbonization process: Fundamentals, main parameter characteristics and possible applications including an effective method of SARS-CoV-2 mitigation in sewage sludge. A review. *Renewable and Sustainable Energy Reviews, 154*, 111873.

d'Espaux, L., Mendez-Perez, D., Li, R., & Keasling, J. D. (2015). Synthetic biology for microbial production of lipid-based biofuels. *Current Opinion in Chemical Biology, 29*, 58–65.

da Silva, J. C. G., Alves, J. L. F., Mumbach, G. D., Andersen, S. L. F., Moreira, R. F. P. M., & Jose, H. J. (2024). Hydrogen-rich syngas production from steam gasification of Brazilian agroindustrial wastes in fixed bed reactor: Kinetics, energy, and gas composition. *Biomass Conversion and Biorefinery, 14*(20), 25901–25924.

de Brito, H. G., de Araújo, B. M. C., Costa, I. O., da Costa, G. L. V., de Oliveira, G. B., de Freitas, F. B. F., et al. (2024). Steam explosion and dilute acid Pretreatments to improve enzymatic hydrolysis and ethanol fermentation of green coconut fibers. *Waste and Biomass Valorization*, 1–13.

de Farias Silva, C. E., & Bertucco, A. (2016). Bioethanol from microalgae and cyanobacteria: A review and technological outlook. *Process Biochemistry, 51*(11), 1833–1842.

de Oliveira, M. C., Bassin, I. D., & Cammarota, M. C. (2022). Microalgae and cyanobacteria biomass pretreatment methods: A comparative analysis of chemical and thermochemical pretreatment methods aimed at methane production. *Fermentation, 8*(10), 497.

de Souza, A. P., Grandis, A., Leite, D. C. C., & Buckeridge, M. S. (2014). Sugarcane as a bioenergy source: History, performance, and perspectives for second-generation bioethanol. *Bioenergy Research, 7*, 24–35.

Díez, L. I., García-Mariaca, A., Llera-Sastresa, E., & Canalís, P. (2024). On the oxy-combustion of blends of coal and agro-waste biomass under dry and wet conditions. *Fuel, 365*, 131265.

Divyabharathi, R., Kalidasan, B., Js, S. S. R., & Chinnasamy, S. (2024). Recent advances in sustainable agro residue utilisation, barriers and remediation for environmental management: Present insights and future challenges. *Industrial Crops and Products, 216*, 118790.

Durán-Valle, C. J., & López-Coca, I. M. (2024). Biochemical and thermochemical conversion technologies for agriculture waste transformation. In *Transforming agriculture residues for sustainable development: From waste to wealth* (pp. 47–84). Springer.

Efremenko, E. N., Nikolskaya, A. B., Lyagin, I. V., Senko, O. V., Makhlis, T. A., Stepanov, N. A., et al. (2012). Production of biofuels from pretreated microalgae biomass by anaerobic fermentation with immobilized clostridium acetobutylicum cells. *Bioresource Technology, 114*, 342–348.

Fakudze, S., & Chen, J. (2023). A critical review on co-hydrothermal carbonization of biomass and fossil-based feedstocks for cleaner solid fuel production: Synergistic effects and environmental benefits. *Chemical Engineering Journal, 457*, 141004.

Fischer, C. R., Klein-Marcuschamer, D., & Stephanopoulos, G. (2008). Selection and optimization of microbial hosts for biofuels production. *Metabolic Engineering, 10*(6), 295–304.

Fryda, L., & Visser, R. (2015). Biochar for soil improvement: Evaluation of biochar from gasification and slow pyrolysis. *Agriculture, 5*(4), 1076–1115.

Gupta, A. P., Upadhyay, P., Sen, T., & Dutta, J. (2023). Agricultural waste as a resource: The lesser travelled road to sustainability. In *Agriculture waste management and bioresource: The circular economy perspective* (pp. 1–20).

Gupta, S., Gupta, G. K., & Mondal, M. K. (2019). Slow pyrolysis of chemically treated walnut shell for valuable products: Effect of process parameters and in-depth product analysis. *Energy, 181*, 665–676.

Ha, G.-S., El-Dalatony, M. M., Kurade, M. B., Salama, E.-S., Basak, B., Kang, D., et al. (2020). Energy-efficient pretreatments for the enhanced conversion of microalgal biomass to biofuels. *Bioresource Technology, 309*, 123333.

Hájek, M., Vávra, A., de Paz Carmona, H., & Kocík, J. (2021). The catalysed transformation of vegetable oils or animal fats to biofuels and bio-lubricants: A review. *Catalysts, 11*(9), 1118.

Hamdi, G. M. H., Abbas, M. N., & Ali, S. A. K. (2024). Bioethanol production from agricultural waste: A review. *Journal of Engineering and Sustainable Development, 28*(02), 233–252.

Huzir, N. M., Aziz, M. M. A., Ismail, S. B., Abdullah, B., Mahmood, N. A. N., Umor, N. A., Muhammad, S. A. F., & a. S. (2018). Agro-industrial waste to biobutanol production: Eco-friendly biofuels for next generation. *Renewable and Sustainable Energy Reviews, 94*, 476–485.

Jang, Y.-S., Park, J. M., Choi, S., Choi, Y. J., Cho, J. H., & Lee, S. Y. (2012). Engineering of microorganisms for the production of biofuels and perspectives based on systems metabolic engineering approaches. *Biotechnology Advances, 30*(5), 989–1000.

Javed, M. R., Noman, M., Shahid, M., Ahmed, T., Khurshid, M., Rashid, M. H., et al. (2019). Current situation of biofuel production and its enhancement by CRISPR/Cas9-mediated genome engineering of microbial cells. *Microbiological Research, 219*, 1–11.

Ji, L.-Q. (2015). An assessment of agricultural residue resources for liquid biofuel production in China. *Renewable and Sustainable Energy Reviews, 44*, 561–575.

Kan, T., Strezov, V., & Evans, T. J. (2016). Lignocellulosic biomass pyrolysis: A review of product properties and effects of pyrolysis parameters. *Renewable and Sustainable Energy Reviews, 57*, 1126–1140.

Kauffman, N., Hayes, D., & Brown, R. (2011). A life cycle assessment of advanced biofuel production from a hectare of corn. *Fuel, 90*(11), 3306–3314.

Khiari, Z. (2022). Sustainable upcycling of fisheries and aquaculture wastes using fish-derived cold-adapted proteases. *Frontiers in Nutrition, 9*, 875697.

Kim, Y. S., Lee, J. J., Kim, T. S., & Sohn, J. L. (2011). Effects of syngas type on the operation and performance of a gas turbine in integrated gasification combined cycle. *Energy Conversion and Management, 52*(5), 2262–2271.

Klassen, V., Blifernez-Klassen, O., Wobbe, L., Schlüter, A., Kruse, O., & Mussgnug, J. H. (2016). Efficiency and biotechnological aspects of biogas production from microalgal substrates. *Journal of Biotechnology, 234*, 7–26.

Konde, K. S., Nagarajan, S., Kumar, V., Patil, S. V., & Ranade, V. V. (2021). Sugarcane bagasse based biorefineries in India: Potential and challenges. *Sustainable Energy & Fuels, 5*(1), 52–78.

Kothari, R., Vashishtha, A., Singh, H. M., Pathak, V. V., Tyagi, V. V., Yadav, B. C., et al. (2020). Assessment of Indian bioenergy policy for sustainable environment and its impact for rural India: Strategic implementation and challenges. *Environmental Technology & Innovation, 20*, 101078.

Kumar, S., & Shukla, S. K. (2016). A review on recent gasification methods for biomethane gas production. *International Journal of Energy Engineering, 6*(1A), 32–43.

Kung, Y., Runguphan, W., & Keasling, J. D. (2012). From fields to fuels: Recent advances in the microbial production of biofuels. *ACS Synthetic Biology, 1*(11), 498–513.

Kuthiala, T., Thakur, K., Sharma, D., Singh, G., Khatri, M., & Arya, S. K. (2022). The eco-friendly approach of cocktail enzyme in agricultural waste treatment: A comprehensive review. *International Journal of Biological Macromolecules, 209*, 1956–1974.

Ledesma, B., & Beltramone, A. (2021). Revalorization of agro-industrial waste as a catalyst source for production of biofuels. *Renewable Energy, 174*, 747–757.

Lü, J., Sheahan, C., & Fu, P. (2011). Metabolic engineering of algae for fourth generation biofuels production. *Energy & Environmental Science, 4*(7), 2451–2466.

Majidian, P., Tabatabaei, M., Zeinolabedini, M., Naghshbandi, M. P., & Chisti, Y. (2018). Metabolic engineering of microorganisms for biofuel production. *Renewable and Sustainable Energy Reviews, 82*, 3863–3885.

Malode, S. J., Prabhu, K. K., Mascarenhas, R. J., Shetti, N. P., & Aminabhavi, T. M. (2021). Recent advances and viability in biofuel production. *Energy Conversion and Management: X, 10*, 100070.

Manu, J., Kailas, T. G., & Madav, V. (2023). Exploring the synergetic effects of rice husk, cashew shell, and cashew husk biomass blends on fluidized bed gasification for enhanced hydrogen production. *Journal of Cleaner Production, 419*, 137991.

Meenakshisundaram, S., Fayeulle, A., Leonard, E., Ceballos, C., & Pauss, A. (2021). Fiber degradation and carbohydrate production by combined biological and chemical/physicochemical pretreatment methods of lignocellulosic biomass—A review. *Bioresource Technology, 331*, 125053.

Mirmohamadsadeghi, S., Karimi, K., Azarbaijani, R., Yeganeh, L. P., Angelidaki, I., Nizami, A.-S., et al. (2021). Pretreatment of lignocelluloses for enhanced biogas production: A review on influencing mechanisms and the importance of microbial diversity. *Renewable and Sustainable Energy Reviews, 135*, 110173.

Mohan, D., Pittman, C. U., Jr., & Steele, P. H. (2006). Pyrolysis of wood/biomass for bio-oil: A critical review. *Energy & Fuels, 20*(3), 848–889.

Mondal, S., Neogi, S., & Chakraborty, S. (2024). Optimization of reactor parameters for amplifying synergy in enzymatic co-hydrolysis and microbial co-fermentation of lignocellulosic agro-residues. *Renewable Energy, 225*, 120281.

Murillo-Alvarado, P. E., & Flores Russell, E. (2022). Optimization approach for bioethanol production from agro-industrial waste. *Frontiers in Energy Research, 10*, 975133.

Mushtaq, Q., Ishtiaq, U., Joly, N., Spalletta, A., & Martin, P. (2024). Harnessing Bacillus subtilis QY5 PP784163 for bioethanol production from potato Peel waste and nutrient recovery for animal feed: Maximizing resource efficiency. *Fermentation, 10*(10), 523.

Mushtaq, Z., Maqbool, R., & Bhat, K. A. (2023). Genetic engineering and fifth-generation biofuels. In *Environmental sustainability of biofuels* (pp. 237–251). Elsevier.

Nair, L. G., Agrawal, K., & Verma, P. (2022). An overview of sustainable approaches for bioenergy production from agro-industrial wastes. *Energy Nexus, 6*, 100086.

Naitam, M. G., Tomar, G. S., Kaushik, R., Singh, S., & Nain, L. (2022). Agro-industrial waste as potential renewable feedstock for biopolymer poly-hydroxyalkanoates (PHA) production. *Enzyme Engineering, 11*, 190–206.

Nimmas, T., Wongsakulphasatch, S., Chanthanumataporn, M., Vacharanukrauh, T., & Assabumrungrat, S. (2024). Thermochemical transformation of CO_2 into high-value products. *Current Opinion in Green and Sustainable Chemistry*, 100911.

Noor, R. S., Shah, A. N., Tahir, M. B., Umair, M., Nawaz, M., Ali, A., et al. (2024). Recent trends and advances in additive-mediated composting technology for agricultural waste resources: A comprehensive review. *ACS Omega, 9*(8), 8632–8653.

Nunta, R., Techapun, C., Sommanee, S., Mahakuntha, C., Porninta, K., Punyodom, W., et al. (2023). Valorization of rice straw, sugarcane bagasse and sweet sorghum bagasse for the production of bioethanol and phenylacetylcarbinol. *Scientific Reports, 13*(1), 727.

Odedina, M. J., Charnnok, B., Saritpongteeraka, K., & Chaiprapat, S. (2017). Effects of size and thermophilic pre-hydrolysis of banana peel during anaerobic digestion, and biomethanation potential of key tropical fruit wastes. *Waste Management, 68*, 128–138.

Ortega, Z., Morón, M., Monzón, M. D., Badalló, P., & Paz, R. (2016). Production of banana fiber yarns for technical textile reinforced composites. *Materials, 9*(5), 370.

Panpatte, D. G., & Jhala, Y. K. (2019). Agricultural waste: a suitable source for biofuel production. In *Prospects of renewable bioprocessing in future energy systems* (pp. 337–355).

Parmar, A., Singh, N. K., Pandey, A., Gnansounou, E., & Madamwar, D. (2011). Cyanobacteria and microalgae: A positive prospect for biofuels. *Bioresource Technology, 102*(22), 10163–10172.

Pattanaik, L., Pattnaik, F., Saxena, D. K., & Naik, S. N. (2019). Biofuels from agricultural wastes. In *Second and third generation of feedstocks* (pp. 103–142). Elsevier.

Paudel, P. P., Kafle, S., Park, S., Kim, S. J., Cho, L., & Kim, D. H. (2024). Advancements in sustainable thermochemical conversion of agricultural crop residues: a systematic review of technical progress, applications, perspectives, and challenges. *Renewable and Sustainable Energy Reviews, 202*, 114723.

Paykani, A., Chehrmonavari, H., Tsolakis, A., Alger, T., Northrop, W. F., & Reitz, R. D. (2022). Synthesis gas as a fuel for internal combustion engines in transportation. *Progress in Energy and Combustion Science, 90*, 100995.

Peralta-Yahya, P. P., Zhang, F., Del Cardayre, S. B., & Keasling, J. D. (2012). Microbial engineering for the production of advanced biofuels. *Nature, 488*(7411), 320–328.

Perera, M., Yan, J., Xu, L., Yang, M., & Yan, Y. (2022). Bioprocess development for biolubricant production using non-edible oils, agro-industrial byproducts and wastes. *Journal of Cleaner Production, 357*, 131956.

Poornima, S., Manikandan, S., Prakash, R., Deena, S. R., Subbaiya, R., Karmegam, N., et al. (2024). Biofuel and biochemical production through biomass transformation using advanced thermochemical and biochemical processes—A review. *Fuel, 372*, 132204.

Popp, J., Harangi-Rákos, M., Gabnai, Z., Balogh, P., Antal, G., & Bai, A. (2016). Biofuels and their co-products as livestock feed: Global economic and environmental implications. *Molecules, 21*(3), 285.

Procentese, A., Raganati, F., Olivieri, G., Russo, M. E., De La Feld, M., & Marzocchella, A. (2019). Agro food wastes and innovative pretreatments to meet biofuel demand in Europe. *Chemical Engineering & Technology, 42*(5), 954–961.

Raheem, A., Prinsen, P., Vuppaladadiyam, A. K., Zhao, M., & Luque, R. (2018). A review on sustainable microalgae based biofuel and bioenergy production: Recent developments. *Journal of Cleaner Production, 181*, 42–59.

Rahimi, Z., Anand, A., & Gautam, S. (2022). An overview on thermochemical conversion and potential evaluation of biofuels derived from agricultural wastes. *Energy Nexus, 7*, 100125.

Raisa, D. M., Sirajuddin, S. N., Syamsu, J. A., & Arsyad, M. (2024). Sustainable agricultural optimization: A review of integrated farming system, resilience, and technology.

Rame, R., Purwanto, P., & Sudarno, S. (2023). Biotechnological approaches in utilizing agro-waste for biofuel production: An extensive review on techniques and challenges. *Bioresource Technology Reports, 24*, 101662.

Râpă, M., Darie-Niță, R. N., & Coman, G. (2024). Valorization of fruit and vegetable waste into sustainable and value-added materials. *Waste, 2*(3), 258–278.

Rathore, S., & Pandey, A. K. (2020). Waste utilization and minimization in food industry. In *Waste to energy: Prospects and applications* (pp. 287–307).

Rauch, R., Hrbek, J., & Hofbauer, H. (2016). Biomass gasification for synthesis gas production and applications of the syngas. In *Advances in bioenergy: The sustainability challenge* (pp. 73–91).

Raud, M., Kikas, T., Sippula, O., & Shurpali, N. J. (2019). Potentials and challenges in lignocellulosic biofuel production technology. *Renewable and Sustainable Energy Reviews, 111*, 44–56.

Robles-Iglesias, R., Naveira-Pazos, C., Fernández-Blanco, C., Veiga, M. C., & Kennes, C. (2023). Factors affecting the optimisation and scale-up of lipid accumulation in oleaginous yeasts for sustainable biofuels production. *Renewable and Sustainable Energy Reviews, 171*, 113043.

Rodionova, M. V., Poudyal, R. S., Tiwari, I., Voloshin, R. A., Zharmukhamedov, S. K., Nam, H. G., et al. (2017). Biofuel production: Challenges and opportunities. *International Journal of Hydrogen Energy, 42*(12), 8450–8461.

Rogeri, R. C., Fuess, L. T., de Araujo, M. N., Eng, F., & do Vale Borges, A., Damianovic, M. H. R. Z., & da Silva, A. J. (2024). Methane production from sugarcane vinasse: The alkalinizing potential of fermentative-sulfidogenic processes in two-stage anaerobic digestion. *Energy Nexus, 14*, 100303.

Sadh, P. K., Duhan, S., & Duhan, J. S. (2018). Agro-industrial wastes and their utilization using solid state fermentation: a review. *Bioresources and Bioprocessing, 5*(1), 1–15.

Salehi, E., Abedi, J., & Harding, T. (2011). Bio-oil from sawdust: Effect of operating parameters on the yield and quality of pyrolysis products. *Energy & Fuels, 25*(9), 4145–4154.

Samantaray, B., Mohapatra, S., Mishra, R. R., Behera, B. C., & Thatoi, H. (2024). Bioethanol production from agro-wastes: A comprehensive review with a focus on pretreatment, enzymatic hydrolysis, and fermentation. *International Journal of Green Energy, 21*(6), 1398–1424.

Sarkar, N., Chakraborty, D., Dutta, R., Agrahari, P., Bharathi, S. D., Singh, A. A., & Jacob, S. (2021). A comprehensive review on oilseed cakes and their potential as a feedstock for integrated biorefinery. *Journal of Advanced Biotechnology and Experimental Therapeutics, 4*(3), 376–387.

Shahzad, H. M. A., Asim, Z., Khan, S. J., Almomani, F., Mahmoud, K. A., Mustafa, M. R. U., & Rasool, K. (2024). Thermochemical and biochemical conversion of agricultural waste for bioenergy production: An updated review. *Discover Environment, 2*(1), 134.

Shanmugam, S., Sekar, M., Sivaramakrishnan, R., Raj, T., Ong, E. S., Rabbani, A. H., et al. (2021). Pretreatment of second and third generation feedstock for enhanced biohythane production: Challenges, recent trends and perspectives. *International Journal of Hydrogen Energy, 46*(20), 11252–11268.

Sharma, P., Vishvakarma, R., Gautam, K., Vimal, A., Gaur, V. K., Farooqui, A., et al. (2022). Valorization of citrus peel waste for the sustainable production of value-added products. *Bioresource Technology, 351*, 127064.

Shen, Y. (2020). A review on hydrothermal carbonization of biomass and plastic wastes to energy products. *Biomass and Bioenergy, 134*, 105479.

Sikarwar, V. S., Zhao, M., Fennell, P. S., Shah, N., & Anthony, E. J. (2017). Progress in biofuel production from gasification. *Progress in Energy and Combustion Science, 61*, 189–248.

Singh, K., Gurung, V., & Gupta, S. (2022a). Exploration of various microbial Systems for the Biofuel Production. *Asian Journal of Biotechnology and Bioresource Technology, 8*(2), 39–49.

Singh, P., Dubey, P., Younis, K., & Yousuf, O. (2024). A review on the valorization of coconut shell waste. *Biomass Conversion and Biorefinery, 14*(7), 8115–8125.

Singh, R., Langyan, S., Rohtagi, B., Darjee, S., Khandelwal, A., Shrivastava, M., et al. (2022b). Production of biofuels options by contribution of effective and suitable enzymes: Technological developments and challenges. *Materials Science for Energy Technologies, 5*, 294–310.

Singhania, R. R., Soccol, C. R., & Pandey, A. (2008). Application of tropical agro-industrial residues as substrate for solid-state fermentation processes. *Current developments in solid-state fermentation* (pp. 412–442).

Srivastava, R. K., Shetti, N. P., Reddy, K. R., Kwon, E. E., Nadagouda, M. N., & Aminabhavi, T. M. (2021). Biomass utilization and production of biofuels from carbon neutral materials. *Environmental Pollution, 276*, 116731.

Stephen, J. L., & Periyasamy, B. (2018). Innovative developments in biofuels production from organic waste materials: A review. *Fuel, 214*, 623–633.

Sun, J., Zhang, L., & Loh, K.-C. (2021). Review and perspectives of enhanced volatile fatty acids production from acidogenic fermentation of lignocellulosic biomass wastes. *Bioresources and Bioprocessing, 8*(1), 68.

Swain, P. K. (2017). Utilisation of agriculture waste products for production of bio-fuels: A novel study. *Materials Today: Proceedings, 4*(11), 11959–11967.

Tabasum, S., Younas, M., Zaeem, M. A., Majeed, I., Majeed, M., Noreen, A., et al. (2019). A review on blending of corn starch with natural and synthetic polymers, and inorganic nanoparticles with mathematical modeling. *International Journal of Biological Macromolecules, 122*, 969–996.

Taherzadeh, M. J., & Karimi, K. (2008). Pretreatment of lignocellulosic wastes to improve ethanol and biogas production: A review. *International Journal of Molecular Sciences, 9*(9), 1621–1651.

Tsegaye, B., Balomajumder, C., & Roy, P. (2019). Microbial delignification and hydrolysis of lignocellulosic biomass to enhance biofuel production: An overview and future prospect. *Bulletin of the National Research Centre, 43*(1), 1–16.

Upadhyay, S. K., Singh, G., Rani, N., Rajput, V. D., Seth, C. S., Dwivedi, P., et al. (2024). Transforming bio-waste into value-added products mediated microbes for enhancing soil health and crop production: Perspective views on circular economy. *Environmental Technology & Innovation, 34*, 103573.

Vásquez, M. C., Silva, E. E., & Castillo, E. F. (2017). Hydrotreatment of vegetable oils: A review of the technologies and its developments for jet biofuel production. *Biomass and Bioenergy, 105*, 197–206.

Verma, M. L., Puri, M., & Barrow, C. J. (2016). Recent trends in nanomaterials immobilised enzymes for biofuel production. *Critical Reviews in Biotechnology, 36*(1), 108–119.

Waheed, Q. M. K., & Williams, P. T. (2013). Hydrogen production from high temperature pyrolysis/steam reforming of waste biomass: Rice husk, sugar cane bagasse, and wheat straw. *Energy & Fuels, 27*(11), 6695–6704.

Wang, S., Sun, X., & Yuan, Q. (2018). Strategies for enhancing microbial tolerance to inhibitors for biofuel production: a review. *Bioresource Technology, 258*, 302–309.

Wojcieszak, D., Przybył, J., Czajkowski, Ł., Majka, J., & Pawłowski, A. (2022). Effects of harvest maturity on the chemical and energetic properties of corn Stover biomass combustion. *Materials, 15*(8), 2831.

Xie, S., Li, Z., Li, H., & Fang, Y. (2024). Integration of carbon capture with heterogeneous catalysis toward methanol production: Chemistry, challenges, and opportunities. *Catalysis Reviews, 66*(4), 1478–1517.

Yadav, P., Anu Kumar Tiwari, S., Kumar, V., Singh, D., Kumar, S., et al. (2024). Sugarcane bagasse: An important lignocellulosic substrate for production of enzymes and biofuels. *Biomass Conversion and Biorefinery, 14*(5), 6111–6142.

Yuliansyah, A. T., Hirajima, T., Kumagai, S., & Sasaki, K. (2010). Production of solid biofuel from agricultural wastes of the palm oil industry by hydrothermal treatment. *Waste and Biomass Valorization, 1*, 395–405.

Zabed, H. M., Akter, S., Yun, J., Zhang, G., Awad, F. N., Qi, X., & Sahu, J. N. (2019). Recent advances in biological pretreatment of microalgae and lignocellulosic biomass for biofuel production. *Renewable and Sustainable Energy Reviews, 105*, 105–128.

Zhao, C., Gai, P., Song, R., Chen, Y., Zhang, J., & Zhu, J.-J. (2017). Nanostructured material-based biofuel cells: Recent advances and future prospects. *Chemical Society Reviews, 46*(5), 1545–1564.

Zhao, P., Shen, Y., Ge, S., Chen, Z., & Yoshikawa, K. (2014). Clean solid biofuel production from high moisture content waste biomass employing hydrothermal treatment. *Applied Energy, 131*, 345–367.

Zinoviev, S., Müller-Langer, F., Das, P., Bertero, N., Fornasiero, P., Kaltschmitt, M., et al. (2010). Next-generation biofuels: Survey of emerging technologies and sustainability issues. *ChemSusChem, 3*(10), 1106–1133.

Chapter 3
Selection, Characterization and Biomass Conversion Approaches for Biofuel Production

Fouziya Parveen, Nafisa Shaheen, Ayush Saxena, and Mohammad Ashfaque

Abstract The mounting problem due to global warming and the increase in the price of conventional fuel have alarmed researchers towards the production of biofuel to meet the future energy demand and reduce the environmental impact. Biofuel is considered as an alternative source of energy. Lignocellulose Biomass, Algal Biomass, Starch Crops, and Oil Crops/Waste Oils are the potential feedstock sources for biofuel production. Biomass to biofuel conversion requires two processes viz, thermochemical conversions such as combustion, gasification, pyrolysis, and hydrothermal liquefaction, and biochemical conversions such as anaerobic digestion, fermentation, and transesterification. The efficiency of biofuel conversion depends on the type of biomass; hence it is vital to understand the biomass composition and its properties. This chapter offers a comprehensive overview of biofuels, focusing on biomass characterization and its conversion approach. Additionally, the chapter examines diverse biofuel production methodologies that elucidate the process of converting biomass into biofuel.

Keywords Lignocellulose · Biofuels · Pyrolysis · Fermentation · Green energy

3.1 Introduction

The growing price of fossil fuel and its harmful environmental impact has influenced researchers towards the production of green energy. The variety of human activities, including transportation, the disposal of waste, and the burning of fossil fuels release CO_2 and deforestation, limiting CO_2 absorption (Dimitriou et al., 2018). On Earth, biomass is one of the most abundant and easily accessible carbon sources. Biomass

F. Parveen · A. Saxena · M. Ashfaque (✉)
Department of Biosciences, Integral University, Lucknow, India
e-mail: mohdashfaquekhan@rediffmail.com; mdashfaque@iul.ac.in

N. Shaheen
Department of Environmental Science, Integral University, Lucknow, India

V. Singh et al. (eds.), *Bioenergy and Value Added Byproducts from Biomass Waste*, Sustainable Environmental Waste Management Strategies,
https://doi.org/10.1007/978-3-032-22849-9_3

or biodegradable products such as waste and residues from agriculture, forestry, and biodegradable fractions of industrial and municipal waste (Saravanan et al., 2022). These biomasses can be further categorized into different generations of biofuel where edible substances like corn, wheat sugarcane, etc. are used in the first generation. The 2nd generation of biofuel utilizes non-food-based biomass which reduces the risk of food security for the current population. Apart from these, the use of algae or seaweed in biofuel production is considered a third generation of biofuel (Zhang & Zhang, 2019), and fourth generation of biofuel is mainly produced from genetically modified microorganisms (GMO's). The 2nd generation of biomass utilization for biofuel can be considered an ideal source for replacing fossil fuel. These biomasses are rich sources of carbohydrate sugar which includes cellulose and hemicellulose and which is the basic constituent of plant cell walls along with lignin that is a phenolic polymer (Saxena et al., 2024). However, the algal biomass is rich in proteins, lipids, and carbohydrates (Singh et al., 2022a). The primary sources of biodiesel production are vegetable oils, other lipid feedstocks, and ethanol from sugar and starch-based feedstocks. However, using food crops for biofuel has increased the global concern about food versus fuel security (Raud et al., 2019). Industries are showing their interest in biomethane production and biogas production for power generation through biomass utilization. The biomass type and operating conditions with integrated technology transform organic matter into biogas (Molino et al., 2013). The biochemical composition and inorganic contents differ in different biomass. Pretreatment of the biomass is important to make biomass easily accessible to the enzyme for hydrolysis and fermentation (Parveen et al., 2024). Biomass hydrolysis and decomposition have significant importance in biofuel production, it requires a variety of different pretreatment methods viz., physical pretreatment, chemical pretreatment, biological pretreatment, and combined pretreatment. These pretreatment methods break the covalently linked lignin with cellulose and hemicellulose which can be beneficial for fermentation (Parveen et al., 2024). The biomasses utilized for this biofuel compete with the food sources such as vegetable oils, beets, sugar cane and cereals), leading to societal and economic problems (Ahorsu et al., 2018). Over the years, the adoption of conversion techniques focused on biomass utilization has increased as a result of the recent fluctuations in fossil fuel prices and the overall shift towards green energy. The development of efficient conversion methods permits the chemical transition to energy and fuels, biomass potential signifies an improved role in global energy demand (Malode et al., 2021). Biomass conversion techniques allow unrefined biomass to be converted into solid, gaseous, or liquid fuels to generate power and heat simultaneously to provide energy services. However, in addition to their relatively low conversion efficiency and the high operational cost preventing the rapid adoption of these technologies, which stops additional use of biomass for direct combustion (Srivastava et al., 2020). Conversely, the drawbacks associated with efficacy during the chemical transition of biomass may be offset by effectively redeveloping the conversion path by using a variety of sub-processes that enhance overall performance. Numerous options for raw materials and finished goods in all industrial processes result in transformation chains with varying conversion rates and total process costs (bin Jumah, 2024) The most strategically significant sustainable fuel source is biofuels which aim to reduce greenhouse gas emissions (GHG), enhancing environment quality.

3.2 Biomass Selection Types

In general, biomass can be described as "organic matter derived from living organisms (Demirbaş, 2001).

These biomasses can be categorized into the following:

- **Herbaceous biomass**: It includes non-woody plant biomass such as switchgrass, miscanthus, corn stover, etc.
- **Deciduous biomass**: It includes woody plant material from trees such as poplar leaves, maple, beech oak, etc.
- **Agricultural biomass**: It includes organic materials from crops and agricultural waste. It is further categorized into (a) Energy Crops: non-food crops, food-based crops, and crop residues, and (b) Non energy crops: straw vegetative grasses, and farm forestry such as sugarcane, sugar beets, corn, and crop residues grain dust, and bagasse respectively.
- **Agroforestry residues**: It includes plant material such as twigs, branches, and leaves.
- **Urban residue**: It includes kitchen waste, yard waste, textile waste, paper and cardboard, etc.

Depending on the variety of biomass utilized for biofuel production it is divided into different generations (Table 3.1). The first generation of biofuels is produced from food or feed-based crops such as barley, wheat, maize, sugar beet, cassava, potatoes, sugarcane, and oilseeds while biodiesel is made from soybean and sunflower, while ethanol is produced by fermenting natural corn using starch-digesting microorganisms like *Saccharomyces cerevisiae* and *Rhizopus* sp. (Srivastava, 2019; Ullah et al., 2018). Converting starch and sucrose to bioethanol via enzymatic hydrolysis is the most commonly used industrial application for a wide range of first-generation biofuels (Kojima & Johnson, 2006). Compared to traditional gasoline, it comes with a higher octane rating and enhances combustion characteristics, enabling machines to run at a higher compression ratio (Zhang et al., 2023). The process of biodiesel production involves esterification or trans-esterification of animal fats or vegetable oils (soybean or palm oil) with alcohol to produce fatty acid methyl esters (FAME), which can be used as a mixed additive or as a diesel alternative (Araújo et al., 2017). The two diesels differ greatly in their structural makeup: biodiesel, a three-carbon ester that burns similarly to diesel, while diesel derived from fossil fuels is a hydrocarbon with 12–20 carbon atoms. On the other hand, biodiesel is cleaner and achieves comparable fuel efficiency to diesel (Solomon & Bailis, 2014). In places like the European Union and portions of Latin America, such as Argentina and Columbia, biodiesel is widely used However, first-generation biofuel is not considered a good choice due to its conflict with the food chain (Solomon & Bailis, 2014).

The non-food-based biomass are mainly utilized for 2nd generation of biofuel which includes, agricultural and forest residues, energy crops that are inexpensive, plentiful, and renewable. They are also known as lignocellulosic biofuels (Yong & Wu, 2022). The primary benefit of producing second-generation biofuels is that, because inedible biomass is utilized, it eliminates the direct food versus fuel rivalry problems that come with first-generation biofuels (Aro, 2016). There are two unique methods for producing second-generation biofuels: thermochemical and biochemical.

Heating the biomass with or without varying oxygen concentrations is known as thermochemical processing. Thus, various compounds can be made by chemical reformation and thermal degradation. Conversely, microbial and enzymatic processes are the foundation of biological conversion technologies, which turn biomass into sugars and transform them into fuels like butanol or ethanol (Canabarro et al.,

Table 3.1 Different generations of biofuels from a variety of sources and their key components

Generation	Biomass type	Key components	Biofuel	References
First generation	Edible crops: corn, sugarcane, wheat, soybean Vegetable oils: palm oil, rapeseed oil Animal fats	Sucrose, glucose, starch (amylose, amylopectin) Triglycerides, fatty acids	Bioethanol, biodiesel, biogas	Hirani et al. (2018)
Second-generation	Lignocellulosic biomass: straw, wood chips, corn stover Agricultural residues: rice husks, wheat straw Dedicated energy crops: switchgrass, miscanthus	Cellulose, hemicellulose, lignin	Cellulosic ethanol, bio-oil, biogas	Sims et al. (2010) and Bhatia et al. (2017)
Third-generation	Microalgae Macroalgae (seaweed)	Proteins, lipids Carbohydrates	Algal biodiesel, biohydrogen	Alam et al. (2015) and Debnath et al. (2021)
Fourth-generation	Engineered crops: high-yield plants, genetically modified algae	Cellulose, hemicellulose, lignin Lipids	Solar biofuels, carbon-negative fuels	Lü et al. (2011), Khan et al. (2021) and Abdullah et al. (2019)

2013). Additionally, Second-generation biofuels are currently less efficient than first-generation bioethanol because of a lower yield of the finished product (Mahapatra et al., 2021).

The third generation of biofuels is produced from algal biomass. Typically, the amount of lipids in algae determines its ability to produce biofuel. The production of biodiesel from algae has drawn a lot of attention due to their (i) high growth rates and lipid content (20–50%), (ii) ability to grow under harsh conditions, (iii) ability to trap CO_2 from the flue gases, and (iv) affordability (Jha et al., 2024) Many algal species, especially those that grow quickly, have been studied. Chlamydomonas can accumulate enormous amounts of lipids over time (Ho et al., 2014). Lipid accumulation explains the rise in lipid concentrations within microalgae cells but ignores total biomass output, hence the relationship between biomass productivity and lipid accumulation is not always absolute. Both the lipid concentration inside cells and the biomass these cells produce are explained by lipid productivity. An indication that is far more pertinent for determining the possible expenses of producing liquid biofuel is biomass productivity (Zhu et al., 2022).

Genetically engineered microorganisms formed by combining fragments from different organisms can be used to modify algal metabolism networks and improve cell productivity (Moravvej et al., 2019). Genetically modified (GM) algal biomass focuses on improving the light-capturing ability by increasing the light penetration, through minimizing photoinhibition and maximizing the photosynthetic efficiency therefore producing more fuel-generating biomass from algae (Abdullah et al., 2019). Using the chloroplast's truncated chlorophyll antenna is one way to improve light penetration into compact microalgae crops. The increase absorption spectrum range of microalgae in photosynthesis GM microalgae can increase their photosynthesis efficiency (Hu et al., 2023). The GM microalgae can manipulate the pigments by decreasing light absorption efficiency and increasing the light penetration into the microalgae culture by shrinking the size of the chlorophyll antenna (Khoo et al., 2023). The lipid or carbohydrate content of microalgae might rise noticeably as a result of metabolic engineering. One of the most alluring elements that can improve the effectiveness of a microalgae biomass is maximization of lipids and carbohydrates (Ahorsu et al., 2018).

3.3 Characterization of Biomass

The characterization of biomass is a crucial step in biomass conversion to biofuels. It is necessary to analyze its chemical composition, including cellulose, hemicellulose, and lignin, as well as its physical characteristics. The compositions of different biomass are described in Table 3.2. The Physical characterization includes Scanning Electron Microscope (SEM), Transmission Electron Microscope (TEM), X-ray diffraction, etc. chemical characterization techniques include FTIR, XPS, and NMR etc. Few examples of these methods have been shown in Fig. 3.1

Table 3.2 Biomass characterization based on proximate and ultimate analysis

Biomass type	Proximate analysis (%)			Ultimate analysis (%)				References
	Cellulose	Hemicellulose	Lignin	C	H	N	S	
Sugar beet	20–24	25–36	1–3	42.4	6.2	1.7	0.2	Pfersich et al. (2020) and Glaser et al. (2024)
Sugarcane bagasse	32–45	25–30	20–25	49.26	5.26	0.43	0.1	Kumar et al. (2021) and Ahmed et al. (2018)
Corn stover	34–38	20–25	15–20	43.28	5.92	1.96	0.66	He et al. (2018) and Chen et al. (2009)
Wheat straw	30–40	20–25	15–20	42.95	5.64	0.76	0.78	Zhan et al. (2023) and Biswas et al. (2017)
Rice husk	35–40	15–20	20–25	41.92	6.36	1.85	0.47	He et al. (2018) and Dagnino et al. (2018)
Soybean hulls	29–51	10–25	1–4	–	–	–	–	Bittencourt et al. (2022)
Oil palm fruit	17–33	43–65	13–37	41.17	7.43	0.84	0.15	Rahman et al. (2020) and Ramachandran et al. (2020)
Algal biomass	7–10	16–20	2–4	44.76	5.07	4.32	0.48	Ververis et al. (2007) and Moorthy et al. (2017)
Seaweed	1.5–34	-	–	30.0	5.5	3.9	2.9	Tsubaki et al. (2020) and Baghel et al. (2021)

3.4 Physical Method of Biomass Characterization

3.4.1 Scanning Electron Microscope (SEM)

The SEM analysis specifically uses scanning electron microscopy, which is a visual technique to observe the morphology of the biomass. The morphology of biomass depends on the molecular composition of the variety of the biomass. In SEM analysis, the biomass is scanned with an electron beam, the electron interacts with the atoms and generates a signal that provides detailed information about the texture of biomass

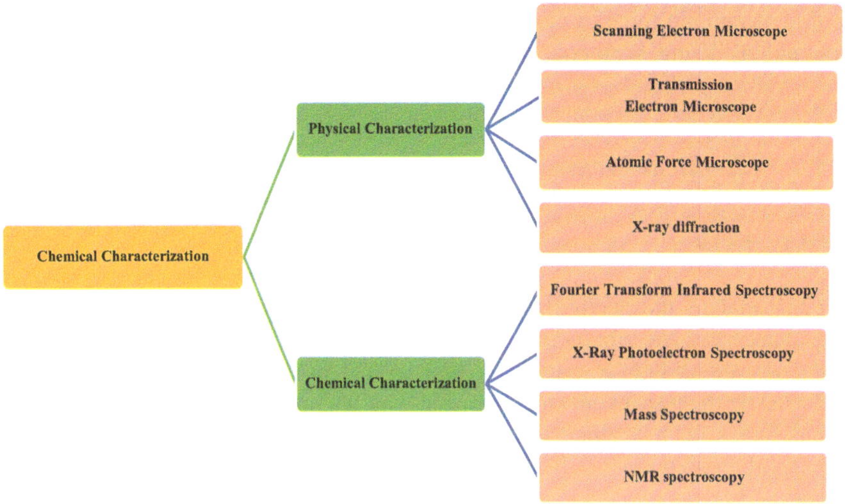

Fig. 3.1 Process of biomass characterization

such as smoothness, and roughness. It is feasible to use different modes at different SEM resolutions for improved characterization (Shah et al., 2018).

3.4.2 Transmission Electron Microscopy (TEM)

This technique also prevails in biomass characterization to study the microstructure and morphology of the biomass at a nanometre scale. In comparison to SEM analysis, TEM provides an accurate result. In this analysis the electron beam passes through the sample to observe the internal structure, however, in SEM analysis it only analyses the surface morphology. It is an important technique to observe the cell wall structure lignification or lignin distribution and hemicellulose deposition (Mayeen et al., 2018).

3.4.3 Atomic Force Microscopy (AFM)

AFM is a highly developed analytical tool that produces three-dimensional images. AFM measures the interaction occure between a surface and a sharp tip (radius of curvature of a few nanometers) that is positioned on a cantilever spring ($0.01\ N/m < k < 150\ N/m$). A high spatial resolution topographical image can be produced by raster-scanning the tip across the surface (Charrier et al., 2018). The interatomic force between the substance and the tip deflects at the flexible beam that connects the tip. The beam deflection is measured by an optical instrument and is proportional to the interatomic force. It can be used to observe a variety of surfaces, including polymer insulating, synthetic, or natural samples (Singh et al., 2021).

3.4.4 Chemical Method

3.4.4.1 FTIR

In FTIR analysis the electromagnetic radiation interacts with the biomass or sample and creates excited vibrational and rotational states in organic molecules, these molecules absorb light at specific frequencies and the generated spectrum data were collected by the instrument and compared with the reference data. Every functional group stands for a distinct infrared spectrum component which can identify the functional group present in the sample. Electromagnetic radiation interacts with the simple in the form of transmission, specular reflection, diffuse reflectance, and attenuated total reflectance (ATR). Since ATR and transmission methods do not require any sample preparation that is why they are frequently utilized in FTIR for lignocellulosic biomass analysis (Zhou et al., 2011). The FTIR analysis is used in comparing raw and processed sample spectra by observing the allocating peaks of the functional groups of each component (cellulose, hemicelluloses, and lignin), it also detected changes that took place in the material during pretreatment. Therefore, the reduction or elimination of a particular peak that is ascribed to a particular percentage can interpreted as the efficiency of treatment analysis of the sample (Mamilla et al., 2019).

3.4.4.2 Mass Spectroscopy (MS)

MS is also an analytical tool used to characterize biomass by providing a detailed analysis of the individual components such as carbohydrates, lignin, and lipids, and molecular structure on the basis of mass-to-charge ratio (Singh et al., 2021). Advanced MS is a combined chromatography technique, offers in-depth understanding of the constituents, impurities, and inhibitors that can affect process optimization and product quality. The MS that are crucial for a thorough investigation such as ICP–MS, GC–MS, LC–MS, and MALDI–MS (Gutierrez Reyes et al., 2025).

3.5 Biomass Conversion Approaches

3.5.1 Bioethanol

Bioethanol is a renewable energy produced by fermenting lignocellulosic biomass such as sugarcane bagasse, rice husk, corn cob, etc. (Table 3.3). It is considered a potential and most commonly produced biofuel in the world (Golušin et al., 2013). The processes for efficient bioethanol production are pretreatment biomass, hydrolysis, fermentation, and distillation.

3.5.2 Pretreatment

An efficient pretreatment process is required due to the complex structure of lignocellulose, which makes hydrolysis challenging. The different pretreatment breaks down the biomass's resistant components increasing the substrate porosity. As a result, it allows cellulases to be exposed to the maximum amount of cellulose surface area to achieve efficient hydrolysis with the least amount of energy and maximum sugar recovery, which can be utilized for bioethanol production (Limayem & Ricke, 2012). Pretreatment encompasses various categories such as physical, chemical, thermochemical, and biological (Jönsson & Martín, 2016).

3.5.3 Hydrolysis

The hydrolysis of holocellulose is an important process in bioethanol production that can be done using chemicals or enzymes. Most bacteria (*Clostridium, Cellulomonas, Bacillus*) and fungi (*Trichoderma, Schizophyllum, Aspergillus*) produce enzymes that can be used for hydrolysis and microbial enzymes are an economic process as compared to chemicals (Sun & Cheng, 2002). The end product of hydrolysis mostly contains reducing sugars (glucose, fructose, xylose, etc.) These reducing sugars can be further exploited for bioethanol production (Rastogi & Shrivastava, 2017).

3.5.4 Fermentation

Fermentation is a biological process in which microorganisms like bacteria, fungus, or yeast transform monomeric sugar into bioethanol (Fig. 3.3). It was observed that the microbial cellulolytic enzyme exhibited the highest activity at 50 °C, which decreases with decreasing temperature, while bacteria require an ideal temperature range of 28–37 °C for ethanol fermentation. Most microorganisms and enzymes

Table 3.3 Types of biofuels and their bioconversion process

Biofuel Type	Source biomass	Conversion approach	Key steps involved	Examples	References
Bioethanol	Sugarcane, corn, lignocellulose	Biochemical conversion	• Pretreatment • Enzymatic hydrolysis • Fermentation	Sugarcane ethanol, cellulosic ethanol	Panahi et al. (2022) and Toor et al. (2020)
Biodiesel	Vegetable oils, animal fats	Chemical conversion—transesterification	• Oil extraction • Reaction with alcohol (methanol/ethanol) • Catalyst addition (base/acid)	Soybean biodiesel, palm oil biodiesel	Veljković et al. (2014) and Singh et al. (2020)
Biogas	Organic waste, MSW, manure	Anaerobic digestion	• Feedstock pre-treatment • Anaerobic fermentation • Gas collection	CO_2., methane, H_2	Jameel et al. (2024) and Kumar et al. (2022)
Bio-oil	Lignocellulosic biomass, agricultural residues	Thermochemical conversion (pyrolysis)	• Rapid heating in low oxygen • Liquid bio-oil collection • By-products (char, gas)	Pyrolysis bio-oil from wood chips, corn stalk, rice husk	Singh et al. (2023)
Biochar	Agricultural and forestry waste	Thermochemical conversion pyrolysis, torrefaction. gasification	• Low-temperature pyrolysis • Solid biochar recovery	Biochar from rice husk, SCB	Yaashikaa et al. (2020)

cannot work effectively under such conditions, and several variables, such as temperature, pH, and the ability to tolerate ethanol in the media, affect the expected theoretical yield (Saxena et al., 2024). However, *Saccharomyces cerevisiae* is the most widely used yeast to convert monomeric sugars into bioethanol (Ahmad et al., 2021).There are several limitations during the fermentation process some of which are mentioned above additionally the hydrolysis of pentoses is the major issue as decreases the ethanol yield. Therefore, recent researchers have focused on making genetically engineered microbes that can utilize pentoses as well as hexoses for efficient bioethanol production. Some examples of these microbes are *Escherichia coli*, *Zymomonas mobilis*, and *Saccharomyces* that are used for genetic medication for efficient ethanol production and are found to be effective in increased bioethanol yield (Saxena et al., 2024).

3.5.5 Distillation

Distillation is the downstream processing step of bioethanol production from biomass. The first-generation bioethanol can reach 12% wt, however, 2nd generation typically only recovers 5% wt of ethanol at the end of fermentation (Branco et al., 2018). For the maximum recovery of ethanol, the crude broth was first passed through a stripper column that concentrated the ethanol up to 20% wt, it was distilled in a rectifier column to 95.0% wt ethanol in water. The 95.0% wt recovered ethanol is referred to as hydrous ethanol, and it is rarely used, unlike pure absolute or anhydrous ethanol (Behera et al., 2022). The ethanol is separated using a variety of distillation methods, including conventional, liquid extraction, fermentation hybrid, azeotropic, absorption, extractive, and membrane separation (Khalid et al., 2019) (Table 3.3).

3.5.5.1 Biodiesel

Biodiesel is methyl ester produced from edible and non-edible oils it is a potential substitute for low-carbon fuel. Biodiesel is a promising renewable fuel that can be combined with conventional petrochemical fuels without compromising engine performance. Combining biodiesel with conventional fuel can lower greenhouse gas emissions while also providing lubrication to the engine. Biodiesel is primarily classified into the first, second, and third generations of biodiesel. The first-generation biodiesel from soybean, rapeseed, and palm oils continues to dominate the global production of biodiesel (Ambat et al., 2018). In earlier days, biodiesel production from edible substrates was quite popular. However, its major drawback was it has increased the risk of scarcity of food supply. Additionally, high cost and limited area for cultivation were also the major concerns that restricted the researchers toward alternate source (Tariq et al., 2012).The second-generation biodiesel production was done by utilizing non-edible substrates such as Neem oil, Jatropha, Karanja oil,

Mahua indica oil, etc. Utilization of such substrate surpasses the drawback of first-generation (Shameer & Ramesh, 2017). Microalgae and waste oils can be used to produce third-generation biodiesel. Third-generation biodiesel's primary benefits include a maximum oil content, a reduced environmental effect, accelerated growth and productivity, ease of access to agricultural land, and less impact on the food supply (Bajhaiya et al., 2010). The methods used for the production of biodiesel include pyrolysis, transesterification, and the supercritical fluid approach, among others (Ibrahim et al., 2022).

3.5.5.2 Transesterification

Biodiesel production involves the transesterification of short-chain alcohol such as methanol or ethanol in the presence of a catalyst (Table 3.3). The most common catalysts for reactions are NaOH and KOH (Avhad & Marchetti, 2016). Transesterification can also be done without using a catalyst using a supercritical process which includes three reversible reactions: the conversion of triglycerides into diglycerides, the conversion of diglycerides into monoglycerides, and then the conversion of monoglycerides into glycerol. Each glyceride that reacts results in an ester (biodiesel) production (Shahid & Jamal, 2011).

$$\text{Triglycerides} + 3\,\text{Alcohol} \xrightarrow{\text{catalyst}} 3\,\text{Alkyl ester} + \text{glycerol}$$

A material that speeds up a reaction without being consumed is called a catalyst. There are two types of catalysts used in biodiesel production (1) Homogenous catalyst and Heterogenous catalyst Different types of catalysts are described in Fig. 3.2.

3.5.5.3 Homogenous Catalyst

Homogenous catalyst, where the catalyst acts in the same phase of the reaction mixture due to this mass transfer resistance is reduced. These catalysts can be acidic basic and enzymatic. The most common basic catalysed used were KOH and NaOH thar are easily dissolved in methanol. Acidic catalysts include HCl, and NaOH however enzymatic catalysts include Lipases (Ajala et al., 2015).The drawback of homogenous catalysts is that they cannot be recycled after use. Additionally, it causes corrosion in the reactor and is quite difficult to separate from the reaction mixture (Borges & Díaz, 2012).

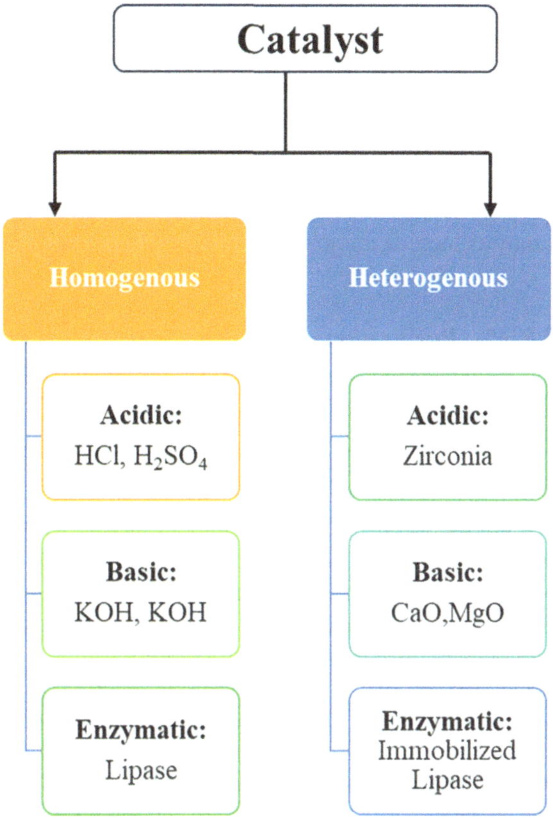

Fig. 3.2 Types of catalysts used in biodiesel production

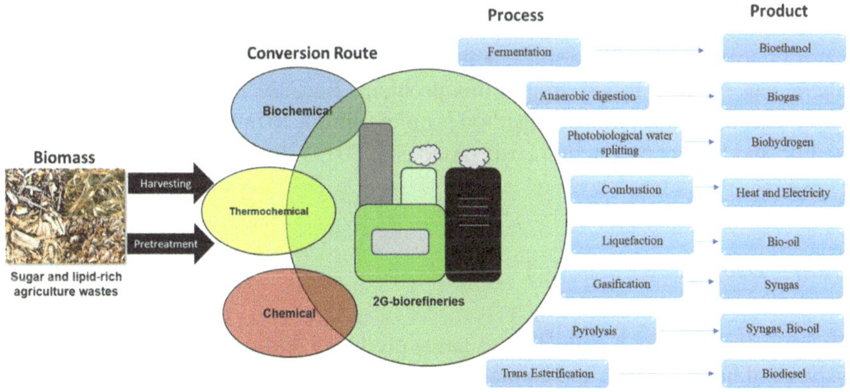

Fig. 3.3 Various methods of biomass utilization for biofuel production

3.5.5.4 Heterogeneous Catalyst

A heterogeneous catalyst is when a catalyst acts in different phases with the reaction mixture. Using a homogenous catalyst in biodiesel production increases its sustainability and reduces the cost of production. This catalyst's recovery is not tedious but can also be regenerated and reused. Acid heterogeneous catalysts include hydrotalcite, and acid basic catalysts include metal oxides such as CaO, MgO and enzymatic catalysts include immobilized lipases (Nasreen et al., 2018).

3.5.5.5 Purification

After Transesterification, purification of biodiesel is necessary to meet quality standards and enable future commercialization. First, in the decanter, the glycerin and biodiesel phases are separated by centrifugation or sedimentation because of their different densities. Evaporation or distillation can be employed to extract the excess alcohol used in the reaction from both phases. The recovered biodiesel after distillation still contains some catalyst, soap, alcohol, free glycerol unreacted mono, di, and triglycerides (Atadashi et al., 2011). The most widely used method for biodiesel purification is washing it with waste. This process is easy to use, effective, and produces highly pure biodiesel. Biodiesel can be mixed with distilled water, deionized water, and an organic solvent (Jesus et al., 2019).

The remaining glycerol, methanol, catalyst, and any soap produced during the process can all be removed because of their water solubility. Finally, it is necessary to dehydrate cleaned biodiesel. Water-based biodiesel purification takes a long time and generates a lot of wastewater that cannot be discarded into the environment due the impurities (Veljković et al., 2014) The cost of producing biodiesel goes up because wastewater effluent needs to be treated. A dry washing purification method that uses solid sorbents like silica, activated carbon, and ion exchange resin, among others, can be used to get around this problem. Because does not requires hence zero wastewater, this purifying method is quick, simple to implement in an industrial facility, and lowers production costs (Atadashi, 2015) (Fig. 3.3).

3.5.6 Biogas

Biogas is the term for a collection of gases produced during the decomposition and disposal of human, plant, and animal waste. These gases are produced in the absence of oxygen in the presence of anaerobic bacteria, particularly methanogenic bacteria (Hagos et al., 2017). This gas must be collected to be used. It is naturally produced in marshes, swamps, and urban waste disposal sites (Table 3.3). The biomass for biogas is divided into five categories: (1) agricultural residues and wastes. (2) Municipal solid waste (MSW), (3) municipal wastewater, (4) animal excrement and (5) industrial solid and liquid waste (Jameel et al., 2024). A variety of gases make up biogas, primarily carbon dioxide (CO_2) and methane (CH_4), with 1–5% of other gases, such as hydrogen (H_2). Bacteria play an important role in the production of these gases during the anaerobic bio-degradation of organic matter. Because of the energy it releases, biogas can be used as a fuel for cooking and heating in any nation. Additionally, biogas can be utilized in an anaerobic digester, which uses a gas engine to transform the gas' energy into heat and power (Fig. 3.3) (Sutaryo, 2012).

3.5.6.1 Biogas Production

In general, the primary step of biogas production is anaerobic digestion of biomass. The microorganisms that are responsible for anaerobic digestion of complex organic material are mainly bacteria, fungi, and archaea. This microorganism produces diverse range or enzyme such as cellulase, xylanase, amylase, protease which can easily hydrolyse the biomass. The biomass degradation process includes the following steps: hydrolysis, acidogenesis, acetogenesis and methanogenesis (Schnürer, 2016). In first, these microorganism converts the complex organic material into soluble monomers and oligomers such as sugars amino acids and fatty acids through microbial or enzymatic degradation. Followed by the second step where it is subsequently converted into short-chain fatty acids, alcohols, hydrogen and carbon dioxide via fermentation. In the third step, the alcohols/acids produced were further degraded through anaerobic oxidation leading to the production of hydrogen, carbon dioxide, and acetate that are used by methanogens for biogas production (Azman et al., 2015).

3.5.7 Biochar

Biochar, a byproduct of the thermochemical conversion of organic carbon-rich biomass through pyrolysis, torrefaction, and gasification of biomass under different. It has a high oxygen content (27–34%) in the form of carboxylic and phenolic acid groups, and the inclusion of a sulfonic group enhances the catalyst's surface molecular absorption and catalytic activity. Because of its remarkable properties, including stability, porosity, recalcitrance potential, and CO_2 sequestration, biochar is a highly fixed carbon material derived from organic sources with numerous applications (Wijitkosum & Jiwnok, 2019). Biochar is a promising method for sequestering carbon (C), generating energy, boosting soil productivity, and enhancing soil and environmental quality. Biochar has the potential in bioenergy production, contributing to economic sustainability, reducing greenhouse gas, and improving soil health. The use of feedstocks such as agricultural, forest, and municipal waste plays an important role in biochar production and its quality (Table 3.3) (Li et al., 2023).

3.5.7.1 Pyrolysis Technology

One of the earliest techniques for processing biomass is pyrolysis, a thermochemical process that involves burning compounds at high temperatures in a closed reactor with little to no oxygen present (Osman et al., 2023). The result is biochar, also known as bio-crude, which is made from various biomasses under higher pressure and temperature (Wei et al., 2022). The main products are categorized as solid (biochar), liquid (bio-oil), and gas (syngas) at temperatures between 350 and 650 °C. Pyrolysis is categorized based on temperature, residence time, particle size, and heating rate as slow pyrolysis and fast pyrolysis (Lachos-Perez et al., 2023).

In slow pyrolysis, a variety of organic and non-organic materials, including heavy crude oil, agricultural and forest residues, algal biomass, etc. are utilized for biochar formation. The slow pyrolysis process involves breaking down biomass at a temperature between 350 and 500 °C, which gives the pyrolysis vapor enough time to settle and raises the secondary cracking level of the biomass. Optimum temperature plays an important role in the quality and yield of biochar (Premchand et al., 2023).

In contrast, fast pyrolysis uses extremely high heating rates (about 1000 °C/min) to reach the pyrolysis temperature of 500 °C, and the vapor's residence period is often shorter than two seconds (Choi et al., 2019). Fast pyrolysis produces pyrolysis fumes and biochar (10–15 weight percent) by breaking down biomass particles quickly. Biochar is the result of this process, and the condensable content in the pyrolysis fumes is quenched and collected in the downstream facilities as a dark-brown liquid known as bio-oil (Fahmy et al., 2020).

3.5.7.2 Gasification

The process of gasification, which typically occurs at 700–1000 °C, produces a gas product by incompletely burning biomass with various gasifying agents, including air, pure oxygen, or steam plus oxygen. In a biomass gasification process, researchers typically start by figuring out how to increase the quantity and quality of syngas while lowering certain pollutants including tar, fly ash, nitrogen oxides, and sulfur dioxide (You et al., 2017). The carbon content of biochar generated by biomass gasification directly impacts its quality. The equivalency ratio (ER), feedstock characteristics, gasifying agent, and pressure are the primary gasified parameters that influence it. The ER value is thought to be the most significant of these parameters influencing the gasification process; based on the physical–chemical characteristics of biomass, the ideal value is between 0.25 and 0.28. Increasing ER typically raises the gasification temperature, which has an impact on the quality of the biochar that is produced (Hansen et al., 2015).

3.5.7.3 Torrefaction

The biomass torrefaction is to produce a "charred" product that can be utilized as fuel or as a soil amendment (Arriola et al., 2020). In an inert atmosphere, biomass is heated directly or indirectly to temperatures between 200 and 300 °C of heating (less than 50 °C/min) and a comparatively extensive residence time (20–120 min) in a standard torrefaction process (Wang et al., 2017). In this process, about 30% of the mass of extremely reactive volatile chemicals is converted to torrefied vapor. The goal of this procedure is to produce torrefied biochar, a dark brown solid fuel with 90% of the initial energy content, yielding an energy density of roughly 1.3% (Ahmed et al., 2023; Singh & Mishra, 2020, 2021; Singh et al., 2016, 2022b; Zakaria et al., 2023).

3.6 Conclusion

The sustainable production of biofuels depends mainly on the variety of biomass. The biomass selection is based on its composition, availability, and compatibility for particular conversion processes. The utilization of different types of biomasses for biofuel production depends on its composition. The researchers are focusing on economic and sustainable methods for biofuel production but its implementation at a commercial scale has various limitations. Biomass can be converted into biofuels like bioethanol, biodiesel, biogas, and bio-oil using a variety of conversion techniques, including as thermochemical (pyrolysis, gasification) and biochemical (fermentation, anaerobic digestion) processes. The efficiency of converting biomass into biofuel has been further improved by developments in pretreatment technologies, catalytic processes, and microbial engineering. All things considered, the energy sector is among the most significant especially biofuel, working in coordination with the agricultural sector by using its byproduct to minimize food dependency for biofuel production. The most widely grown cereal crops in the world, including corn, rice, and wheat where their byproducts, can be consumed, additionally, corn stover, rice straw, and wheat straw, can also be utilized for biofuel production. In the end, biomass-derived biofuel is more economical and sustainable.

Environmental impact assessment, effective processing methods, and ongoing research and innovation in biomass choices are critical to developing a sustainable bioeconomy. Reducing reliance on fossil fuels, cutting greenhouse gas emissions, and promoting global energy security can all be achieved by combining the production of biofuel with the ideas of the circular economy.

Acknowledgements Authors are thankful to the Integral University for providing necessary support.

References

Abdullah, B., Muhammad, S. A. F. S., Shokravi, Z., Ismail, S., Kassim, K. A., Mahmood, A. N., & Aziz, M. M. A. (2019). Fourth generation biofuel: A review on risks and mitigation strategies. *Renewable and Sustainable Energy Reviews, 107*, 37–50.

Ahmad, A., Naqvi, S. A., Jaskani, M. J., Waseem, M., Ali, E., Khan, I. A., Faisal Manzoor, M., Siddeeg, A., & Aadil, R. M. (2021). Efficient utilization of date palm waste for the bioethanol production through Saccharomyces cerevisiae strain. *Food Science & Nutrition, 9*, 2066–2074.

Ahmed, G., Jamal, F., Tiwari, R. K., Singh, V., Rai, S. N., Chaturvedi, S. K., Pandey, K., Singh, S. K., Kumar, A., Narayan, S., & Vamanu, E. (2023). Arsenic exposure to mouse visceral leishmaniasis model through their drinking water linked to the disease exacerbation via modulation in host protective immunity: a preclinical study. *Scientific Reports*, *13*(1). https://doi.org/10.1038/s41598-023-48642-z

Ahmed, N., Zeeshan, M., Iqbal, N., Farooq, M. Z., & Shah, S. A. (2018). Investigation on bio-oil yield and quality with scrap tire addition in sugarcane bagasse pyrolysis. *Journal of Cleaner Production, 196*, 927–934.

Ahorsu, R., Medina, F., & Constantí, M. (2018). Significance and challenges of biomass as a suitable feedstock for bioenergy and biochemical production: A review. *Energies, 11*, 3366.

Ajala, O. E., Aberuagba, F., Odetoye, T. E., & Ajala, A. M. (2015). Biodiesel: Sustainable energy replacement to petroleum-based diesel fuel–a review. *ChemBioEng Rev, 2*, 145–156.

Alam, F., Mobin, S., & Chowdhury, H. (2015). Third generation biofuel from algae. *Procedia Engineering, 105*, 763–768.

Ambat, I., Srivastava, V., & Sillanpää, M. (2018). Recent advancement in biodiesel production methodologies using various feedstock: A review. *Renewable and Sustainable Energy Reviews, 90*, 356–369.

Araújo, K., Mahajan, D., Kerr, R., & da Silva, M. (2017). Global biofuels at the crossroads: An overview of technical, policy, and investment complexities in the sustainability of biofuel development. *Agriculture, 7*, 32.

Aro, E.-M. (2016). From first generation biofuels to advanced solar biofuels. *Ambio, 45*, 24–31.

Arriola, E., Chen, W.-H., Chih, Y.-K., De Luna, M. D., & Show, P. L. (2020). Impact of post-torrefaction process on biochar formation from wood pellets and self-heating phenomena for production safety. *Energy, 207*, 118324.

Atadashi, I. M. (2015). Purification of crude biodiesel using dry washing and membrane technologies. *Alexandria Engineering Journal, 54*, 1265–1272.

Atadashi, I. M., Aroua, M. K., Aziz, A. R. A., & Sulaiman, N. M. N. (2011). Refining technologies for the purification of crude biodiesel. *Applied Energy, 88*, 4239–4251.

Avhad, M. R., & Marchetti, J. M. (2016). Innovation in solid heterogeneous catalysis for the generation of economically viable and ecofriendly biodiesel: A review. *Catal Rev, 58*, 157–208.

Azman, S., Khadem, A. F., Van Lier, J. B., Zeeman, G., & Plugge, C. M. (2015). Presence and role of anaerobic hydrolytic microbes in conversion of lignocellulosic biomass for biogas production. *Critical Reviews in Environment Science and Technology, 45*, 2523–2564.

Baghel, R. S., Reddy, C. R. K., & Singh, R. P. (2021). Seaweed-based cellulose: Applications, and future perspectives. *Carbohydrate Polymers, 267*, 118241.

Bajhaiya, A. K., Mandotra, S. K., Suseela, M. R., Toppo, K., & Ranade, S. (2010). Algal biodiesel: The next generation biofuel for India. *Asian J Exp Biol Sci, 4*, 728–739.

Behera, S. S., Saranraj, P., & Ray, R. C. (2022) Microbial bioethanol fermentation technologies—Recent trends and future prospects. *Biofuels and Biorefining*, 75–108.

Bhatia, S. K., Kim, S.-H., Yoon, J.-J., & Yang, Y.-H. (2017). Current status and strategies for second generation biofuel production using microbial systems. *Energy Conversion and Management, 148*, 1142–1156.

bin Jumah, A. (2024). A comprehensive review of production, applications, and the path to a sustainable energy future with hydrogen. *RSC Advances, 14*, 26400–26423.

Biswas, B., Pandey, N., Bisht, Y., Singh, R., Kumar, J., & Bhaskar, T. (2017). Pyrolysis of agricultural biomass residues: Comparative study of corn cob, wheat straw, rice straw and rice husk. *Bioresource Technology, 237*, 57–63.
Bittencourt, G. A., de Souza Vandenberghe, L. P., Valladares-Diestra, K. K., & Soccol, C. R. (2022). Soybean hull valorization for sugar production through the optimization of citric acid pretreatment and enzymatic hydrolysis. *Industrial Crops and Products, 186*, 115178.
Borges, M. E., & Díaz, L. (2012). Recent developments on heterogeneous catalysts for biodiesel production by oil esterification and transesterification reactions: A review. *Renewable and Sustainable Energy Reviews, 16*, 2839–2849.
Branco, R. H. R., Serafim, L. S., & Xavier, A. M. R. B. (2018). Second generation bioethanol production: On the use of pulp and paper industry wastes as feedstock. *Fermentation, 5*, 4.
Canabarro, N., Soares, J. F., Anchieta, C. G., Kelling, C. S., & Mazutti, M. A. (2013). Thermochemical processes for biofuels production from biomass. *Sustainable Chemical Processes, 1*, 1–10.
Charrier, A. M., Lereu, A. L., Farahi, R. H., Davison, B. H., & Passian, A. (2018). Nanometrology of biomass for bioenergy: The role of atomic force microscopy and spectroscopy in plant cell characterization. *Frontiers in Energy Research, 6*, 11.
Chen, M., Zhao, J., & Xia, L. (2009). Comparison of four different chemical pretreatments of corn stover for enhancing enzymatic digestibility. *Biomass and Bioenergy, 33*, 1381–1385.
Choi, J. H., Kim, S.-S., Kim, J., & Woo, H. C. (2019). Fast pyrolysis of fermentation residue derived from *Saccharina japonica* for a hybrid biological and thermal process. *Energy, 170*, 239–249.
Dagnino, E. P., Felissia, F. E., Chamorro, E., & Area, M. C. (2018). Studies on lignin extraction from rice husk by a soda-ethanol treatment: Kinetics, separation, and characterization of products. *Chemical Engineering Research and Design, 129*, 209–216.
de Jesus, S. S., Ferreira, G. F., Maciel, M. R. W., & Maciel Filho, R. (2019). Biodiesel purification by column chromatography and liquid-liquid extraction using green solvents. *Fuel, 235*, 1123–1130.
Debnath, C., Bandyopadhyay, T. K., Bhunia, B., Mishra, U., Narayanasamy, S., & Muthuraj, M. (2021). Microalgae: Sustainable resource of carbohydrates in third-generation biofuel production. *Renewable and Sustainable Energy Reviews, 150*, 111464.
Demirbaş, A. (2001). Biomass resource facilities and biomass conversion processing for fuels and chemicals. *Energy Conversation and Management, 42*, 1357–1378.
Dimitriou, I., Goldingay, H., & Bridgwater, A. V. (2018). Techno-economic and uncertainty analysis of biomass to liquid (BTL) systems for transport fuel production. *Renewable and Sustainable Energy Reviews, 88*, 160–175.
Fahmy, T. Y. A., Fahmy, Y., Mobarak, F., El-Sakhawy, M., & Abou-Zeid, R. E. (2020). Biomass pyrolysis: Past, present, and future. *Environment, Development and Sustainability, 22*, 17–32.
Glaser, S. J., Abdelaziz, O. Y., Demoitié, C., Galbe, M., Pyo, S.-H., Jensen, J. P., & Hatti-Kaul, R. (2024). Fractionation of sugar beet pulp polysaccharides into component sugars and pre-feasibility analysis for further valorisation. *Biomass Conversion and Biorefinery, 14*, 3575–3588.
Golušin, M., Dodić, S., & Popov, S. (2013). Strategic priorities of sustainable energy development. *IU Sustainable Energy Management*, 243–333
Gutierrez Reyes, C. D., Onigbinde, S., Amarasekara, A. S., Adeniyi, M., Solomon, J., Gonzalez Ponce, H. A., & Mechref, Y. (2025). MS-based characterization of biomass-derived materials: activated carbons and solvent liquefaction products. *Polymers (Basel), 17*, 258.
Hagos, K., Zong, J., Li, D., Liu, C., & Lu, X. (2017). Anaerobic co-digestion process for biogas production: Progress, challenges and perspectives. *Renewable and Sustainable Energy Reviews, 76*, 1485–1496.
Hansen, V., Müller-Stöver, D., Ahrenfeldt, J., Holm, J. K., Henriksen, U. B., & Hauggaard-Nielsen, H. (2015). Gasification biochar as a valuable by-product for carbon sequestration and soil amendment. *Biomass and Bioenergy, 72*, 300–308.

He, X., Liu, Z., Niu, W., Yang, L., Zhou, T., Qin, D., Niu, Z., & Yuan, Q. (2018). Effects of pyrolysis temperature on the physicochemical properties of gas and biochar obtained from pyrolysis of crop residues. *Energy, 143*, 746–756.

Hirani, A. H., Javed, N., Asif, M., Basu, S. K., & Kumar, A. (2018). A review on first-and second-generation biofuel productions. *Biofuels: Greenhouse Gas Mitigation and Global Warming*, 141–154 (next Generation Biofuels Role in Biotechnology).

Ho, S.-H., Nakanishi, A., Ye, X., Chang, J.-S., Hara, K., Hasunuma, T., & Kondo, A. (2014). Optimizing biodiesel production in marine *Chlamydomonas* sp. JSC4 through metabolic profiling and an innovative salinity-gradient strategy. *Biotechnology for Biofuels, 7*, 1–16.

Hu, J., Wang, D., Chen, H., & Wang, Q. (2023). Advances in genetic engineering in improving photosynthesis and microalgal productivity. *International Journal of Molecular Sciences, 24*, 1898.

Ibrahim, N. A., Rashid, U., Hazmi, B., Moser, B. R., Alharthi, F. A., Rokhum, S. L., & Ngamcharussrivichai, C. (2022). Biodiesel production from waste cooking oil using magnetic bifunctional calcium and iron oxide nanocatalysts derived from empty fruit bunch. *Fuel, 317*, 123525.

Jameel, M. K., Mustafa, M. A., Ahmed, H. S., Jassim Mohammed, A., Ghazy, H., Shakir, M. N., Lawas, A. M., Khudhur Mohammed, S., Idan, A. H., & Mahmoud, Z. H. (2024). Biogas: Production, properties, applications, economic and challenges: A review. *Results in Chemistry*, 101549

Jha, S., Singh, R., Pandey, B. K., Tiwari, A. K., Shukla, S., Dikshit, A., & Bhardwaj, A. K. (2024). Recent aspects of algal biomass for sustainable fuel production: A review. *Discover Sustainability, 5*, 300.

Jönsson, L. J., & Martín, C. (2016). Pretreatment of lignocellulose: Formation of inhibitory by-products and strategies for minimizing their effects. *Bioresource Technology, 199*, 103–112.

Khalid, A., Aslam, M., Qyyum, M. A., Faisal, A., Khan, A. L., Ahmed, F., Lee, M., Kim, J., Jang, N., & Chang, I. S. (2019). Membrane separation processes for dehydration of bioethanol from fermentation broths: Recent developments, challenges, and prospects. *Renewable and Sustainable Energy Reviews, 105*, 427–443.

Khan, N., Sudhakar, K., & Mamat, R. (2021). Role of biofuels in energy transition, green economy and carbon neutrality. *Sustainability, 13*, 12374.

Khoo, K. S., Ahmad, I., Chew, K. W., Iwamoto, K., Bhatnagar, A., & Show, P. L. (2023). Enhanced microalgal lipid production for biofuel using different strategies including genetic modification of microalgae: A review. *Progress in Energy and Combustion Science, 96*, 101071.

Kojima, M., & Johnson, T. (2006). *Potential for biofuels for transport in developing countries*. World Bank.

Kumar, A., Kumar, V., & Singh, B. (2021). Cellulosic and hemicellulosic fractions of sugarcane bagasse: Potential, challenges and future perspective. *International Journal of Biological Macromolecules, 169*, 564–582.

Kumar, R., Kumar, A., & Pal, A. (2022). Overview of hydrogen production from biogas reforming: Technological advancement. *International Journal of Hydrogen Energy, 47*, 34831–34855.

Lachos-Perez, D., Martins-Vieira, J. C., Missau, J., Anshu, K., Siakpebru, O. K., Thengane, S. K., Morais, A. R. C., Tanabe, E. H., & Bertuol, D. A. (2023). Review on biomass pyrolysis with a focus on bio-oil upgrading techniques. *Analytica, 4*, 182–205.

Li, Y., Gupta, R., Zhang, Q., & You, S. (2023). Review of biochar production via crop residue pyrolysis: Development and perspectives. *Bioresource Technology, 369*, 128423.

Limayem, A., & Ricke, S. C. (2012). Lignocellulosic biomass for bioethanol production: Current perspectives, potential issues and future prospects. *Progress in Energy and Combustion Science, 38*, 449–467.

Lü, J., Sheahan, C., & Fu, P. (2011). Metabolic engineering of algae for fourth generation biofuels production. *Energy & Environmental Science, 4*, 2451–2466.

Mahapatra, S., Kumar, D., Singh, B., & Sachan, P. K. (2021). Biofuels and their sources of production: A review on cleaner sustainable alternative against conventional fuel, in the framework of the food and energy nexus. *Energy Nexus, 4*, 100036.

Mamilla, J. L., Novak, U., Grilc, M., & Likozar, B. (2019). Natural deep eutectic solvents (DES) for fractionation of waste lignocellulosic biomass and its cascade conversion to value-added bio-based chemicals. *Biomass and bioenergy, 120*, 417–425.

Malode, S. J., Prabhu, K. K., Mascarenhas, R. J., Shetti, N. P., & Aminabhavi, T. M. (2021). Recent advances and viability in biofuel production. *Energy Conversation and Management X, 10*, 100070.

Mayeen, A., Shaji, L. K., Nair, A. K., & Kalarikkal, N. (2018). Morphological characterization of nanomaterials. In *Characterization of nanomaterials* (pp. 335–364). Elsevier.

Molino, A., Migliori, M., Ding, Y., Bikson, B., Giordano, G., & Braccio, G. (2013). Biogas upgrading via membrane process: Modelling of pilot plant scale and the end uses for the grid injection. *Fuel, 107*, 585–592.

Moorthy, R. K., Premalatha, M., & Arumugam, M. (2017). Batch sedimentation studies for freshwater green alga *Scenedesmus abundans* using combination of flocculants. *Frontiers in Chemistry, 5*, 37.

Moravvej, Z., Makarem, M. A., & Rahimpour, M. R. (2019). The fourth generation of biofuel. In *Second third generation feeding* (pp. 557–597). Elsevier.

Nasreen, S., Nafees, M., Qureshi, L. A., Asad, M. S., Sadiq, A., & Ali, S. D. (2018). Review of catalytic transesterification methods for biodiesel production. *Biofuels: State of Development, 6*, 93–119.

Osman, A. I., Farghali, M., Ihara, I., Elgarahy, A. M., Ayyad, A., Mehta, N., Ng, K. H., Abd El-Monaem, E. M., Eltaweil, A. S., & Hosny, M. (2023). Materials, fuels, upgrading, economy, and life cycle assessment of the pyrolysis of algal and lignocellulosic biomass: A review. *Environmental Chemistry Letters, 21*, 1419–1476.

Panahi, H. K. S., Dehhaghi, M., Guillemin, G. J., Gupta, V. K., Lam, S. S., Aghbashlo, M., & Tabatabaei, M. (2022). Bioethanol production from food wastes rich in carbohydrates. *Current Opinion in Food Science, 43*, 71–81.

Parveen, F., Saxena, A., Hussain, A., Giri, B. S., & Ashfaque, M. (2024). Facile fractionation of lignocellulosic biomass: A review on promising green technology of deep eutectic solvent for bioethanol production. *Biomass Conversion and Biorefinery*, 1–20.

Pfersich, J., Arauzo, P. J., Lucian, M., Modugno, P., Titirici, M.-M., Fiori, L., & Kruse, A. (2020). Hydrothermal conversion of spent sugar beets into high-value platform molecules. *Molecules, 25*, 3914.

Premchand, P., Demichelis, F., Chiaramonti, D., Bensaid, S., & Fino, D. (2023). Biochar production from slow pyrolysis of biomass under CO_2 atmosphere: A review on the effect of CO_2 medium on biochar production, characterisation, and environmental applications. *Journal of Environmental Chemical Engineering, 11*, 110009.

Rahman, N. A., Hasan, M. F., Nyakuma, B. B., Rahman, M. R. A., Rahman, M. M., Osman, S. A., Bosiran, M. A., & Abdullah, T. A. T. (2020). Characteristics of densified oil palm empty fruit bunches blended with carbon nanotube particles. *Annales de Chimie-Science des Materiaux, 44*.

Ramachandran, V., Ismail, F. S., Noor, M. J. M. M., Akhir, F. N. M. D., Zakaria, Z., & Hara, H. (2020). Extraction and intensive conversion of lignocellulose from oil palm solid waste into lignin monomer by the combination of hydrothermal pretreatment and biological treatment. *Bioresource Technology Reports, 11*, 100456.

Rastogi, M., & Shrivastava, S. (2017). Recent advances in second generation bioethanol production: An insight to pretreatment, saccharification and fermentation processes. *Renewable and Sustainable Energy Reviews, 80*, 330–340.

Raud, M., Kikas, T., Sippula, O., & Shurpali, N. J. (2019). Potentials and challenges in lignocellulosic biofuel production technology. *Renewable and Sustainable Energy Reviews, 111*, 44–56.

Saravanan, A., Kumar, P. S., Jeevanantham, S., Karishma, S., & Vo, D.-V.N. (2022). Recent advances and sustainable development of biofuels production from lignocellulosic biomass. *Bioresource Technology, 344*, 126203.

Saxena, A., Parveen, F., Hussain, A., Khubaib, M., & Ashfaque, M. (2024). Second-generation biorefineries: single platform for the conversion of lignocellulosic wastes to environmentally important biofuels. *Environmental Science and Pollution Research*, 1–32

Schnürer, A. (2016). Biogas production: microbiology and technology. *Anaerobes Biotechnology*, 195–234.

Shah, M. A., Khan, M. N. S., & Kumar, V. (2018). Biomass residue characterization for their potential application as biofuels. *Journal of Thermal Analysis and Calorimetry, 134*, 2137–2145.

Shahid, E. M., & Jamal, Y. (2011). Production of biodiesel: A technical review. *Renewable and Sustainable Energy Reviews, 15*, 4732–4745.

Shameer, P. M., & Ramesh, K. (2017). Green technology and performance consequences of an eco-friendly substance on a 4-stroke diesel engine at standard injection timing and compression ratio. *Journal of Mechanical Science and Technology, 31*, 1497–1507.

Sims, R. E. H., Mabee, W., Saddler, J. N., & Taylor, M. (2010). An overview of second generation biofuel technologies. *Bioresource Technology, 101*, 1570–1580.

Singh, V., Verma, M., Singh, N., Singh, J., Kaur, P. K., Singh, N., & Mishra, V. (2021). Advancement on biomass classification, analytical methods for characterization, and its economic importance. *Bioenergy Research: Biomass Waste to Energy*, 249–272.

Singh, J., Bhonsle, A. K., & Atray, N. (2022). Life-cycle analysis of a hydrocarbon biorefinery. In *Hydrocarb. Biorefinery* (pp 387–408). Elsevier.

Singh, D., Sharma, D., Soni, S. L., Sharma, S., Sharma, P. K., & Jhalani, A. (2020). A review on feedstocks, production processes, and yield for different generations of biodiesel. *Fuel, 262*, 116553.

Singh, N., Singh, V., & Singh, M. P. (2022b). Microbial degradation of lignocellulosic biomass for bioenergy production: A metagenomic-based approach. *Biocatalysis and Biotransformation, 41*(1), 15–25. https://doi.org/10.1080/10242422.2022.2056451

Singh, S., Pant, K. K., & Krishania, M. (2023). Current perspective for bio-oil production from agricultural residues in commercialization aspect: A review. *Journal of Analytical and Applied Pyrolysis, 175*, 106160.

Singh, V., & Mishra, V. (2020). Coronavirus Disease 2019 (COVID-19): Current situation and therapeutic options. *Coronaviruses, 2*(4), 481–491. https://doi.org/10.2174/266679670199920 1005211854

Singh, V., & Mishra, V. (2021). Sustainable reduction of Cr (VI) and its elemental mapping on chitosan coated citrus limetta peels biomass in synthetic wastewater. *Separation Science and Technology, 57*(10), 1609–1626. https://doi.org/10.1080/01496395.2021.1993921

Singh, V., Singh, M. P., Verma, V., Singh, P., Srivastava, R., & Singh, A. K. (2016). Characteristics of cold adapted enzyme and its comparison with mesophilic and thermophilic counterpart. *Cellular and Molecular Biology, 62*, 144.

Solomon, B. D., & Bailis, R. (2014). *Sustainable development of biofuels in Latin America and the Caribbean*. Springer.

Srivastava, R. K. (2019). Bio-energy production by contribution of effective and suitable microbial system. *Materials Science for Energy Technologies, 2*, 308–318.

Srivastava, R. K., Shetti, N. P., Reddy, K. R., & Aminabhavi, T. M. (2020). Biofuels, biodiesel and biohydrogen production using bioprocesses. A review. *Environmental Chemistry Letters, 18*, 1049–1072.

Sun, Y., & Cheng, J. (2002). Hydrolysis of lignocellulosic materials for ethanol production: A review. *Bioresource Technology, 83*, 1–11.

Sutaryo, S. (2012). Optimisation and inhibition of anaerobic digestion of livestock manure.

Tariq, M., Ali, S., & Khalid, N. (2012). Activity of homogeneous and heterogeneous catalysts, spectroscopic and chromatographic characterization of biodiesel: A review. *Renewable and Sustainable Energy Reviews, 16*, 6303–6316.

Toor, M., Kumar, S. S., Malyan, S. K., Bishnoi, N. R., Mathimani, T., Rajendran, K., & Pugazhendhi, A. (2020). An overview on bioethanol production from lignocellulosic feedstocks. *Chemosphere, 242*, 125080.

Tsubaki, S., Nishimura, H., Imai, T., Onda, A., & Hiraoka, M. (2020). Probing rapid carbon fixation in fast-growing seaweed Ulva meridionalis using stable isotope 13C-labelling. *Science and Reports, 10*, 20399.

Ullah, K., Sharma, V. K., Ahmad, M., Lv, P., Krahl, J., & Wang, Z. (2018). The insight views of advanced technologies and its application in bio-origin fuel synthesis from lignocellulose biomasses waste, a review. *Renewable and Sustainable Energy Reviews, 82*, 3992–4008.

Veljković, V. B., Stamenković, O. S., & Tasić, M. B. (2014). The wastewater treatment in the biodiesel production with alkali-catalyzed transesterification. *Renewable and Sustainable Energy Reviews, 32*, 40–60.

Ververis, C., Georghiou, K., Danielidis, D., Hatzinikolaou, D. G., Santas, P., Santas, R., & Corleti, V. (2007). Cellulose, hemicelluloses, lignin and ash content of some organic materials and their suitability for use as paper pulp supplements. *Bioresource Technology, 98*, 296–301.

Wang, L., Barta-Rajnai, E., Skreiberg, Ø., Khalil, R., Czégény, Z., Jakab, E., Barta, Z., & Grønli, M. (2017). Impact of torrefaction on woody biomass properties. *Energy Procedia, 105*, 1149–1154.

Wei, R., Li, H., Chen, Y., Hu, Y., Long, H., Li, J., & Xu, C. C. (2022). Environmental issues related to bioenergy. In *Comprehensive renewable energy* (pp. 92–106). Elsevier.

Wijitkosum, S., & Jiwnok, P. (2019). Elemental composition of biochar obtained from agricultural waste for soil amendment and carbon sequestration. *Applied Sciences, 9*, 3980.

Yaashikaa, P. R., Kumar, P. S., Varjani, S., & Saravanan, A. (2020). A critical review on the biochar production techniques, characterization, stability and applications for circular bioeconomy. *Biotechnol Reports, 28*, e00570.

Yong, K. J., & Wu, T. Y. (2022). Second-generation bioenergy from oilseed crop residues: Recent technologies, techno-economic assessments and policies. *Energy Conversion and Management, 267*, 115869.

You, S., Ok, Y. S., Chen, S. S., Tsang, D. C. W., Kwon, E. E., Lee, J., & Wang, C.-H. (2017). A critical review on sustainable biochar system through gasification: Energy and environmental applications. *Bioresource Technology, 246*, 242–253.

Zakaria, M. R., Farid, M. A. A., Andou, Y., Ramli, I., & Hassan, M. A. (2023). Production of biochar and activated carbon from oil palm biomass: Current status, prospects, and challenges. *Industrial Crops and Products, 199*, 116767.

Zhan, Q., Lin, Q., Wu, Y., Liu, Y., Wang, X., & Ren, J. (2023). A fractionation strategy of cellulose, hemicellulose, and lignin from wheat straw via the biphasic pretreatment for biomass valorization. *Bioresource Technology, 376*, 128887.

Zhang, J., & Zhang, X. (2019). The thermochemical conversion of biomass into biofuels. *Biomass, Biopolymer-Based Materials and Bioenergy*, 327–368 (Elsevier)

Zhang, B., Chen, Y., Jiang, Y., Lu, W., & Liu, W. (2023). Effect of compression ratio and Miller cycle on performance of methanol engine under medium and low loads. *Fuel, 351*, 128985.

Zhu, Z., Sun, J., Fa, Y., Liu, X., & Lindblad, P. (2022). Enhancing microalgal lipid accumulation for biofuel production. *Frontiers in Microbiology, 13*, 1024441.

Zhou, G., Taylor, G., & Polle, A. (2011). FTIR-ATR-based prediction and modelling of lignin and energy contents reveals independent intra-specific variation of these traits in bioenergy poplars. *Plant methods, 7*(1), 9.

Chapter 4
Microbial-Based Bioenergy Production: Sustainable Biofuels from Organic Waste and Biomass

Akhtar Hussain, Irum, Gyanendra Tripathi, Ayush Saxena, Madhvi Singh, Alvina Farooqui and Mohammad Ashfaque

Abstract Microbial bioenergy production offers a sustainable approach to fuel generation by utilizing the metabolic capabilities of microorganisms to convert organic waste and biomass into biofuels. This process not only reduces dependence on fossil fuels but also provides an environmentally friendly alternative for waste management. Key biofuels including bioethanol, biodiesel, biogas, and advanced biofuels such as butanol and biohydrogen are synthesized through microbial fermentation, anaerobic digestion, and electrosynthesis. These processes rely on diverse microbial pathways, including enzymatic fermentation, lipid biosynthesis, and methanogenesis, which utilize feedstocks such as lignocellulosic biomass, agricultural residues, and industrial waste. Recent breakthroughs in genetic engineering, synthetic biology, and metabolic engineering have enhanced microbial efficiency and increased biofuel yields. Technologies such as omics-driven pathway optimization and microbial fuel cells further improve bioenergy production. However, several challenges persist, including scalability limitations, high pretreatment costs, and environmental concerns such as water and nutrient demands. Additionally, regulatory policies and land-use competition influence the feasibility of large-scale commercialization. To overcome these barriers, ongoing research and the adoption of innovative strategies such as synthetic microbial consortia and integrated biorefineries are essential. These approaches enhance efficiency by enabling multiple product streams and optimizing microbial interactions. Aligning microbial bioenergy production with global sustainability goals, such as greenhouse gas emission reduction and circular economy principles, highlights its role in addressing climate change. With continuous technological advancements and supportive policies, microbial bioenergy has the potential to drive the transition toward a cleaner and sustainable energy landscape.

A. Hussain · A. Saxena · M. Singh · M. Ashfaque (✉)
Department of Biosciences, Integral University, Lucknow, India
e-mail: mohdashfaquekhan@rediffmail.com

Irum · G. Tripathi · A. Farooqui
Department of Bioengineering, Integral University, Lucknow, India

V. Singh et al. (eds.), *Bioenergy and Value Added Byproducts from Biomass Waste*, Sustainable Environmental Waste Management Strategies,
https://doi.org/10.1007/978-3-032-22849-9_4

Keywords Microbes · Biomass · Bioenergy · Biowaste

4.1 Introduction

The global demand for energy has experienced a important increase, attributed to an growing population alongside the intensified impact of transportation, industrialization, and enhanced living values. Consequently, the incessantly enhance energy demands may be addressed through the utilization of biofuels as a viable sustainable alternative, which is characterized by a minimal environmental impact (Yadav et al., 2020). Numerous research endeavours are ongoing globally, and the IEA (International Energy Agency) forecasts that production of biofuel levels will increase by 26% by the year 2024–25. Empirical studies suggest that biofuels exhibit performance metrics comparable to those of conventional fossil fuels, and their sustained application has the potential to alleviate the pressures associated with fossil fuel depletion. Biofuels are classified as liquid or gaseous fuels mainly derived from biomass. Some advantages emerge when considering these alternatives in lieu of traditional fuels, including the abundant accessibility of biomass for energy generation, reduced CO_2 emissions during combustion, heightened environmental accountability, economic benefits, and natural biodegradability (Chandra et al., 2021). Agricultural residues constitute a substantial biomass resource found across the terrestrial landscape, predominantly emanating from diverse sectors such as food processing, residential areas, commercial entities, kitchens, cafeterias, restaurants, and markets for fresh produce (Sarkar et al., 2018).

The emission of methane, a greenhouse gas that has a harmful effect 23 times more intense than CO_2 in worsening environment change. Waste biomass produced from food processing, encompassing grains, vegetables, and fruits, is abundant in proteins, carbohydrates, and lipids. The substantial organic composition, coupled with elevated chemical oxygen demand (COD) and biological oxygen demand (BOD) levels present in these wastes, renders them particularly conducive to microbial growth. The accumulation of these wastes contributes to environmental pollution and generates Odors that may pose significant ecological and health dangers. But the challenges posed by these ecofriendly issues are viewed as propitious for the exploitation of Agro-industrial wastes as substrates for generation of bioenergy. Biofuels such as biodiesel, bioethanol, biohydrogen and biogas, can be synthesized from these waste materials, contingent upon their specific composition and characteristics. Biofuels originating from biomass not only facilitate the generation of renewable energy but also maintain minimal or negligible carbon footprints (Kiehbadroudinezhad et al., 2023). The construction of bio-based refineries aspires to reduce environmental pollution in contemporary contexts and is anticipated to alleviate climate change effects over time. It also aims to promote a "zero-waste society" through sustainable bioenergy production, thereby diminishing global dependence on fossil fuels (Sarkar et al., 2018). Moreover, advancements in biotechnology present an additional advantage, offering a comprehensive and reliable framework for screening and analysis that enhances future energy production applicability (Nair et al., 2022).

This chapter highlights, above all, the conversion of waste into energy by vaporizing Agro-waste in bioenergy production while also pinpointing the ever-growing demand for scarce natural resources, among other processes involved in energy generation. The study further elaborates on these already mentioned processes and other processes adopted to produce biofuels from agro-industrial waste.

4.2 Agro-industrial Wastes Are a Worthwhile Area of Biofuel Production

Agro-industrial waste includes all the waste generated from various agricultural processing sectors, farms, and livestock; hence, it is highly effective for biofuel generation. Mainly their internal composition consists of carbohydrate polymers such as cellulose (C5) and hemicellulose (C6), starch, but also include fibers, lipids, proteins, and other organic materials (Chen et al., 2019; Sindhu et al., 2019). A high amount of organic matter found in these wastes can pose a threat to the environment unless handled properly. The variation in composition of the agro-industrial waste can vastly change the biofuel productivity. For example, rice and vegetable wastes contribute to enhancement in carbohydrate composition, whereas meat and lipids add to lipid and protein content (Achmon et al., 2019). Some of these include seeds (pomace), leaves, stalk, skin, and etc., left behind from vegetables and fruits. Another important waste during cold storage is vegetables with high potassium content (Achmon et al., 2019; Li et al., 2018). They are known to be rich sources of mono, di, polysaccharides, and starch which show the possibilities of being used as raw materials for commercial alcohol production (Tiwari & Pal, 2021).

4.2.1 Forests Waste

Forests have been another bounteous source of fuel, timber, coal, and charcoal, unlimited, depending on factors such as their natural state or the fact that they are to be managed. All residues that remain after cutting the stem wood, such as roots, branches, and leaves, add up to a valuable source of LCB. Approximately 4.7 Gt of produced woody biomass is annually around the world (Tripathi et al., 2019).

4.2.2 Agricultural Waste

Crops are very abundant natural resources because of their ease of collection and storage. sugarcane bagasse, Wheat straw, groundnut, corn cobs, cotton stalks, rice husks, and coconut shells are some of the items included in the residue classification. They have the potential of being converted into second-generation biofuels. In the year 2018, India generated nearly 684 million tonnes of agricultural residues derived from several harvests, which were later put into industrial applications, energy generation, and animal feed (Usmani et al., 2021).

4.2.3 Agro-energy Crops

Agricultural enables energy production from bioenergy farms through practices such as energy crop farming. These energy crops are specially bred for their ability to generate both liquid and gaseous biofuel; they have also proliferated very rapidly. They are planted basically for biomass, the raw material for biorefinery operations (Fig. 4.1).

4.2.4 Animal Waste

The combustible nature of this carbon-rich material makes it suitable for many other uses as a plentiful fuel. Dung cakes cooked on stoves become a source of biogas when it is a part of a household operation in rural settings.

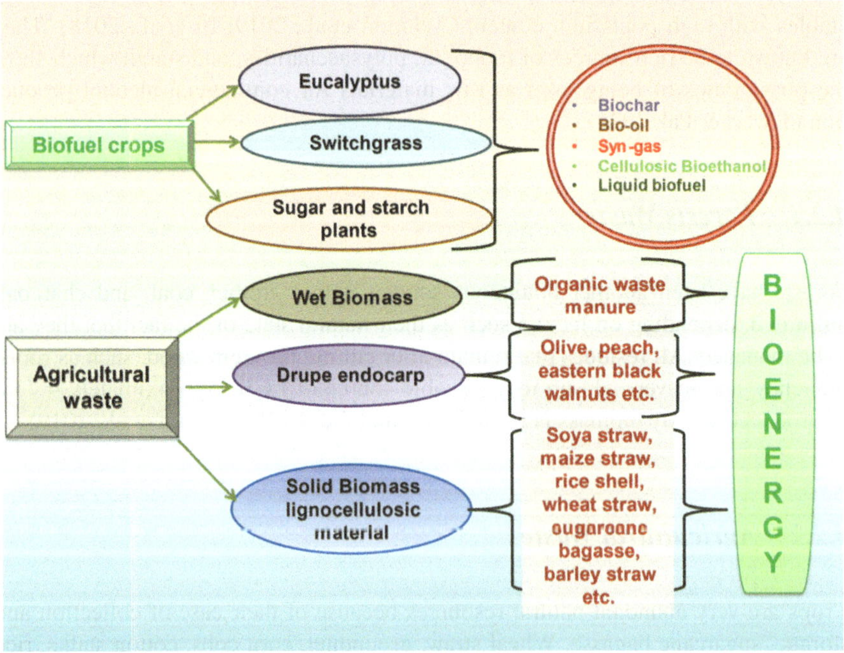

Fig. 4.1 Different sources of (LCB) lignocellulosic biomass

4.2.5 Urban Waste

Waste can be further categorized as municipal solid waste or liquid waste, respectively, since municipal solid waste consists of refuse coming from the households, kitchens, and gardens, while liquid waste is coming from home wastewater and effluents (wastewater) (Retrieved on 10 February 2021 from https://www.mnre.gov.in/wasteto-energy/current-status.) (Hussain et al., 2024a, 2024b).

4.2.6 Agro-industrial Waste

Biofuels can be manufactured from waste resources coming from many sectors of industries, including: effluents from the glucose and starch production, paper and pulp industry by-products, and waste from distilleries, tanneries, and palm oil production. The raw material of LCB in the pulp and paper industry is lignocellulosic biomass. Biomass has industrial applications mainly to extract cellulose and lignin. According to data retrieved on February 10, 2021, from https://www.mnre.gov.in/wasteto-energy/current-status, (Hussain et al., 2024a, 2024b). India total potential for energy generation from (industrial and urban) biological waste was approximately 5691 MW.

4.3 Advantages and Disadvantages of Types of Wastes

The composite referred to as lignocellulosic biomass periodically includes three major polymers: cellulose, hemicellulose, and lignin. It may also possess certain traces of pectin, protein, extractives, inorganic minerals, and phenolic compounds. Different polymeric components of LCB exhibit different composition. LCB is one of the major streams of perennial herbaceous plants, woody crops, and other plant wastes primarily sourced from cultivated plants and considered by-products. Hemicellulose surrounds the cellulose in a matrix, while lignin encloses and shields the other components. Altogether, those 3 form a covalently bonded and cross-linked composite structure (Fig. 4.1).

4.4 Raw Materials

Raw materials must be present for the economy to work properly and make the process efficient. The diversity of the living world is an abundant reservoir of energy-maintained energy-mainly expressed as hydrocarbons, whereby hydrocarbon conversion to biofuel production is shown in Table 4.1. Depending on

Table 4.1 Classification of biomass feedstock (Marichamy, 2024; Sikarwar et al., 2017)

Types of biomasses	Sources of examples
Organic wastes	Industrial organic, MSW (municipal solid waste)
Algae	Eukaryotic algae, prokaryotic algae, kelps
Food crops	Oil crops, grains
Lichen	Fruticose lichens, Crustose lichens, foliose lichens,
Bio-renewable waste	Urban organic waste, mill wood waste, agricultural waste, crop residues
Agro-energy crops	Switchgrass, oilseed crops, sugar crops, miscanthus, forage crops, short-rotation woody crops,
Forest crops	Trees, wood, logging, residue, sawdust, bark, shrubs, and wood residues
Sugar waste crops	Molasses, sugar cane, sugar beets, sorghum
Aquatic plants	Water hyacinth, algae, reed, rushes and water weed
Mosses	Polytrichales, bryophyta
Landfill	(Liquid and inert) waste, hazardous waste, non-hazardous waste

requirements, feedstocks can be distinguished based on the origin of feedstock production. Feedstocks can be turned efficiently into biofuels, and the availability of such materials in many forms offers a basis for various production technologies of biofuels.

4.5 Classification of Biofuels

Biofuels can be categorized on the raw materials' origin into four generations: 1st, 2nd, 3rd, and 4th (Khoo et al., 2023). The 1st generation is derived from sugar, starch, fats, and vegetable oils that were usually converted to FAME or bio-alcohols ($n = 1$–4). The 2nd one depends on carbon negativity, mainly influenced by the concentration of CO, with reliance on lignocellulosic biomass (Shahid et al., 2021). The production of 3rd and 4th generation biofuels employs blue-green algae and algae technologies among others to convert lipids into biofuels. In the case of 3rd generation, algae are utilized for biofuel production, while for 4th generation metabolically-engineering algae from oxygenic photosynthetic microorganisms to form artificial carbon reservoirs are employed (Geraldi et al., 2023). These may also be considered chemically synthesized and biologically synthesized biofuels. Chemical synthesis refers to biofuels generated with the help of a catalyst under variable conditions such as pH, temperature, and pressure. Though not as environmental-friendly, the chemical methods are preferred over biological ones, as they yield a little more than those of biological origin. In chemical production, the use of catalysts instead of various physical processes allows, in this case, a more specific formation of the product desired and a lower cost for separation Generally, complex, and hard-to-separate forms of products mean greater costs are incurred with downstream processing. The biological methods utilize lignocellulosic biomass that undergoes microbial treatments with, fungi, bacteria, algae halophilic and archaea. Other biological processes apply biocatalysts (enzymes) to generate biofuels: cellulases, xylanases, laccase, LiP (lignin peroxidase), LPMO (lytic polysaccharide monooxygenase), multicopper oxidases, MnP (manganese peroxidase), CBDH (cellobiose dehydrogenase), and lipases (Kanakdande & Mane, 2023). Besides, this is the cleanest way of biofuels' production, using enzymes since it deals with some problems connected with chemical process approaches. Some advantages for the production process through enzymatic reactions are: specific production of product is possible under defined pH, temperature, and pressure conditions. That means, from the point of industrial applicability, the use of enzymes from extremophiles, which can perform a wider range of functions over pH and temperature conditions ranging from 4 to 80 degrees Celsius, even under pressure, will add value. So far, all biological processes had more production costs than all chemical processes, and biological process costs may improve by the deeper understanding of microbial pathways, efficiency of enzyme actions, and substrate usage strategies. New biological strains have shown remarkable promise for breaking existing barriers (Fig. 4.2).

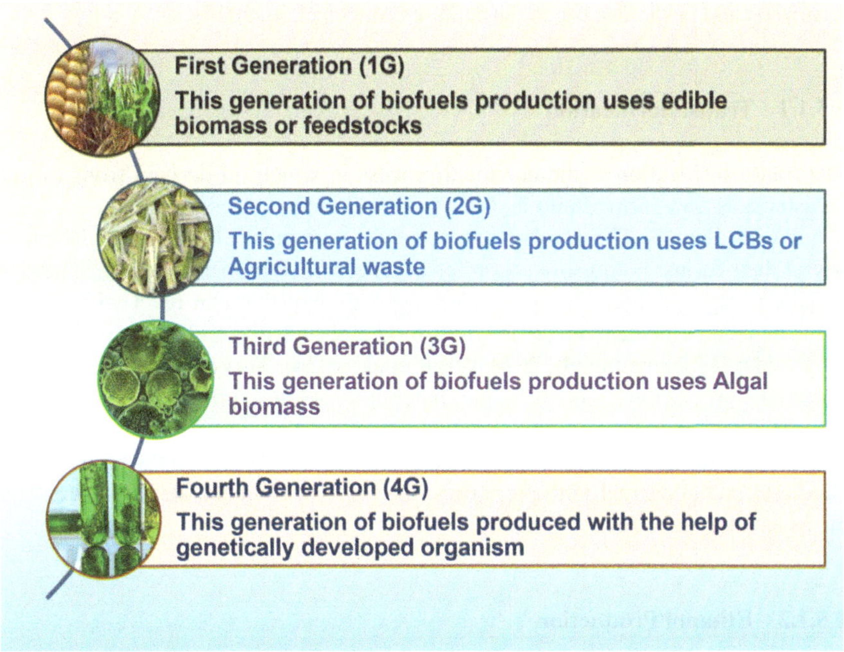

Fig. 4.2 Different generations of biofuels

4.5.1 First-Generation Biofuels

This bio refinery has been taken seriously by the researchers, as conventional energy sources are in short supply and prices for crude oil rise with each passing day. The generation one (1G) bio refinery processes food crops including corn, sugarcane, rapeseed, and cassava into biofuels (Kaparaju & Mukherjee, 2025) The biofuels produced from such feedstocks are mainly biodiesel, ethanol, and biogas. Biodiesel is used to replace conventional diesel in a process that is formed by the transesterification of fats and vegetable oils. With slight modifications to the properties of biodiesel, one can get an even more efficient substitute for diesel. Biogenic ethanol, also commonly referred to as flexi-fuel, is an effective substitute for gasoline (Singh, 2024). It is produced through the fermentation of sugar- or starch-based materials, then blended in with ethanol to create ethyl tert-butyl ether (ETBE), and uses a gasoline blend perfectly well. In addition, biogas and biomethane can be generated by anaerobic digestion of household or municipal waste. Even though the feedstocks can be the same, food crops used to make biodiesel, bioethanol, and biogas pose challenges because of the rapidly increasing demand for edible oils (Chen et al., 2024a, 2024b). Besides food-and-fuel conflicts, there are problems connected with the high-water usage of these first-generation biofuel systems (Fig. 4.1), high fertilizers, and pesticides use in agriculture (Ameen et al., 2024).

4.5.1.1 Transesterification

The transesterification is the chemical process in which oil derived from various feedstocks is transformed into FAME (fatty acid methyl esters) using the reactions of oil with alcohol. The alcohol is a reactive component along with catalysts of any of their forms: homogeneous or heterogeneous in a proper condition. The conversion process includes the use of the right alcohol that can react with triglyceride molecules present in oil to yield biodiesel and glycerol, a very important by-product (Mohamed et al., 2024; Nisar et al., 2024). This can be done with oils obtained from various sources: algae, halophilic isolates, animal fats, and vegetable oils. Transesterification is a reversible reaction: therefore, it requires the continuous mixing of the reactants so that the esters (biodiesel) could be produced. Biodiesel is a good fuel with properties like those of ordinary diesel, hence suitable as an alternative fuel.

4.5.1.2 Ethanol Production

Sugar, starch, and cellulose are main sources of ethanol from agricultural crops. Bio alcohol derived from LCB and food crops is referred to as bio-ethanol, which can be generated through efficient biochemical pathways (Periyasamy et al., 2023). Starch is composed of polymeric units of D-glucopyranose, which are first converted to glucose molecules with the help of the enzyme "amylase." This process converts starch to maltodextrin through liquefaction. The dextrin's and oligosaccharides are hydrolyzed under the action of the enzymes pullulanase and glucoamylase. These enzymes act on the substrate, degrading it into mainly glucose and maltose. The fermentation is then carried out by yeast at a temperature of about 30 °C to produce ethanol (Saeid et al., 2024).

4.5.1.3 Fermentation

Fermentation is a process in which the biotrophic and enzyme activity of many microorganisms converts organic compounds into simpler constituents. Glucose conversion to ethanol involves fermentative as an intermediate pyruvate (Ray, 2024). This procedure can be divided into anaerobic and aerobic, depending on whether O_2 is a requirement for the microorganisms involved. Depending upon whether the fermentative substrate is from butyrate, ethanol from glycerol mixed with hydrogen obtained from senegalensis, a range of microorganisms can perform the fermentation, like as bacteria-*Zymomonas mobilis*, and halophilic bacteria-*Halanaerobium saccharolyticum* yeast-*Saccharomyces cerevisiae*, (Chauhan, 2020). Most of eukaryotic and prokaryotic organisms follow the (EMP) Embden-Meyerhof-Parnas, generally called the glycolysis pathway, but *Zymomonas mobilis* follows the Entner-Doudoroff (ED) pathway. *Z. mobilis* achieved theoretical yield of ethanol up to 97%, and its ethanol yield is approximately 2.6-fold higher than *S. cerevisiae* (Asefi et al., 2024). Besides, *Z. mobilis* has been recently optimally engineered to use a range of C6 and C5 sugars from lignocellulosic substrates, such as glucose, D-xylose and D-mannose, (Narisetty et al., 2022). Also, microorganism Clostridia could produce hydrolytic complexes enabling the use of both types of sugars. Many of such microorganisms are much better at fermenting mixed or individual C6 and C5 combination sugars (Wang & Yin, 2021).

4.5.1.4 Anaerobic Fermentation

The process happens under anaerobic conditions to develop biogas. Microorganisms in environments devoid of oxygen decompose organic biomass into methane (65–75%) and CO_2 (Mishra et al., 2024a, 2024b). From an economic perspective, anaerobic digestion represents a cost-effective methodology and produces biofertilizers for agronomic applications. Initially, the classical method is mainly producing gaseous fuels for domestic use and electricity generation, while new technologies are increasingly devoted to liquid fuels to help reduce the separation cost (Lakzian et al., 2024). The SSF process together with anaerobic digestion of birch wood using *Saccharomyces cerevisiae* is where sugars are converted to ethanol, with the remaining stillage used for biomethane production (Fan et al., 2021).

4.5.1.5 Whole-Crop Utilization

Whole-crop utilization by turning the total plant into a commercially viable product; for example, Jatropha oil seed contains 29–39% oil content. Utilization of whole-crop, changed portions of the crop are processed separately. The oil is chemically modified to generate biodiesel, which can then be further processed into oleochemicals; the remaining de-oiled cake, however, will be applied in gasification to produce syngas. Finally, syngas will be further transformed into other fuel types (Rodríguez et al., 2024).

4.5.2 Second-Generation Biofuels

The second-generation biofuels are produced from non-edible feed stocks such as wheat straw, rice straw and cane bagasse agricultural residue. They also use organic waste obtained from municipal solid waste as another input source. Again, these biofuels do not compete directly with food., which leads to a lower greenhouse gas emissions yield in the production of these products; thus, less land is required to manufacture raw materials for them. The combination of feedstocks can include a heterogeneous one. 2G feedstock is more complex, which has increased the risk of operation technology compared to the previous generation and increased the cost of process development. A comprehensive understanding of these processes, of the choice of route, requires advanced technical knowledge and process modification (Chauhan, 2020). Plant biomass is made of plant cells and consists mainly of about 75% polysaccharides, thus a considerable potential source of carbon (Jacobson et al., 2025). Lignocellulosic biomass conversion into advanced biofuels includes either hydrolysis before fermentation into bioethanol and biobutanol or gasification for producing biodiesel, dimethyl ether, and synthetic natural gas (Fig. 4.1). The raw materials that are used to manufacture advanced biofuels involve rotating crops like poplar and Eucalyptus, Miscanthus, switchgrass, and agricultural waste crops (Cecchi et al., 2021). Advanced bioethanol could substitute gasoline; sugars from hydrolysis could be transformed further into biomass-to-liquids processes or Fischer- Tropsch diesel thereby making it feasible to entirely replace conventional diesel. In this scenario, LCB is converted to syngas by using specific techniques for gasification. It is then used to produce liquid hydrocarbons, such as kerosene, diesel, or bio-synthetic natural gas, or to produce (DME) dimethyl ether or methane (Cecchi et al., 2021). Second-generation biofuels can follow two different strategic pathways. The first pathway, which involves thermo-chemical conversion, is the route whereby biomass is converted through the heat degradation and chemical reformation under different oxygen concentration. The second path refers to biochemical conversion, whereby microbial agents and catalysts or biocatalysts effectuate the conversion. Normally, thermochemical conversion is a more convincing method of biofuels production as it converts the whole organic matter into biofuels, while biochemical conversion is limited in that it can only process polysaccharides into biofuels (Osman et al., 2021).

4.5.3 Physical Process

4.5.3.1 Mechanical Extraction

In mechanical extraction modes, crude oil has generally been obtained from oilseeds in a mechanical process by the application of screw press-based external mechanical pressure. However, a slight amount of oil in the de-oiled cake is extracted simultaneously with the recovered oil, which amounts to about 15–20% of the total oil content. In some methods, the oil is first pre-pressed before solvent extraction; this means oil recovery efficiency of 30–40% (Vadke, 2023). On the other hand, with full pressing, much higher pressures of 9.5 MPa are used to maximize oil content extraction.

4.5.3.2 Briquetting

A various biomass resources, characterized as second-generation fuels, include agricultural and forestry residues, although their inconsistent bulk characteristics pose difficulties for direct biofuel application (Mukherjee et al., 2025). This dilemma, however, is mitigated through densification mechanisms that transform the material into a more uniform and compact configuration. Those densification strategies include pressing and maceration, or a combination of both. In the pressing process, the material's density is increased by application of mechanical pressure for densification during the first stages of compression; with further increased pressure, density of the material affected negatively, until it reaches density of water, whereas this fills neither have dependent on pressure in maceration but rather employ methods like cutting, grinding, and pulverization to slag the bigger materials like tree branches very efficiently. Those models were more effective than ultrafine grinding techniques (Pan, 2024).

4.5.3.3 Distillation

The raw materials are crushed and subsequently subjected to steam distillation. In this process, the vapours are separately condensed into liquid products. An advanced distillation technique called molecular distillation is used when other conventional methods are not sufficiently effective to obtain a fragrance (Pal et al., 2024).

4.5.3.4 Thermochemical Conversion

Thermochemical conversion consists of heating biomass followed by its chemical transformation to biofuels (Poornima et al., 2024). It presents gasification, combustion, liquefaction, and pyrolysis used in O_2-depleted or noncontrolled O_2 situations to produce syngas, mainly of H_2 and CO. The syngas is then further treated to obtain gaseous or liquid (Rey et al., 2024).

4.5.3.5 Combustion

The chemical reactions generally take place in the air whereby the fuels element reacts with O_2, producing by-products including CO_2 and H_2O, and energy released as heat (Sivabalan et al., 2021). Although combustion generally produces less product, it is by far the commonest technique of converting biomass to energy.

4.5.3.6 Gasification

Gasification is a high-temperature pyrolysis that yields a mixed gas called producer gas (H_2, CO_2, CO, N_2, and CH_4). Regarding biomass, methanol, ammonia, and ethylene can be made using gasification. For the last 30 years, gasification has been used for the biofuel generation process of lignocellulosic biomass. This involves multiple operations simultaneous with each other: drying, pyrolysis, and partial oxidation (Amenaghawon et al., 2021). The Production of syngas from the gasification procedure may attain (non-catalytic and catalytic) modes. Non-catalytic syngas production usually requires high temperatures in the range of 1000–1300 °C while non-catalytic methods function at around 899–900 °C due to developments in catalysis (Makaryan et al., 2023). A crucial area of gasification is the control of H_2/CO ratios to maximize various product concentrations (Siddiqui et al., 2022).

4.5.3.7 Bio Methanol

As biomass is converted into producer gas, oxygen plays a vital role, producing a mix of gases consisting of CO, CO_2, H_2, CH_4, and N_2, which ably combine to form syngas. The produced syngas may be applied in stationary power generation. The gas mixture is then reduced to 300 °C. Thereafter, it is admitted into a shift reactor under predetermined operating conditions, where the H_2/CO ratio is adjusted to 2.1, ideal for methanol synthesis (Aydin et al., 2024).

$$CO + 2H_2 \leftrightarrow CH_3OH$$

$$CO_2 + 3H_2 \leftrightarrow CH_3OH + H_2O$$

4.5.3.8 Methane

The process of syngas preparation stays the same. In this respect, when in the process of tar decomposition with a catalyst, methane needs a higher H_2/CO ratio of 3.1. The process takes place at a pressure of 31 bar and within 300 and 600 °C, ensuring optimal performance of these reactions and avoiding unwanted side reactions (Minyukova& Dokuchits, 2023).

$$CO + 3H_2 \leftrightarrow CH_4 + H_2O$$

4.5.3.9 Bio-ethanol Production

Biomass is a natural source for generating power with steam; SO_2 is combined with steam in an explosion at about 206 °C for a very rapid time of about 5 min, and after that, steam recovery is carried out by means of product formation. Flashing occurs, and the vapour produced therein is reduced to retrieve furfural. Upon vapor from recovery, liquid is obtained with the help of more H_2O for simultaneous fermentation. Saccharification, the conversion of biomass into the simplest sugars, occurs at innocuous low temperatures of up to 40 °C (Dong, 2019). saccharification delivers the simple sugars of hexoses and pentoses, which undergo fermentation to bioethanol. The ethanol and methanol production is done through this methodology. Biomass composition involves cellulose and hemicellulose hydrolysed to produce ethanol as a starting material; followed by methanol production, in where the leftover components, for example, lignin along with the residual hydrolysed biomass, are applied (Devi et al., 2022). Other methodologies include hydrolysis of biomass subsequently augmented through hydrogenation using catalyst, for example, lignin combined with the residual hydrolysed biomass for methanol production (Devi et al., 2022). Alternatively, biomass is subjected to hydrolysis combined with hydrogenation by means of a catalyst; for instance, Ru at a temperature of 160 °C and pressure of 50 bara. One of the main advantages of the method is that the stability of the sugars permits for the production of alcohol (Shao et al., 2023).

4.5.3.10 Liquefaction

Biomass solid is treated with alkali, glycerine, propanol, butanol, or direct liquefaction (da Costa et al., 2023). This methodology leads to a high-viscosity product that is immiscible in water; in this case, solvents, carbon monoxide (reducing gases), hydrogen, or a catalyst will be required within biomass. In the process of thermochemical conversion, LCB goes into the formation of liquid oils via direct liquefaction (Muh et al., 2021).

4.5.3.11 Pyrolysis of Biomass

The basic thermochemical procedure changes biomass into a hydrocarbon-rich gas mixture under controlled or inert atmosphere conditions with the help of an external furnace because the very organic material is converted into a notable compound, in addition to other liquid by-products like acetic acid, acetone, and methanol; plus, solid by-products are also formed into charcoal, and gases produced are generally non-condensable gases when heated to 480 °C in the presence of air (Blasi et al., 2023). About 70% of the production of bio-oil is industrial-scale manufacture with no preliminary processing of the biomass economically feasible, taking into consideration (Das, 2022) Thereupon, bio-oil becomes corrosive to equipment, with its quality having to be improved to reduce oxygen concentration, while hydrogenation also allows for alkali removal, and oil can tolerate catalytic cracking. Concerning lignocellulosic biomass, wood, pyrolysis works as a zonal mechanism involving thermally induced decomposition of the hemicellulose, cellulose, and lignin (Kashif et al., 2025) Also, in essence, pyrolysis consists of five stages (Fig. 4.3) (Sun et al., 2022).

Fast Pyrolysis process is performed in a higher temperature range of 560–960 °C, achieving rapid heating rates in a one -time period of 0.5–10 s. During fast pyrolysis, bio-mass devolatilizes to vapours, aerosol, and char. In the next cooling stage and consequent condensation of vapour, some aerosols were converted to a

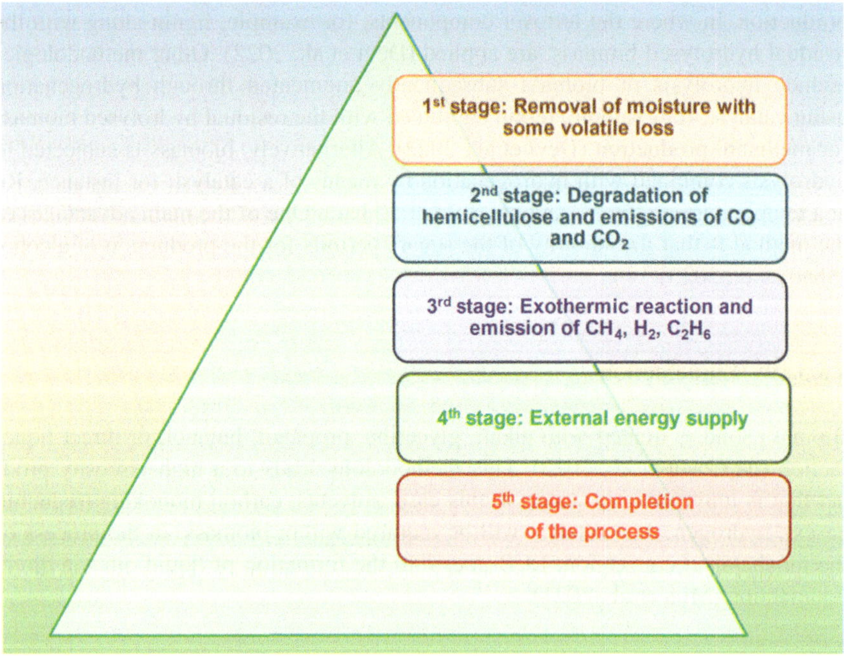

Fig. 4.3 Pyrolysis consists of five distinct stages

dark liquid, which gave 59–74% bio-oil, 16–26% char, and 9–14% non-condensable gases (Mati et al., 2022).

Flash pyrolysis operates at temperature ranges higher than fast pyrolysis, such as 777–1027 °C, with a very fast heating rate above 723 °C/s and particle sizes smaller than 0.2 mm. The bio-oil from flash pyrolysis is combined with char to form a bio slurry mixture. This bio slurry is then fed into the gasifier at 26 bar pressure and a temperature of 654–954 °C for conversion into syngas, with 70% conversion efficiency. The bio-oil from flash pyrolysis is utilized for power generation in engines and turbines (Panchasara & Ashwath, 2021). The biomass syngas feeds into power generation of different biofuels, usually with contamination from particulates, tars, nitrogen sulphate, or some trace element. Such impurities need to be cleaned before entering the catalytic reactor system (Sikarwar et al., 2017). While second-generation(2G) biofuel does not compete with food feedstocks as much as first-generation (1G) biofuel does, fibre crops as raw material contend against food crops. To some extent, pesticide and fertilizer use remains yet another grave concern for the production of second-generation biofuels.

4.5.4 Third-Generation Biofuels

Algae have been used to produce the third-generation (3G) biofuel. Algae include microalgae, macroalgae from seaweed, and blue-green algae (BGA) or cyanobacteria. There are over 800,000 types of algae on the planet; they grow quickly, within a month, and require less nutrition for their growth-and the bonus part is that the economic production comes from wastewater (Bowyer et al., 2018). The oil production potential of algae was found to be 100-fold greater on an area basis than that of soybeans and canola oils (Selvam et al., 2024). The optimum control of the growth of algae should be done at a temperature of 20–30 °C because growth continues, although slowly, when the temperature is less than 16 °C, and if the temperature goes beyond 35 °C, then it is detrimental for the growth of algae. Microalgae can, indeed, be cultivated from wastewater treatment facilities (Mohsenpour et al., 2021). The major limitation about this system is that the various parameters cannot be regulated.

4.5.4.1 Photo Bioreactor (PBRs)

A closed system offers better internal control in modulating the growth conditions of pure algal cultures; for example, it can control critical factors, including H_2O, pH levels, temperature, CO_2, light intensity, and mixing density (Zhu et al., 2022). Algae are cultivated in photobioreactors (PBRs), usually in glass or plastic tubes, and supplied with light at selective wavelengths to allow optimal algae growth. Light usually penetrates 7.6–10.2 cm into the PBRs under turbidity induced by algal growth (Zhu et al., 2022). Various conditions in PBRs have been optimized, including the volume-to-surface area ratio for optimal light penetration, spatial distribution of light inside the reactor, the rate of transfer CO_2, and monitoring of growth in the reactor to assist smooth operation of the PBRs. All installations and running costs can be considerable; however, product yields, especially oil, reportedly are between 40 and 55% (Adamu et al., 2023). Low operational costs for open systems allow their effective integration with closed PBR types for both cost and product yield optimization (Zhang et al., 2024). Algae are unparalleled sources of biofuels, among them biodiesel, ethanol, and petroleum. Algae are reported to be 602 times more productive than crops, such as corn and switchgrasses (Zhang et al., 2024). One of the advantages attributed to algae is their high productivity and comparatively less land required for cultivation. Oil extracted from algae undergoes the transesterification process to produce biodiesel, both chemically and biologically. In addition to that, biological production uses a lipase enzyme that transfers lipids into biodiesel at room temperature, so the energy demands for biodiesel production are considerably lowered.

4.5.5 Fourth-Generation Biofuels

The emergence of novel synthetic biological instruments has heralded the inception of fourth-generation biofuels. This contemporary iteration harbours the potential for significant advancements in photobiological and electrobiological biofuels (Bajpai, 2022). The synthetic biology associated with algae and cyanobacteria involves the engineering of innovative and promising strains through the application of recombinant metals such as cobalt, iron, and manganese. Furthermore, artificial photosynthesis emulates the molecular mechanisms engaged in this process, employing catalytically active enzymes as biocatalysts, including hydrogenase, to facilitate hydrogen production through dark fermentation (Ren et al., 2022). In a straightforward procedure, nanotechnology promotes hydrogen generation through a non-molecular mechanism by means of light-driven catalysis across diverse surfaces. These surfaces may comprise metals, semiconductors, or nanostructured carbon-based materials. Nevertheless, the constraints associated with the non-molecular mechanism stem from the utilization of rare and expensive metals as catalysts. The interaction between molecular and non-molecular mechanisms introduces complexity to the process, culminating in a diminished comprehension. The challenges engendered by the non-molecular mechanism predominantly arise from the dependency on rare metals as catalysts, which inflates expenses.

4.5.5.1 Microbial Conversion

The production of biofuels persistently presents challenges, as the production costs surpass those of the raw materials. Within the realm of bioprocessing, the microbial conversion phase transmutes LCB into biofuel through action of microorganism, devoid of enzymatic involvement. This methodology has garnered heightened interest among scholars in comparison to first-generation biofuel production, particularly with respect to cost-effectiveness (Hussain et al., 2024a, 2024b). The nascent trend in bioprocessing emphasizes the potential for production of ethanol via microorganism pathways or their combinations, primarily leveraging the C6 formation alongside other constituents (C5and lignin) of the pre-treated biomass, thereby achieving a substantial of ethanol yield. Many aerobic fungi can directly decompose pre-treated feedstocks into CO_2, yielding only a minimal quantity of ethanol. Nonetheless, the electrons which will be evolved will be transferred into the respiratory chain for oxidative phosphorylation; the metabolic pathway will not use substrate-level phosphorylation for ethanol production. The general principle states that 2 mol of alcohol are produced from every mole of hexose sugar. Yeast such as *Saccharomyces cerevisiae* use the glycolytic pathway to produce alcohol from sugars. All industrial processes are executed at a higher pH and temperature levels so that they can efficiently utilize consortia of strong fungal strains such as Sarocladium strictum along with thermophiles like Halomonas to improve the degradation of wheat straw (Irum et al., 2024).

4.5.5.2 Enzymatic Conversion to Biofuel

The generation of biofuels from lignocellulosic biomass has presented a novel opportunity to address global energy demands, yet it simultaneously poses a substantial challenge concerning global warming. The production of biologically-derived energy, referred to as "bioenergy," constitutes a paramount concern within biocatalysts (which employ various enzymes to facilitate biomass conversion) (Amesho et al., 2024). The complex architecture of LCB is initially hydrolyzed into its simplest elements employing several hydrolysing enzyme mixtures. Nevertheless, the deployment of enzyme mixtures has exhibited maximum conversion efficiency in the formation of industrial products derived from LCB conversion. An enzyme consortium represents another strategy employed to dismantle recalcitrant substrates into valuable products (Chen et al., 2024a, 2024b). This enzymatic consortium transformation necessitates extensive investigations into diverse enzymes and their interactions to operate in a synergistic manner. To this point, Sprizyme, a commercial enzymatic formulation produced by Novozymes, is accessible in three distinct variants in the market for the comprehensive conversion of carbohydrates. Moreover, Novozymes is dedicated to the process of converting cellulose into ethanol. Enzymatic mixtures facilitate the hydrolysis of cell wall structures owing to their specific catalytic actions. Furthermore, the product generated by one enzyme functions as the substrate for another enzyme, thereby allowing individual enzymes to exert their effects on the available substrates. During the interactions between enzymes and substrates, produced enzyme inhibitors, which can hinder the hydrolysis process. Consequently, an in-depth analysis of enzyme mechanisms can significantly improve the conversion efficiency of LCB. Each substrate within cellulosic biomass possesses distinct characteristics and attributes that require tailored conditions for its conversion into the simplest form.

Cellulases a collective of three enzymes exists that work synergistically on C6 to completely change it into glucose. Enzymes include β-glucosidase exoglycanase, and endoglucanase (Yaverino et al., 2024).

Endoglucanase enzyme indiscriminately hydrolyses β 1–4 glycosidic linkages at the amorphous regions, producing 2-carbon units of cellobiose from the cellulose polymer (Sharma et al., 2024).

Exoglucanase commonly referred to as cellobiohydrolase, avicelase, and exocellulase, this enzyme cleaves the terminal ends of the crystalline regions of cellulose, generating cellobiose as a by-product.

β-glucosidase enzyme catalyses the hydrolysis of the 2-carbon chain molecule (cellobiose) into glucose. these three enzymes are collectively termed cellulolytic enzymes, predominantly synthesized by *Trichoderma reesei.* Additionally, various thermophilic and potent strains, such as thermophilic clostridia and Thermo aner bacterium, are also recognized for their cellulase production (Ghosh, 2024).

4.5.5.3 Xylanases

Xylan is a pentosan polymer, which is a constituent of hemicellulose, made up of 5C chain units. Xylanases are enzymes acting only on the Xylan units linked to lignin and cellulose (Mishra et al., 2024a, 2024b). The lignocellulosic biomass xylose removal improves access to other hydrolysing enzymes or chemicals and, thus, improves cellulose saccharification. Many microbial agents cause Xylan degradation at thermophilic temperatures, such as *Trichoderma reesei* or *Humicola insoles* (Pradeep et al., 2022).

4.5.5.4 Lignolytic Enzymes

These enzymes degrade lignin and use both phenolic and non-phenolic groups of lignin as a substrate, including LiP (lignin peroxidase), MnP (manganese peroxidase), laccases, VP (versatile peroxidase), and dye-oxidizing enzymes DyP (A and B). For LiP, the mechanism occurs on the surfaces of (non-phenolic and phenolic) aromatic components, whereas MnP requires phenolic substrates that are oxidized with the usage of Mn^{3+}, and its characteristics exhibit the hybrid role between those of LiP and MnP (Jose et al., 2022). These enzymes that degrade lignin play a vital role in the pre-treatment stage of lignocellulosic biomass by removing the lignin and thereby exposing sites for the attack by cellulolytic enzymes.

4.5.5.5 Cellobiose Dehydrogenase (CBDH)

This enzyme functions as an oxidation agent for many sugars, such as lactose, glucose, and cellobiose. In the course of this reaction, CBDH oxidizes the substrate and reduces the flavin on the other hand, into flavin adenine dinucleotide (Roy et al., 2024). The second function of CBDH is to activate cellulases by eliminating product inhibition caused by cellobiose. There is another enzyme that falls under this category, the lytic polysaccharide monooxygenase copper (LPMO), which, in order to cleave the C1, C4, or both glycosidic bonds of C6 requires an external electron donor. First, LPMO takes a hit at the C-H bond of primary alcohols, then oxygen-dependent breakdown follows ultimately forming a carbohydrate (Lindley 2022).

4.6 Effect of Surfactant on Enzymatic Hydrolysis

Surfactants greatly assist in reducing the time for hydrolysis and the amount of enzyme needed for hydrolysis (Rokhati et al., 2022). Lignocellulosic biomass surfactants hydrophobic ally interacts with its lignin component to yield almost twice the amount due to enhanced surface modification or disruption; they also stabilize enzymes (Vasile & Baican, 2023). Non-ionic surfactants, like as Tween (80, 20) and polyethylene glycol (PEG), promote enzymatic hydrolysis to a significantly greater extent than their ionic counterparts, as they increase the surface area available for hydrolysis (Singh et al., 2021) (Table 4.2).

4.7 Lignin Strategy to Biofuel

4.7.1 Lignin Structure

The lignin matrix is heterogeneous and is composed of structural elements. Some typical ones are p-coumaryl alcohol, coniferyl alcohol, and sinapyl alcohol of aromatic nature. Lignin is an aromatic polymer which consists of three different monomers such as guaiacyl (G, from coniferyl alcohol), p-hydroxy-phenyl (H, from coumaryl alcohol), and syringyl (S, from sinapyl alcohol), each conferring unique properties on lignin (Hussain et al., 2024a, 2024b) The differences in monolignol composition impart different properties to woods. The G/S-Units ratio is mostly characteristic for softwood lignin, while the hardwood composition has a variant ratio of G/S/H (Liu, 2024). The chemical treatment of lignin can transform its aromatic structure into hydrocarbon (HC) fuels and yield products with similar properties to gasoline and diesel (Busca, 2021).

Table 4.2 Advantages and disadvantages from biofuels in terms of GHG emissions (Elfasakhany et al., 2019)

Biofuel	Advantage	Disadvantage
Natural gases	Low PM, NO_x methane performance improved by adding hydrogen	PAHs, formaldehyde
Biogas, compressed natural gas (CNG), liquefied petroleum gas (LPG), and hybrids	GHG reductions with hybrids like compressed natural gas (CNG), liquefied petroleum gas (LPG)	
Ethanol	Reduction CO, 1,3 butadiene, aromatics HC, particulate matter (PM), NO_x	Increase acetaldehyde emissions
Hydrogen	Zero emissions	
Synthetic fuel	Polycyclic aromatic hydrocarbons (PAHs), and mutagenicity, PM, NO_x and aldehyde, 1,3 butadiene, benzene	
DME	Clean burning	Challenging as a vehicle fuel
FAME	Reduction in PM	Enhance in NO_x and aldehyde emissions

4.7.2 Lignin Valorization

In a cellulosic bio refineries, the component that is cellulosic is 45–60% and is utilized for the manufacture of ethanol. About 20% of the rest will produce lignin with more emphasis on the rendering of lignin into biofuel and value-added by-products (Yogalakshmi et al., 2023). Processing entails subjecting lignin to depolymerised incorporation of hydrogen as reducing agent resulting to formations of aromatic products poor in oxygen but very stable and will give rise to diverse conditions under which different lignin monomers might interact with each other, which will enable the formation of desirable products (Seidi et al., 2025). The interaction barrier is overcome by different pretreatment methods for lignocellulosic biomass. Hydrogenation, mild in nature and using a catalyst appropriate for the case, will allow the formation of desired products in this very complex reaction. An in situ catalytic characterization indicates that guaiacol conversion is accomplished under hydrogenation conditions with a selectivity to cyclohexanol of 99% and some 94% in 7 h, 220 °C, and 2 MPa initial pressure using the right ratio of water/methanol/lignin as feedstock (20:5:0.5). The highly selective hydrogenation performed on a mild condition applied over a stepwise precipitation with a catalyst (Ni/SiO_2) successfully provides conditions for carrying out the selective hydrogenation for 7 h, at 120 °C, and 2 MPa, for complete conversion into guaiacol (Awan, 2019). Hydrogenation using wet impregnation affords a good alumina-doped palladium catalyst, whereby 3 weights % of the catalyst converts

4-ethylphenyl(cyclohexanol) during a reaction in an aqueous phase over 6 h at 60 °C (Awan, 2019). Together with chemical catalysts, biological catalysts such as lignolytic enzymes are also being researched for lignin Valorization in view of biofuel production in the coming future. Lignin-derived value-added HC fuels can be transformed into C7-C18 hydrocarbons, very suitable for jet fuel applications (Karim et al., 2022). The proposition whereby lignin is converted into biofuel thus entails the degradation of lignin to make the O_2/C and H/C ratios coincide with those of the conventional fuels. On HDO technology, lignin monomers increase the H/C ratio, while simultaneously decreasing the O/C ratio. Herein, the monomeric lignin units are involved in CO bond cleavage, while the oxygen removal occurs via high-pressure hydrogen treatment, forming water vapor and reducing in O/C ratio. Saturation of aromatic carbon double bond by hydrogen occurs in the presence of a catalyst (palladium and activated carbon) the consists of advances in microbial bioenergy technology (Miskolczi et al., 2023).

4.8 Technological Advances in Microbial Bioenergy Production

Microbial bioenergy production has witnessed great advancement as a result of the increasing global demand for sustainable energy. In essence, it employs microorganisms to convert organic matter into usable energy, thereby providing renewable and environment-friendly energy alternative to fossil fuels (Gueboudji, 2024). new technological developments have made great advancements in efficiency, scalability, and cost-effectiveness of microbial bioenergy systems, such that, now, they stand as one of the most promising alternatives to energy challenges.

4.8.1 Synthetic Biology and Genetic Engineering

Synthetic biology have changed the process of microbial bioenergy production. Scientists can modify microbial genomes through optimization of metabolic pathways for production improvement. Engineered strains of Escherichia coli and *Saccharomyces cerevisiae* are generated to enhance the production of large quantities of ethanol, butanol and other biofuels (Joshi et al., 2022). Tools like CRISPR-Cas9 enable precise genetic modifications allowing introduction of new pathways or enhancement of existing ones. All these have considerably boosted the efficiency of microbial bioenergy systems.

4.8.2 Metabolic Engineering

Metabolic engineering seeks to reconstitute the metabolism of microorganisms to improve the quality and quantity of energy-rich compound production. Enzymes and regulatory networks can thus be manipulated toward pathways that favor the production of biofuels. An example would be the metabolic engineering of oleaginous yeast species such as Yarrowia lipolytica for increased biodiesel production (Sartaj et al., 2023). It is the same metabolic engineering that has begun to make it possible to use some non-conventional substrates for enhanced sustainability in bioenergy production, conspicuously including lignocellulosic biomass and industrial waste.

4.8.3 Microbial Electrochemical Systems

MES (Microbial electrochemical systems), which comprise MFCs (microbial fuel cells) and MECs (microbial electrolysis cells), represent a main milestone in bioenergy technology. MFCs convert organic matter into electricity by means of electroactive bacteria; hydrogen gas is produced through microbial-driven water electrolysis in MECs (Albarracin-Arias et al., 2021). Advances in electrode materials, like those with graphene-based anodes and cathodes, have greatly enabled power output and efficiency improvement in these systems. Integration with wastewater treatment plants has shown the feasibility of MES for simultaneous energy generation and environmental remediation.

4.8.4 Omics Technologies

Omics technologies-genomics, transcriptomics, proteomics, and metabolomics-have revolutionized our understanding of microbial bioenergy production. They provide insights into microbial physiology and allow the identification of key genes, proteins, and metabolites involved in biofuel synthesis. For instance, omics studies have discovered additional pathways of methane production in methanogenic archaea from which biogas production systems could be made more efficient (Wirth et al., 2023).

4.8.5 Integration with Renewable Energy Systems

Microbial bioenergy production systems with different renewable energy systems-solar and wind type-this approach is becoming more attractive. For example, excess electricity from solar panels could drive microbial electrolysis cells for increased hydrogen production (Zhao & Yuan, 2023). microbial bioenergy systems with carbon-capture technologies would pave the way for the production of carbon–neutral or carbon-negative biofuels. Such integrative frameworks demonstrate that microbial bioenergy has the potential to engender a sustainable energy economy.

4.9 Environmental and Economic Benefits

Reduction in Greenhouse Gas Emissions the utilization of microbial-based bioenergy produce offers a promising pathway toward reducing GHG (greenhouse gas) emissions. Unlike fossil fuels, which release significant amounts of CO_2 and other pollutants, microbial biofuels derived from organic waste and biomass contribute to a more sustainable carbon cycle. Anaerobic digestion and microbial fermentation efficiently convert agricultural residues, food waste, and sewage sludge into biofuels such as biogas, bioethanol, and biodiesel (Sharma et al., 2024). This process not only mitigates methane emissions from landfills but also provides a cleaner alternative to conventional fossil fuels, thereby playing a crucial role in combating climate change (Gupta et al., 2022). Waste Management and Circular Economy Integration In addition to emission reduction, microbial bioenergy production enhances waste management and integrates seamlessly into the circular economy model. Organic waste, which would otherwise contribute to environmental pollution, is repurposed to generate valuable energy resources. Microbial consortia break down complex organic matter, optimizing biofuel yield while minimizing residual waste (Sinduja et al., 2023). The digestate byproduct from anaerobic digestion can also be used as a nutrient-rich biofertilizer, creating a closed-loop system that supports sustainable agriculture and reduces dependency on synthetic fertilizers (Lamolinara et al., 2022).

Cost Comparison with Fossil Fuels and Traditional Biofuels from an economic perspective, microbial biofuels present a cost-effective alternative to fossil fuels and traditional biofuels. Advances in bioprocessing technologies, such as genetic engineering of microbes and improved bioreactor designs, have enhanced yield efficiency and reduced production costs (Joshi et al., 2022). Compared to conventional biofuels derived from crops like corn and sugarcane, microbial biofuels utilize non-food biomass, reducing competition with food supply and lowering feedstock costs (Mahapatra et al., 2021). As fossil fuel prices fluctuate and carbon pricing mechanisms expand, microbial bioenergy becomes increasingly competitive in global energy markets. Regional Case Studies Demonstrating Economic Viability several regional case studies highlight the economic viability of microbial biofuel production. In Sweden, biogas plants utilizing microbial digestion of municipal waste have demonstrated profitability while reducing landfill waste (Gustafsson et al., 2024). Similarly, India's large-scale implementation of bioethanol production from agricultural residues has improved rural economies and energy security (Arora et al., 2024).

4.10 Challenges and Limitations

4.10.1 Feedstock Heterogeneity and Pretreatment Challenges

The variability in organic waste and biomass composition—ranging from lignocellulosic agricultural residues to lipid-rich food waste—poses a significant barrier to efficient microbial conversion. Pretreatment processes, such as thermal, chemical, or enzymatic methods, are often required to break down complex polymers (e.g., lignin) into fermentable substrates. However, these processes are energy-intensive and can generate inhibitory byproducts (e.g., furans, phenolic compounds) that reduce microbial activity. Recent studies highlight the need for tailored pretreatment strategies to optimize feedstock compatibility with microbial consortia (Kumar et al., 2023).

4.10.2 Microbial Conversion Efficiency and Metabolic Limitations

Microbial strains used in biofuel production (e.g., Clostridium, Saccharomyces) often exhibit suboptimal substrate utilization rates or sensitivity to environmental stressors (pH, temperature). Metabolic bottlenecks, such as carbon flux inefficiencies or competition between pathways for biofuel synthesis (e.g., ethanol vs. butanol), further limit yields. Advances in synthetic biology, including CRISPR-based genome editing, aim to engineer robust microbial chassis with enhanced metabolic capabilities, but scalability remains a hurdle (Zhang et al., 2022).

4.10.3 Scaling Up from Lab to Industrial Systems

While lab-scale bioreactors achieve promising biofuel yields, industrial-scale operations face challenges like mass transfer limitations, inconsistent mixing, and microbial contamination. Large-scale systems also struggle to maintain sterility and optimal conditions for diverse microbial communities. Recent pilot-scale studies emphasize the need for modular reactor designs and real-time monitoring technologies to improve process control (Paul and Sarkar 2026).

4.10.4 Economic Viability and Resource Competition

High capital and operational costs for bioreactors, pretreatment, and downstream purification hinder commercialization. Additionally, competition for feedstocks between bioenergy production and other industries (e.g., animal feed, composting) raises ethical and logistical concerns. A 2023 techno-economic analysis underscored the importance of integrating waste Valorization into circular economy frameworks to offset costs (Soni and Soni 2025).

4.10.5 Environmental and Social Trade-Offs

Although microbial bioenergy reduces reliance on fossil fuels, indirect environmental impacts—such as land-use change for biomass cultivation or greenhouse gas emissions from waste transport—require careful assessment. Social acceptance of waste-derived biofuels also varies, with public scepticism about Odor, safety, and scalability. Life-cycle assessments (LCAs) stress the need for transparent sustainability metrics to guide policy (Iqbal et al., 2025)

4.10.6 Policy and Regulatory Gaps

Inconsistent global policies on waste management and bioenergy incentives create uncertainty for investors. For instance, the lack of standardized carbon credit systems for microbial biofuels discourages market adoption. Recent reports by the International Energy Agency (IEA, 2019) advocate for harmonized regulations and subsidies to align bioenergy production with net-zero targets, highlighting the role of public–private partnerships in bridging gaps.

4.11 Conclusion

Microbial bioenergy production holds significant promise for a sustainable energy future by harnessing the ability of microbes to convert organic waste and biomass into biofuels. Through advanced processes such as fermentation, anaerobic digestion, and microbial electrosynthesis, renewable fuels like bioethanol, biodiesel, and biogas can be efficiently generated. The integration of cutting edge technologies, including genetic engineering and synthetic biology, has further optimized microbial efficiency, leading to improved biofuel yields and cost-effectiveness. Additionally, utilizing diverse biomass feedstocks—such as agricultural residues and industrial waste offers an environmentally friendly approach to waste management while reducing reliance on fossil fuels. Despite its potential, microbial bioenergy production faces several challenges, including technical and economic constraints. such as yield efficiency, scalability, and high pretreatment costs hinder widespread adoption. Environmental concerns such as water and nutrient demands, along with land-use competition must also be carefully addressed.

Furthermore, policy and regulatory frameworks play a crucial role in shaping the feasibility and expansion of microbial biofuels. Overcoming these barriers requires ongoing research, technological advancements, and supportive policies to drive large-scale commercialization. In advance, innovations in microbial bioenergy such as integrated biorefineries and synthetic microbial ecosystems will be instrumental in creating a more sustainable and economically viable biofuel industry. Aligning microbial bioenergy production with global sustainability goals, including greenhouse gas reduction and the promotion of circular economy principles, underscores its importance in combating climate change. By leveraging microbial consortia and synthetic biology, the future of bioenergy appears increasingly promising, paving the way for a cleaner, more sustainable energy landscape.

Acknowledgements Authors are thankful to the Integral University, Lucknow for necessary research support.

References

Achmon, Y., Claypool, J. T., Pace, S., Simmons, B. A., Singer, S. W., & Simmons, C. W. (2019). Assessment of biogas production and microbial ecology in a high solid anaerobic digestion of major California food processing residues. *Bioresource Technology Reports, 5*, 1–11.

Adamu, H., Bello, U., Yuguda, A. U., Tafida, U. I., Jalam, A. M., Sabo, A., & Qamar, M. (2023). Production processes, techno-economic and policy challenges of bioenergy production from fruit and vegetable wastes. *Renewable and Sustainable Energy Reviews, 186*, 113686.

Albarracin-Arias, J. A., Yu, C. P., Maeda, T., Quintero, W. V., & Sanchez-Torres, V. (2021). Microbial community dynamics and electricity generation in MFCs inoculated with POME sludges and pure electrogenic culture. *International Journal of Hydrogen Energy, 46*(74), 36903–36916.

Ameen, M., Mahmood, A., Shehzad, A. N., Zia, M. A., & Javaid, M. M. (2024). Sorghum's potential unleashed: A comprehensive exploration of bio-energy production strategies and innovations. *Bioresource Technology Reports*, 101906.

Amenaghawon, A. N., Anyalewechi, C. L., Okieimen, C. O., & Kusuma, H. S. (2021). Biomass pyrolysis technologies for value-added products: a state-of-the-art review. *Environment, Development and Sustainability*, 1–55.

Amesho, K. T., Kadhila, T., Shangdiar, S., Shihomeka, S. P., Shopati, A. K., Edoun, E. I., & Mohan, C. (2024). Bioeconomy for sustainable environment management practices. In *Bioeconomy for sustainability* (pp. 41–63). Springer Nature Singapore.

Arora, J., Joshi, A., & Ray, R. C. (2024). Transforming agriculture residues for sustainable development.

Asefi, S., Nouri, H., Pourmohammadi, G., & Moghimi, H. (2024). Comprehensive network of stress-induced responses in *Zymomonas mobilis* during bioethanol production: From physiological and molecular responses to the effects of system metabolic engineering. *Microbial Cell Factories, 23*(1), 180.

Awan, I. Z. (2019). Heterogeneous catalysts for the valorisation of renewable feedstocks by depolymerisation of lignins.

Aydin, S. (Ed.). (2024). *Anaerobic digestion: Biotechnology for environmental sustainability*. BoD—Books on Demand.

Bajpai, P. (2022). *Fourth Generation Biofuels*. Springer.

Blasi, A., Verardi, A., Lopresto, C. G., Siciliano, S., & Sangiorgio, P. (2023). Lignocellulosic agricultural waste valorization to obtain valuable products: An overview. *Recycling, 8*(4), 61.

Bowyer, J., Howe, J., Levins, R. A., Groot, H., Fernholz, K., Pepke, E., & Henderson, C. (2018). Third gen eration biofuels implications for wood-derived fuels. Available online: http://www.dovetailinc.org/report_pdfs/2018/dovetail3gbiofuel0218.pdf. Accessed 28 Feb 2019

Busca, G. (2021). Production of gasolines and monocyclic aromatic hydrocarbons: From fossil raw materials to green processes. *Energies, 14*(13), 4061.

Cecchi, T., De Carolis, C., Cecchi, T., & de Carolis, C. (2021). Current state of art of the usual food waste valorization. In *Biobased products from food sector waste: Bioplastics, biocomposites, and biocascading* (pp. 79–105).

Chandra, M. S., Srinivasulu, M., Yadav, P. S., Ramesh, B., Narasimha, G., & Chandrasekhar, T. (2021). Value added products from agriculture, paper and food waste: A source of bioenergy production. *Bioenergy Research: Commercial Opportunities & Challenges*, 91–126.

Chauhan, A. K. (2020). Biofuel: Types and process overview. *Nanomaterials in Biofuels Research*, 1–28.

Chen, B., Cai, J., Chen, X., Wu, D., & Pan, Y. (2024a). A review on oil shale in-situ mining technologies: Opportunities and challenges. *Oil Shale, 41*(1).

Chen, J., Cui, L., Yan, Y., Ji, X., Fei, Q., & Ma, Y. (2024b). In situ preparation of oriented microbial consortium-based compound enzyme strengthens food waste disintegration and anaerobic digestion: Performance, mechanism, microbial communities and global metabolic pathways. *Chemical Engineering Journal, 486*, 150208.

Chen, J., Fan, Y., Jiaqiang, E., Cao, W., Zhang, F., Gong, J., et al. (2019). Effects analysis on the gasification kinetic characteristics of food waste in supercritical water. *Fuel, 241*, 94–104.

da Costa, A. A. F., de Oliveira, A. D. N., Esposito, R., Auvigne, A., Len, C., Luque, R., et al. (2023). Glycerol and microwave-assisted catalysis: Recent progress in batch and flow devices. *Sustainable Energy & Fuels, 7*(8), 1768–1792.

Das, P. (2022). Pyrolytic bio-oil—Production and applications. In *Advanced biofuel technologies* (pp. 243–304). Elsevier.

Devi, A., Bajar, S., Kour, H., Kothari, R., Pant, D., & Singh, A. (2022). Lignocellulosic biomass valorization for bioethanol production: A circular bioeconomy approach. *Bioenergy Research, 15*(4), 1820–1841.

Dong, C. (2019). Key features of thermochemical pretreatment induced lignin structure to saccharification and fermentation.

Elfasakhany, A. (2019). Biofuels in automobiles: advantages and disadvantages: a review. *Current Alternative Energy, 3*(1), 27–33.

Fan, M., Zhao, C., Huang, X., Zhang, H., & Xie, J. (2021). Enhanced digestibility and fermentability of sugarcane bagasse in biofuel production by surfactant-assisted dilute acid pretreatment. *Industrial Crops and Products, 172*, 114006.

Geraldi, A., Syarafa, H. S., Febriana, E., Setianingsih, P. N. M., Tamamia, E., Al Hanif, A. K. R., et al. (2023). Application of genetic engineering for increasing biofuel yield from microalgae: a review. In *AIP conference proceedings* (Vol. 2679, No. 1). AIP Publishing.

Ghosh, A. (2024). Hydrolases from extremophiles: Exploring alternative strategies for low-cost biofuel production. In *Biofuels* (pp. 440–463). CRC Press.

Gueboudji, Z. (2024). Waste valorization techniques. In *Generation of energy from municipal solid waste: Circular economy and sustainability* (pp. 29–52). Springer Nature Switzerland.

Gupta, J., Ghosh, P., Kumari, M., & Thakur, I. S. (2022). Solid waste landfill sites for the mitigation of greenhouse gases. In *Biomass, biofuels, biochemicals* (pp. 315–340). Elsevier.

Gustafsson, M., Meneghetti, R., Souza Marques, F., Trim, H., Dong, R., Al Saedi, T., et al. (2024). A perspective on the state of the biogas industry from selected member countries. In *IEA bioenergy task* (Vol. 37, No. 2024, p. 2).

Hussain, A., Parveen, F., Saxena, A., & Ashfaque, M. (2024a). A review of nanotechnology in enzyme cascade to address challenges in pre-treating biomass. *International Journal of Biological Macromolecules*, 132466.

Hussain, A., Saxena, A., Irum, Farooqui, A., & Ashfaque, M. (2024b). Environment of lignocellulosic waste to biofuel. In *Agricultural biomass nanocatalysts for green energy applications* (pp. 19–43). Springer Nature Singapore.

IEA. (2019). Renewables 2019: Analysis and forecast to 2024. *International Energy Agency, 204*. https://www.iea.org/reports/renewables-2019

Iqbal, K., Mishra, A., & Sreedharan, S. M. (2025). The quintessence of algal biomass in bioplastic production: insightful advancement and sustainable use. *Bioresources and Bioprocessing, 12*(1), 91.

Irum, Asif, M., Dubey, P., Tripathi, G., Khujamshukurov, N. A., Farooqui, A., & Mishra, V. (2024). Lignocellulosic waste to biofuel-paddy straw to bioethanol: advancement in technology. In *Paddy straw waste for biorefinery applications* (pp. 87–107). Springer Nature Singapore.

Jacobson, T., Edwards, M., Qiande, M., Robert, M., Moncrieff, J., & Voiniciuc, C. (2025). Intracellular mannanases sustain matrix polysaccharide biosynthesis. *bioRxiv*, 2025-02.

Jose, A., Thurakkal, L., & Porel, M. (2022). Structure, properties, and functions of manganese peroxidase for enzymatic pretreatment of waste Biomass. In *Enzymes in the valorization of waste* (pp. 113–135). CRC Press.

Joshi, A., Verma, K. K., D Rajput, V., Minkina, T., & Arora, J. (2022). Recent advances in metabolic engineering of microorganisms for advancing lignocellulose-derived biofuels. *Bioengineered, 13*(4), 8135–8163.

Kanakdande, A. P., & Mane, R. S. (2023). Nanotechnology for bioenergy and biofuel production. In *Nanomaterials for sustainable development: Opportunities and future perspectives* (pp. 283–296). Springer Nature Singapore.

Kaparaju, P., & Mukherjee, T. (2025). Characteristic Requirements of Biomass for Biorefineries.

Karim, A., Islam, M. A., Nayeem, A., & Yousuf, A. (2022). Conversion pathways for biomass-derived aviation fuels. In *Sustainable alternatives for aviation fuels* (pp. 1–25). Elsevier.

Kashif, M., Sabri, M. A., Aresta, M., Dibenedetto, A., & Dumeignil, F. (2025). Sustainable synergy: Unleashing the potential of biomass in integrated biorefineries. *Sustainable Energy & Fuels, 9*(2), 338–400.

Khoo, K. S., Ahmad, I., Chew, K. W., Iwamoto, K., Bhatnagar, A., & Show, P. L. (2023). Enhanced microalgal lipid production for biofuel using different strategies including genetic modification of microalgae: A review. *Progress in Energy and Combustion Science, 96*, 101071.

Kiehbadroudinezhad, M., Merabet, A., Ghenai, C., Abo-Khalil, A. G., & Salameh, T. (2023). The role of biofuels for sustainable MicrogridsF: A path towards carbon neutrality and the green economy. *Heliyon, 9*(2).

Kumar, A., et al. (2023). *Bioresource Technology, 384*, Article 129295.

Lakzian, E., Yazdani, S., Salmani, F., Mahian, O., Kim, H. D., Ghalambaz, M., et al. (2024). Supersonic separation towards sustainable gas removal and carbon capture. *Progress in Energy and Combustion Science, 103*, 101158.

Lamolinara, B., Pérez-Martínez, A., Guardado-Yordi, E., Fiallos, C. G., Diéguez-Santana, K., & Ruiz-Mercado, G. J. (2022). Anaerobic digestate management, environmental impacts, and techno-economic challenges. *Waste Management, 140*, 14–30.

Li, Y., Xu, F., Li, Y., Lu, J., Li, S., Shah, A., et al. (2018). Reactor performance and energy analysis of solid state anaerobic co-digestion of dairy manure with corn stover and tomato residues. *Waste Management, 73*, 130–139.

Lindley, P. J. (2022). *Substrate-induced perturbations to the active site electronic structure of AA10 lytic polysaccharide monooxygenases* (Doctoral dissertation, University of York).

Liu, R. (2024). Heterogeneity of softwood kraft lignin: Effects on fractionation and nanoparticles formation.

Mahapatra, S., Kumar, D., Singh, B., & Sachan, P. K. (2021). Biofuels and their sources of production: A review on cleaner sustainable alternative against conventional fuel, in the framework of the food and energy nexus. *Energy Nexus, 4*, 100036.

Makaryan, I. A., Salgansky, E. A., Arutyunov, V. S., & Sedov, I. V. (2023). Non-Catalytic partial oxidation of hydrocarbon gases to syngas and hydrogen: A systematic review. *Energies, 16*(6), 2916.

Marichamy, M. K. (2024). Process optimization in biofuel production using different biomass. In *Biomass-based clean technologies for sustainable development: sustainable resources of energy production and utilization* (pp. 135–159). Springer Nature Singapore.

Mati, A., Buffi, M., Dell'Orco, S., Lombardi, G., Ruiz Ramiro, P. M., Kersten, S. R., & Chiaramonti, D. (2022). Fractional condensation of fast pyrolysis bio-oil to improve biocrude quality towards alternative fuels production. *Applied Sciences, 12*(10), 4822.

Minyukova, T. P., & Dokuchits, E. V. (2023). Hydrogen for CO_2 processing in heterogeneous catalytic reactions. *International Journal of Hydrogen Energy, 48*(59), 22462–22483.

Mishra, S., Banerjee, A., Chattaraj, S., Samantaray, A., Panigrahi, S., Bauri, K. P., & Thatoi, H. (2024a). Microbial process in anaerobic digestion of food wastes for biogas production: A review. *Systems Microbiology and Biomanufacturing*, 1–19.

Mishra, R. K., Gariya, B., Savvasere, P., Dhir, D., Kumar, P., & Mohanty, K. (2024b). Thermocatalytic pyrolysis of waste areca nut into renewable fuel and value-added chemicals. *ACS Omega, 9*(24), 25779–25792.

Miskolczi, N., Gao, N., & Quan, C. (2023). Pyrolysis-gasification of biomass and municipal plastic waste using transition metal modified catalyst to investigate the effect of contaminants. *Journal of the Energy Institute, 108*, 101233.

Mohamed, A., Shaaban, D., Ismail, E., Hesham, K., Soliman, M., Mohamed, E. T., et al. (2024). Application of green technology for Biofuel production. البحوث التطبيقية في العلوم والانسانيات, *1*(1), 125–140.

Mohsenpour, S. F., Hennige, S., Willoughby, N., Adeloye, A., & Gutierrez, T. (2021). Integrating micro-algae into wastewater treatment: A review. *Science of the Total Environment, 752*, 142168.

Muh, E., Tabet, F., & Amara, S. (2021). Biomass conversion to fuels and value-added chemicals: A comprehensive review of the thermochemical processes. *Current Alternative Energy, 4*(1), 3–25.

Mukherjee, S., Chander, T., Pradhan, S., Prasad, L., & Mehariya, S. (2025). Physico-chemical characterization of agricultural and forest biomass.

Nair, L. G., Agrawal, K., & Verma, P. (2022). An overview of sustainable approaches for bioenergy production from agro-industrial wastes. *Energy Nexus, 6*, 100086.

Narisetty, V., Cox, R., Bommareddy, R., Agrawal, D., Ahmad, E., Pant, K. K., et al. (2022). Valorisation of xylose to renewable fuels and chemicals, an essential step in augmenting the commercial viability of lignocellulosic biorefineries. *Sustainable Energy & Fuels, 6*(1), 29–65.

Nisar, A., Hashum, K., Bashir, M., Mubeen, N., Younus, S., Mehmood, S., et al. (2024). Advancing sustainable biofuel production from agricultural residues: A comprehensive mini-review. *Sustainable Chemical Engineering*, 115–128.

Osman, A. I., Mehta, N., Elgarahy, A. M., Al-Hinai, A., Al-Muhtaseb, A. A. H., & Rooney, D. W. (2021). Conversion of biomass to biofuels and life cycle assessment: A review. *Environmental Chemistry Letters, 19*, 4075–4118.

Pal, D. B., Yadav, S., Jana, S. K., & Tiwari, A. K. (2024). Waste biomass conversion in useful various chemicals. In *Sustainable clean energy production using waste biomass: sustainable energy production and utilization* (pp. 75–100). Springer Nature Singapore.

Pan, C. (2024). Branch Series Flotation. In *The ECPH encyclopedia of mining and metallurgy* (pp. 180–180). Springer Nature Singapore.

Panchasara, H., & Ashwath, N. (2021). Effects of pyrolysis bio-oils on fuel atomisation—A review. *Energies, 14*(4), 794.

Paul, D., & Sarkar, N. (2026). Scaling up microbial drug production: strategies to overcome key challenges. In *Bioinformatics, AI, and Machine Learning in Microbial Drug Development* (pp. 283–322). Academic Press.

Periyasamy, S., Isabel, J. B., Kavitha, S., Karthik, V., Mohamed, B. A., Gizaw, D. G., & Aminabhavi, T. M. (2023). Recent advances in consolidated bioprocessing for conversion of lignocellulosic biomass into bioethanol—A review. *Chemical Engineering Journal, 453*, 139783.

Poornima, S., Manikandan, S., Prakash, R., Deena, S. R., Subbaiya, R., Karmegam, N., et al. (2024). Biofuel and biochemical production through biomass transformation using advanced thermochemical and biochemical processes—A review. *Fuel, 372*, 132204.

Pradeep, N. S., & Edison, L. K. (Eds.). (2022). *Microbial beta glucanases: Molecular structure, functions and applications*. Springer Nature.

Ray, R. V. (2024). Effects of pathogens and disease on plant physiology. In *Agrios' Plant Pathology* (pp. 63–92). Academic Press.

Ren, Y., Si, B., Liu, Z., Jiang, W., & Zhang, Y. (2022). Promoting dark fermentation for biohydrogen production: Potential roles of iron-based additives. *International Journal of Hydrogen Energy, 47*(3), 1499–1515.

Rey, J. R. C., Longo, A., Rijo, B., Pedrero, C. M., Tarelho, L. A. C., Brito, P. S. D., & Nobre, C. (2024). A review of cleaning technologies for biomass-derived syngas. *Fuel, 377*, 132776.

Rodríguez, R. P., Díaz-Domínguez, Y., Tobío-Pérez, I., Ortiz-Alvarez, M., & Hernández, J. S. (2024). Production of biodiesel from *Jatropha curcas* oil. In *The production of biodiesel and related fuel additives* (pp. 103–153). Bentham Science Publishers.

Rokhati, N., Kusworo, T. D., Prasetyaningrum, A., Hamada, N. A., Utomo, D. P., & Riyanto, T. (2022). Effect of surfactant HLB value on enzymatic hydrolysis of chitosan. *ChemEngineering, 6*(1), 17.

Roy, P., Maturano, J., Hasdemir, H., Lopez, A., Xu, F., Hellman, J., et al. (2024). Elucidating the mechanism of metabolism of cannabichromene by human cytochrome P450s. *Journal of Natural Products, 87*(4), 639–651.

Saeid, A., Akter, F., Ali, M. A., & Rahman, M. H. (2024). Enzymatic modification of starch: Amylases and pullulanase. In *Advanced Research in Starch* (pp. 95–115). Springer Nature Singapore.

Sarkar, O., Butti, S. K., & Mohan, S. V. (2018). Acidogenic biorefinery: Food waste valorization to biogas and platform chemicals. *Waste Biorefinery*, 203–218.

Sartaj, K., Prasad, R., Matsakas, L., & Patel, A. (2023). Transforming recalcitrant wastes into biodiesel by oleaginous yeast: An insight into the metabolic pathways and multi-omics landscape. *Chemical Engineering Journal, 474*, 145625.

Seidi, F., Liu, Y., Huang, Y., Xiao, H., & Crespy, D. (2025). Chemistry of lignin and condensed tannins as aromatic biopolymers. *Chemical Society Reviews*.

Selvam, J., Mal, J., Singh, S., Yadav, A., Giri, B. S., Pandey, A., & Sinha, R. (2024). Bioprospecting marine microalgae as sustainable bio-factories for value-added compounds. *Algal Research, 79*, 103444.

Shahid, M. K., Batool, A., Kashif, A., Nawaz, M. H., Aslam, M., Iqbal, N., & Choi, Y. (2021). Biofuels and biorefineries: Development, application and future perspectives emphasizing the environmental and economic aspects. *Journal of Environmental Management, 297*, 113268.

Shao, X., Yang, S., Fan, L., & Yuan, Y. (2023). Sugar alcohol phase change materials for low-to-medium temperature thermal energy storage: A comprehensive review. *Journal of Energy Storage, 68*, 107848.

Sharma, P., Bano, A., Singh, S. P., Atkinson, J. D., Lam, S. S., Iqbal, H. M., & Tong, Y. W. (2024). Biotransformation of food waste into biogas and hydrogen fuel—A review. *International Journal of Hydrogen Energy, 52*, 46–60.

Siddiqui, M. Z., Sheraz, M., Toor, U. A., Anus, A., Mahmood, A., Haseeb, M., et al. (2022). Recent approaches on the optimization of biomass gasification process parameters for product H_2 and syngas ratio: a review. *Environment, Development and Sustainability*, 1–29.

Sikarwar, V. S., Zhao, M., Fennell, P. S., Shah, N., & Anthony, E. J. (2017). Progress in biofuel production from gasification. *Progress in Energy and Combustion Science, 61*, 189–248.

Sindhu, R., Gnansounou, E., Rebello, S., Binod, P., Varjani, S., Thakur, I. S., et al. (2019). Conversion of food and kitchen waste to value-added products. *Journal of Environmental Management, 241*, 619–630.

Sinduja, M., John, J. E., Suganthi, R., Ragul, S., Balaganesh, B., Mathiyarasi, K., et al. (2023). Microbiota in sustainable degradation of organic waste and its utilisation in agricultural industry. In *Microbiomes for the management of agricultural sustainability* (pp. 29–57). Springer Nature Switzerland.

Singh, R. (2024). *Domestically produced bioethanol contributing to India's Pursuit of energy self-reliance and de-carbonization of the road transport sector* (No. 2024-28-0126). SAE Technical Paper.

Singh, N., Gupta, R. P., Puri, S. K., & Mathur, A. S. (2021). Bioethanol production from pre-treated whole slurry rice straw by thermophilic co-culture. *Fuel, 301*, 121074.

Sivabalan, K., Hassan, S., Ya, H., & Pasupuleti, J. (2021). A review on the characteristic of biomass and classification of bioenergy through direct combustion and gasification as an alternative power supply. In *Journal of physics: Conference series* (Vol. 1831, No. 1, p. 012033). IOP Publishing.

Soni, S. K., & Soni, R. (2025). Techno-economic and environmental assessment. In *Green Biorefinery Solutions: Transforming Biodegradable Waste into Resources* (pp. 291–314). Singapore: Springer Nature Singapore.

Sun, S., Xu, D., Wei, Y., Zhi, Y., Jiang, G., & Guo, Y. (2022). Influence laws of operating parameters on coal pyrolysis characteristics. *Journal of Analytical and Applied Pyrolysis, 167*, 105684.

Tiwari, A. K., & Pal, D. B. (2021). Bio-processing: Biomass to commercial alcohol. *Bioenergy Research: Biomass Waste to Energy*, 149–168.

Tripathi N, Hills CD, Singh RS, Atkinson CJ (2019) Biomass waste utilisation in low-carbon products: Harnessing a major potential resource. *NPJ Climate and Atmospheric Science, 2*, 1–10. https://doi.org/10.1038/s41612-019-0093-5.

Usmani, Z., Sharma, M., Awasthi, A. K., Lukk, T., Tuohy, M. G., Gong, L., et al. (2021). Lignocellulosic biorefineries: The current state of challenges and strategies for efficient commercialization. *Renewable and Sustainable Energy Reviews, 148*, 111258.

Vadke, V. S. (2023). *Principles of vegetable oil extraction*. CRC Press.

Vasile, C., & Baican, M. (2023). Lignins as promising renewable biopolymers and bioactive compounds for high-performance materials. *Polymers, 15*(15), 3177.

Wang, J., & Yin, Y. (2021). Clostridium species for fermentative hydrogen production: An overview. *International Journal of Hydrogen Energy, 46*(70), 34599–34625.

Wirth, R., Bagi, Z., Shetty, P., Szuhaj, M., Cheung, T. T. S., Kovács, K. L., & Maróti, G. (2023). Inter-kingdom interactions and stability of methanogens revealed by machine-learning guided multi-omics analysis of industrial-scale biogas plants. *The ISME Journal, 17*(8), 1326–1339.

Yadav, A. N., Rastegari, A. A., Yadav, N., & Gaur, R. (Eds.). (2020). *Biofuels production-sustainability and advances in microbial bioresources*. Springer.

Yaverino-Gutiérrez, M. A., Wong, A. Y. C. H., Ibarra-Muñoz, L. A., Chávez, A. C. F., Sosa-Martínez, J. D., Tagle-Pedroza, A. S., et al. (2024). Perspectives and progress in bioethanol processing and social economic impacts. *Sustainability, 16*(2), 608.

Yogalakshmi, K. N., Mohamed Usman, T. M., Kavitha, S., Sachdeva, S., Thakur, S., Adish Kumar, S., & Rajesh Banu, J. (2023). Lignocellulosic biorefinery technologies: a perception into recent advances in biomass fractionation, biorefineries, economic hurdles and market outlook. *Fermentation, 9*(3), 238.

Zhang, Q., Guan, Y., Zhang, Z., Dong, S., Yuan, T., Ruan, Z., & Chen, M. (2024). Sustainable microalgae cultivation: A comprehensive review of open and enclosed systems for biofuel and high value compound production. In *E3S web of conferences* (Vol. 577, p. 01008). EDP Sciences.

Zhang, Y., et al. (2022). *Nature Communications, 13*(1), 5102.

Zhao, H., & Yuan, Z. Y. (2023). Progress and perspectives for solar-driven water electrolysis to produce green hydrogen. *Advanced Energy Materials, 13*(16), 2300254.

Zhu, C., Ji, Y., Du, X., Kong, F., Chi, Z., & Zhao, Y. (2022). A smart and precise mixing strategy for efficient and cost-effective microalgae production in open ponds. *Science of the Total Environment, 852*, 158515.

Chapter 5
Recent Updates on Biofuel Production from Food Waste

Lipsa Dehal and Gurinder Kaur Walia

Abstract With a 3.3 billion tonne carbon footprint, food waste (FW) is a major worldwide problem that is mostly caused by poor supply chain management, storage problems, and transportation challenges. In accordance with the SDGs, acidogenic processes such as dark fermentation (DF), anaerobic digestion (AD), and its combination (DF-AD) can valorize FW and generate renewable biofuels like biohydrogen, hydrochar, biochar, bioethanol etc. FW's high moisture content, organic matter, and biodegradability make it the perfect substrate for acidogenic reactions. However, energy output, conversion efficiency, and cost effectiveness all play a role in the selection of FW valorisation paths. In order to give a full knowledge of energy recovery and yield from all acidogenic processes, this study examines comparative studies on the production of biogas, biohydrogen, biohythane etc. from FW via acidogenic processes, with an emphasis on energy yield, energy recovery, and environmental and economic effect. Here we also examine the synergistic impacts of AD and HTC (hydrothermal carbonization) procedures, current developments in digestate slurry management, as well as to choose particular microbes and genetically modify anaerobic bacteria to accelerate the AD process. Studies on pre-treatments and co-digestion as methods to increase biofuel yield received special attention. Finally, a prospective integrated bio-thermo-chemical process is suggested for maximal energy recovery, which includes digestate management, the manufacturing of biofertilizer, and the valuation of food waste to energy vectors (Biohydrogen, Biomethane, and hydro-char).

Keywords Biofuel · Anaerobic digestion · Food waste · Fermentation · Bio-thermo-chemical process

---

L. Dehal (✉) · G. K. Walia
Department of Zoology and Environmental Sciences, Punjabi University, Patiala, Punjab, India
e-mail: lipsa_rs25@pbi.ac.in

V. Singh et al. (eds.), *Bioenergy and Value Added Byproducts from Biomass Waste*, Sustainable Environmental Waste Management Strategies,
https://doi.org/10.1007/978-3-032-22849-9_5

5.1 Introduction

About 44% of the two billion tons of municipal solid waste generated each year is food waste, which is frequently disposed of in landfills in most countries (Sharma & Jain, 2020). With 9.1 billion people on the planet by 2050, the amount of food waste generated will increase, resulting in more waste and pollution (Truong et al., 2019). In landfills, solid food waste decomposes and contaminates water that enters the landfill before it drains as leachate, which exacerbates water pollution issues (Julkipli et al., 2024). Additionally, it emits greenhouse gases, produces unpleasant odors, and serves as a habitat for different pathogen vectors. When flushed down the drain, liquid food processing waste, including leftover cooking oil, can also clog pipes and hinder sewage. To reduce food waste landfilling and its effects on the environment, effective food waste management is necessary. We must aim to feed more people while preserving what we already have, not to increase food production. Between 2012 and 2050, the demand to increase production outputs by 35–50% globally will be fueled by both population and economic growth. Future food availability will be challenging if food scarcity continues. Thus, biodiversity loss, environmental degradation, water scarcity, and climate change are the outcomes of the excessive strain on resources and significant environmental harm (Warming, 2023). Food waste also contributes to the annual release of approximately 3.3 billion tons of carbon dioxide into the atmosphere through the carbon emissions of these waste products (Bhatia et al., 2023).

Food waste is produced at every stage of the distribution network, including manufacture, sales, restaurants, and households. Animal byproducts, mixed catering waste, agricultural residue, and inedible food waste are the four types of food scraps (Georganas et al., 2020). Excessive food waste is produced as a result of food deterioration, surplus supply, and waste shelf life. The disposal of food waste poses a number of difficulties, such as its perishable and heterogeneous nature, high moisture content, and an unpleasant odor brought on by nitrogen and sulfur compounds. Landfilling, composting, and incineration are popular techniques for getting rid of food waste. Studies have concentrated on new technologies for food waste disintegration because these methods are unable to produce any beneficial goods. Just a small portion of this trash is converted into energy by incineration or recycled into another form, like compost. The disposal of food waste poses a number of difficulties, such as its perishable and heterogeneous nature, high moisture content, and an unpleasant odor brought on by nitrogen and sulfur compounds (Sarker et al., 2024). Landfilling, composting, and incineration are popular techniques for getting rid of food waste. Studies have concentrated on new technologies for food waste disintegration because these methods are unable to produce any beneficial goods. Just a small portion of this trash is converted into energy by incineration or recycled into another form, like compost. For biofuel and resource recovery of food waste, aerobic digestion, food brewing, bioethanol fermentation, anaerobic composting, and biological processes are frequently used (Pour & Makkawi, 2021). The conversion of food waste into inexpensive but highly valuable products like biofuel, biodiesel, biochar, biofertilizers, organic acid, and enzymes is one example of how to reuse this waste in an environmentally friendly and sustainable way (Varjani et al., 2021). Contrarily, food waste includes valuable substances as lipids, carbohydrates, amino acids, and other chemicals (Luo et al., 2019). A by-product from one business could be used totally by another company or wind up in the final product of another industry. The bioeconomy will help to reduce emissions and reliance on fossil fuels by producing bioenergy, biofuels, biobased products, food, feed, and other products using more ecologically friendly materials, like biomass (Food Waste) (Scarlat et al., 2015). Numerous high-value products (biofuel, biochar/ash, and water) can be produced through value-adding using integrated pyrolysis–gasification processes with negligible or no environmental impact.

In order to lessen the influence on the environment, waste management procedures have historically included treatment, reduction, and preventive actions. Stricter waste management laws and regulations are required due to the increased interest in sustainability on a global scale. The current study provides comprehensive information regarding the current status of food waste production while taking into account recent developments in food waste management. Utilizing the potential and related sustainability benefits of food waste for the creation of value-added products will be crucial since the study offers a comprehensive picture of how food waste contributes to the circular bioeconomy.

5.2 Food Waste Management

Food waste encompasses a wide range of by-products or waste products generated during various stages of food production, such as the procurement, processing, and handling of fruits, vegetables, cereals, edible and uncooked raw materials from wet markets, as well as leftover food from homes and restaurants. Food loss and waste are problems for producers, retailers, and consumers worldwide. 59 kg of food are thought to be wasted for every 195 kg of food produced. The food waste generation rate is the amount of food waste produced daily or annually by an individual. With 180–200 kg of garbage per capita per year respectively, it can be expressed in kilograms per person (Kaza et al., 2018).

About 17% of the world's food output may be lost, with families accounting for 61% of this loss, food service for 26%, and retail for 13%, according to the United Nations Environment Programme's Food loss Index Report 2021 (Agarwal et al., 2021). Agriculture is the main pillar of the Indian economy. A significant amount of food waste is generated annually, primarily due to post-harvest processes like drying, fanning, threshing, packing, and shipping, and food waste at big, subsidized marketplaces for distribution to retailers and merchants. India's rate of food production is much lower than its rate of food waste (Munesue et al., 2015). The Ministry of Agriculture of India's Crop Statistics reports that 623 million tons of gross crop leftovers are produced each year from a variety of crops in various Indian states. The largest volumes of garbage were generated by India's rice, sugarcane, and wheat production, which produced 110, 199, and 139 million tonnes of waste annually, respectively (Singh & Kalamdhad, 2021).

"Halve global food waste per capita at the retail and consumer level by 2030," according to the UN's 2019 Sustainable Development Goals (SDGs). In order to reduce food loss and jointly contribute to the achievement of the SDGs, the FAO and UNEP use two distinct indices to track progress towards these targets: the food loss index and the food waste index. Both developed and developing countries should impose stringent laws and regulations in order to achieve the SDGs. Slum expansion, urbanization, and population growth are all significantly impacted by globalization, which further exacerbates the food crisis. Food security is still a problem in areas of poverty, while being essential for sustainable development (Chisoro et al., 2023). For example, South Korea reduced the amount of food waste produced by expanding food waste recycling to recover biofuels and value-added products (59%) and implementing a scheme dubbed "Pay as You Throw" (FAO, 2019). Food waste accounted for 15.2% (40.7 million tons) of all municipal solid waste (MSW) generated in the United States in 2017, according to the Environmental Protection Agency. For 30.6, 2.6, and 7.5 million tonnes of food waste, respectively, techniques such as landfilling, composting, and energy recovery were employed (Kannah et al., 2020). Implementing a successful approach is urgently needed since food waste negatively impacts the economy, ecology, food availability, and nutrition. Food waste is a worldwide problem, but it also presents an opportunity to create goods with extra value. The idea of food waste valorization has been presented to illustrate how food waste could be converted into capital.

5.3 Technologies in Biofuel Production from Food Waste

Biofuel production from food waste relies on various advanced technologies that convert organic materials into renewable energy. These technologies include fermentation, anaerobic digestion, and emerging processes like integrated anaerobic digestion- dark fermentation, hydrothermal carbonization as summarized in Table 5.1. Each method is tailored to efficiently process different types of food waste, transforming them into bioethanol, biogas, or biodiesel, while also contributing to waste reduction and sustainable energy generation.

5.3.1 Incineration

Thermal or thermochemical convergence occurs during incineration. The process of thermochemical conversion involves a high-temperature chemical reformation that necessitates the bonding and reforming of organic matter to gas, highly oxidized bio-oil (liquid), and biochar (solid). When a catalyst is present, organic pollutants

Table 5.1 Technological innovations in biofuel production from food waste

Technology	Process	Recent Innovations	Products	References
Incineration	Thermochemical conversion that involves a high-temperature chemical reformation	More energy-efficient incineration plants or combined heat and power (CHP) systems	Bio-oil and Biochar production, converted to heat and electricity (energy)	Shukla et al. (2024) and Mattson et al. (2024)
Pyrolysis	Controlled burning of biomass without an oxygen source	New reactor designs for more efficient conversion	Bio-oils, carbon rich biochar and gaseous products	Burezq and Davidson (2023) and Saxena et al. (2024)
Anaerobic digestion (AD)	Food waste microbiological degradation without oxygen	Improved reactor designs and the creation of more effective microorganisms for complicated waste	Biogas (methane and CO_2)	Abubakar et al. (2022) and Wu et al. (2024)
Dark fermentation (DF)	Yeast ferments food waste to produce bioethanol from carbohydrates	Genetically modified organisms (GMOs) used to increase the fermentation	Convert sugar or starch to bioethanol or biofuel and hydrogen	Hassan et al. (2024) and Jain et al. (2024)
Integrated anaerobic digestion and dark fermentation (AD-DF)	Combining anaerobic digestion and dark fermentation	Microbial strains (GMOs), reactor designs, operational conditions, nutrient management, and hydrogen recovery systems improved	Biogas, hydrogen, organic acids	Srisowmeya et al. (2020) and Subramanian and Suresh (2024)
Photofermentation	Photosynthetic bacteria, like purple non-sulfur bacteria (PNSB), use light as an energy source to process food waste	Genetically engineered bacteria used, optimizing light absorption systems, and enhancing reactor designs also	Clean fuel (hydrogen), organic acids	Tiang et al. (2020) and Gupta et al. (2024)
Hydrothermal carbonization (HTC)	Apply heat and pressure, simulating natural coal formation under controlled conditions	Effective reaction conditions, Improvement in feedstock preprocessing, and enhancing energy efficiency	Biochar, bio-oils, hydrochar	Le et al. (2022) and Yan et al. (2024)

are oxidized to carbon dioxide and water during incineration, which takes place at a lower temperature and uses less fuel than thermal incineration (Shukla et al., 2024).

At temperatures above 800 °C, the biomass is oxidized to produce heat and steam, which is then used to generate power (Mattson et al., 2024). The combustion process can be used to treat any kind of biomass, albeit its effectiveness is dependent on the biomass's moisture level. Ash produced by incineration can be used as a filler in building materials or dumped in a landfill. It contributes to air pollution by generating harmful gases and particulate matter, which negatively impacts the use of combustion as a management strategy for food waste disposal (Khan et al., 2024).

When moisture-containing food waste is burned with organic biomass, dioxins are produced, which can shorten the incinerator's lifespan and provide a secondary environmental risk. Therefore, in order to avoid uncontrolled emissions and the production of energy, it is recommended that fuel gas management be incorporated into the design of an incinerator plant. Residents who live close to incinerators are more likely to be exposed to pollutants through their skin, food, or drink, or by breathing in polluted air (Astuti & Rauf, 2024). Compared to direct incineration, these techniques are more effective when a pretreatment step is added before burning. Drying and crushing/shredding are possible pretreatment steps.

5.3.2 Pyrolysis

Biomass may be effectively converted into biofuel using pyrolysis. Pyrolysis, the controlled burning of biomass without an oxygen source, produces carbon-rich biochar (Burezq & Davidson, 2023). Wood scraps, *Ipomoea carnea*, tea factory waste, and microalgae are just a few of the biomass sources that are increasingly being pyrolyzed to produce fuel bio oils (Saxena et al., 2024). Pyrolysis is divided into slow and fast varieties according to temperature and operation rate. Slow pyrolysis is distinguished by its exceptionally long vapour residence time (5–30 min), slow heating rates (between 10 and 20 °C/min), at least 400 °C, and increased charcoal generation in comparison to gaseous and liquid products (Moya et al., 2024). This is the conventional process that has been used for many years to turn wood into charcoal. Therefore, the formation of solid carbonaceous char benefits from delayed pyrolysis. The flash/fast pyrolysis takes place at a high temperature (750–1050 °C), a quick heating value (> 700 °C/s), and a relatively brief gas residence time (0.5 s). The method is hindered by low thermal stability, high temperature, corrosiveness, and the presence of particles in the bio-oil, even though pyrolysis can provide a very high bio-oil yield (60–75%), 20–25% biochar, and 10–20% gaseous products (Selvarajoo et al., 2022). Therefore, quick pyrolysis of biomass to produce bio-oil avoids the production of unwanted char (Table 5.1).

5.3.3 Anaerobic Digestion (AD)

Biogas, which is primarily made up of methane and carbon dioxide with traces of hydrogen and ammonia, is created when organic materials are digested anaerobically (AD). It also contains a nutrient-rich liquid fertilizer. Generally speaking, organic matter's protein and carbohydrate components convert to biogas more quickly than lipids (Abubakar et al., 2022). Based on temperature, anaerobic digestion processes are commonly divided into two categories, such as thermophilic and mesophilic. The organic matter is broken down by mesophiles and thermophiles into carbon dioxide and methane at average incubation temperatures of 55 °C and 37 °C, respectively (Wu et al., 2024).

Excessive salt concentration systems suffer from chemical inhibitors like ammonium ion and ammonia formation, and nitrogen-rich protein components produce free ammonia during the anaerobic absorption of food waste. To increase energy renewal and anaerobic degradation efficiency, Aworanti et al. (2023) suggested a two-stage anaerobic degradation process for methane fermentation combined with the production of hydrogen or ethanol.

5.3.4 Dark Fermentation (DF)

With over 90% of overall production, steam reforming of hydrocarbons and coal gasification are now the most common methods for creating hydrogen. Dependency on these nonrenewable resources has significantly increased as a result of this reliance on fossil fuels. This problem must be solved by switching to renewable energy-based hydrogen production methods, such as biological hydrogen synthesis, which produces a mixture of hydrogen and carbon dioxide with an energy content that is directly proportional to the hydrogen content (LHV 120 MJ/kg) (Hassan et al., 2024).

Microbial hydrogen generation involves a number of processes, including dark fermentation (DF), photofermentation (PF), and direct and indirect biophotolysis of water (Jain et al., 2024). Because of its low cost, simplicity of use, and abundance of renewable substrates, DF has become known as a very viable technique. Compared to traditional methods like steam reforming natural gas or electrolyzing water, the synthesis of hydrogen gas by DF has several benefits. These advantages include low energy use, reduced greenhouse gas emissions, and the utilization of renewable organic waste as a feedstock (Ahmad et al., 2024).

In the absence of light, the biological phenomenon known as DF of FW converts organic compounds into bio-hydrogen (here defined as the mixture of hydrogen and carbon dioxide) and other useful byproducts. The aforementioned process involves the breakdown of complex organic compounds into simpler components, such as hydrogen gas, acetate, and carbon dioxide, by the action of anaerobic microbes. Other byproducts that can be obtained from this process include VFAs, alcohols, and organic acids, and these byproducts have the potential to be used in the industrial sector, particularly in the areas of chemical residues and bioenergy generation. Pyruvate is first produced during the fermentation of carbohydrates in the DF process. Nicotinamide adenine dinucleotide (NADH) and ATP are produced as energy sources by this metabolic process (Villanueva-Galindo et al., 2023). Ferredoxin reduction is the mechanism by which pyruvate is transformed into acetyl-CoA. A metalloenzyme known as hydrogenase or nitrogenase uses both NADH and reduced ferredoxin to convert H^+ to hydrogen in the case that acetyl-CoA is converted to acetate. This results in a maximum theoretical production of 4 mol hydrogen/mol hexose sugar (Gbiete et al., 2024).

Though it is important to note that the acetate pathway produces 3.3 mol hydrogen/mol pentose sugar, the previously described procedure can also be used with C-5 sugars. Because it efficiently combines protons and electrons, the enzyme hydrogenase is essential to the anaerobic fermentation process, which produces hydrogen. DF is frequently carried out in controlled settings with meticulously regulated substrate concentration, pH, and temperature. The method shows promise for application in trash treatment and the creation of renewable energy. However, in order to achieve the best hydrogen yields, some challenges must be overcome, such as low hydrogen yields and the existence of microbial competition. The presence of organisms that consume hydrogen, such as hydrogenotrophic methanogens, sulphate-reducing bacteria, nitrate-reducing bacteria, and propionate manufacturers, can be blamed for the poor hydrogen yields. The total amount of bio-hydrogen produced decreases as a result of these organisms (Xue et al., 2024). Despite these challenges, the DF method has the ability to significantly reduce greenhouse gas emissions and divert garbage from landfills.

5.3.5 Integrated Anaerobic Digestion and Dark Fermentation (AD-DF)

A biological process called anaerobic digestion (AD) breaks down biomass in an oxygen-free atmosphere creating a gas phase that is high in methane. The four stages of AD are methanogenesis, acetogenesis, acidogenesis, and hydrolysis (Subramanian & Suresh, 2024). For the system to function as efficiently as possible, syntrophy between each species is necessary. Complex polymers in organic biomass are broken down by hydrolysis into smaller parts that microbes can access. In order to break down proteins, lipids, and carbohydrates into sugars, long-chain fatty acids (LCFAs), and amino acids, hydrolytic bacteria release enzymes. The rate-determining phase may be hydrolysis, which is best at pH 5–7 and temperatures between 30 and 50 °C. Because lignin is recalcitrant, the presence of lignocellulosic biomass slows down the process (Chatterjee & Mazumder, 2019).

Acidogenic bacteria convert hydrolysis products into intermediate volatile fatty acids (VFAs), including butyric, propionic, and acetic acids, as well as trace amounts of lactic acid and ethanol, during the acidogenesis process. The VFAs ratios are impacted by process variables such as feedstock, temperature, and pH. However, excessive VFA generation may result in inhibition and, eventually, process failure. One consequence of amino acid breakdown in protein-rich waste that can obstruct AD is ammonia. During acetogenesis, VFAs are transformed into carbon dioxide, hydrogen, and acetic acid. Acetic acid is used by methanogens, which inhibits acetogenic microorganisms and maintains conditions favorable for acetogenesis. Acidogenesis and β-oxidation cause glycerol and LCFAs to undergo acetogenesis, which yields acetate (Wang et al., 2023).

Hydrogen partial pressure, which shouldn't be higher than 10^{-4} atm, impedes the endothermic process. The existence of methanogens, which use hydrogen as an electron donor during the methanogenesis process to create methane through a hydrogenotrophic pathway, prevents this situation. Acetoclastic methanogens, which transform acetate via an energy-gaining process, also create the energy carrier (Magdalena et al., 2020). These microbes only survive on a small variety of substrates and are extremely sensitive to oxygen. They need a lower redox potential and a higher pH. In batch reactors, digestion stops when biogas production stops after roughly 40 days, whereas methanogenesis normally takes 5–16 days.

The most common biological mechanism for producing hydrogen is dark fermentation (DF), which includes the first three phases of AD (hydrolysis, acidogenesis, and acetogenesis). Biomasses and waste biomass are used as substrates by facultative or obligate anaerobes in this process. Hydrogen, carbon dioxide, and VFAs such propionic, butyric, and acetic acids are the byproducts of DF. Since 4 mol of hydrogen may be produced from 1 mol of glucose, the acetate route should be optimized to produce the maximum hydrogen yield (Sahota et al., 2024). Since DF doesn't require a sterile environment and its operations are straightforward and low energy consumption, it is a cost-effective method when compared to other photosynthetic technologies. Because of its versatility, DF can treat a variety of waste biomasses; however, its technical readiness level (TRL) is only 5. The process's bottlenecks should be examined in order to optimize it, as this suggests that it is not yet prepared for industrial-scale implementation to replace conventional fossil fuel-based systems (Karka et al., 2021).

Managing VFAs is a major difficulty in dark fermentation. Once acetogens and acidogens produce a consistent amount of VFAs, their buildup lowers pH, which stops microorganisms from growing. Although VFAs are released into the medium to keep cells stable, they can accumulate and cause cell wall lysis by entering the cytoplasm of biocatalysts. Due to pollution concerns, the liquid residue of DF should not be released into the environment because of its high acid concentration. Although this VFA-rich substrate is appropriate for an AD phase, methanogenesis cannot begin when their concentration exceeds 8000 mg/L (Morison, 2021).

To take advantage of the waste biomass, a two-stage setup that uses DF to make hydrogen and then AD of solid residue containing VFAs to produce CH4 should be taken into consideration. Two-stage anaerobic digestion (TSDA) is the term for this procedure. As illustrated in Fig. 5.1, TSAD distinguishes between acidogenesis-acetogenesis and methanogenesis because they need distinct ideal operating parameters. Utilizing two reactors maximizes gas output and makes process parameter

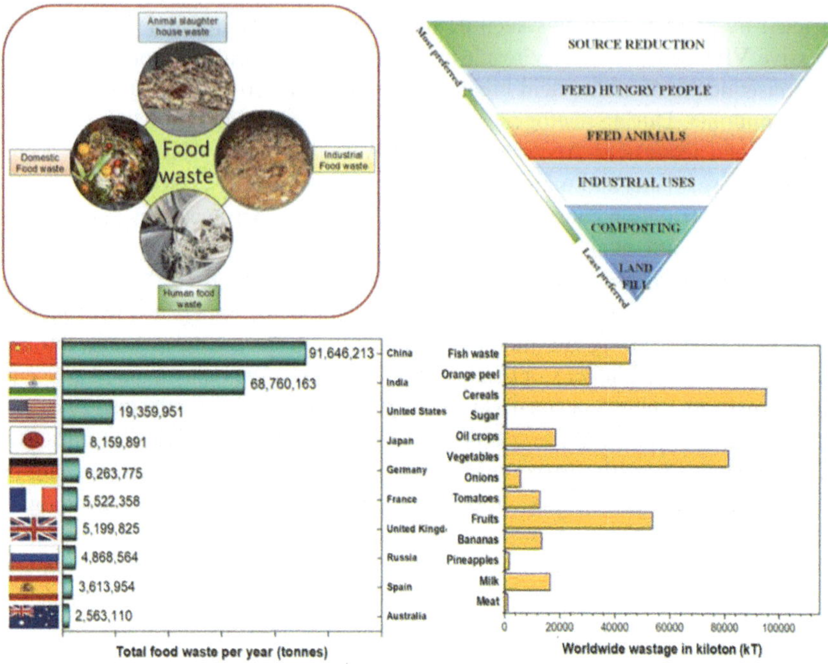

Fig. 5.1 Food waste generation in various countries (source files UNEP food waste index report 2021) (Lahiri et al., 2023)

management easier. By guaranteeing the utilization of carbon-based biomass, this upgrade generates two energy carriers (hydrogen and CH4). Taking into account the energy recovery from methane, AD uses glucose to achieve an efficiency of 84%, while TSAD uses glucose to achieve an efficiency of 86.3–89%, depending on the glucose degradation pathway. Compared to AD, TSAD exhibits greater energy recovery (Srisowmeya et al., 2020).

5.3.6 Photofermentation

A biological process known as photofermentation produces carbon dioxide, hydrogen, and cell energy (ATP) when organic matter ferments in the presence of photosynthetic non-sulfur bacteria (Tiang et al., 2020). The photosynthetic non-sulfur bacteria may finish converting the substrate into hydrogen energy and other products in anaerobic conditions, such as those seen in sunshine or artificial light. Because of its controlled uniformity, intensity, and low heat output, artificial light is the preferable choice even though sunlight is freely available but greatly influenced by outside environmental conditions (such as clouds and warmth). Some bacteria, including *Rhodovulum sulfidophilum, Rhodobacter capsulatus, Rhodospirillum rubrum,* and *Rhodopseudomonas palustris*, are employed in photofermentation (Gupta et al., 2024). Photofermentation produced 870.26 mL hydrogen/g of total volatile fatty acids (TVFA), while dark fermentation produced 226.24 mL hydrogen/g of TS, according to a study comparing the two processing methods (Policastro et al., 2024). The kind of substrate, bacterial strain, temperature, pH, and light intensity all have an impact on how well photo fermentative hydrogen generation works.

5.3.7 Hydrothermal Carbonization (HTC)

By combining high temperatures (170–250 °C) and autogenous pressure (2–10 MPa) in a brief amount of time (5–240 min) and without oxygen, hydrothermal carbonization (HTC), a wet thermochemical conversion process, can turn biomass with high moisture content, like FW, into valuable products (Le et al., 2022). The natural moisture in the feedstock is used as a solvent in the HTC process. In addition to a liquid fraction (oils after extraction) and gaseous products, this process primarily produces hydrochar, a carbon-rich product that is comparable to the biochar produced in slow pyrolysis and can be used as fuel (Yan et al., 2024). HTC has a competitive edge since it uses less energy than other thermal processes because of its low processing temperature and lack of biomass-drying pretreatment stages. When compared to torrefaction treatment, HTC processing used half as much energy to treat moist agroindustrial waste. Furthermore, the processing conditions aid in the inactivation of organic pollutants, the removal of pathogens, and a significant decrease in offensive odors. In order to create hydrochar, the HTC process goes through a number of concurrent chemical reactions, such as hydrolysis, dehydration, decarboxylation, condensation, polymerization, and aromatization (Sharma et al., 2020a, 2020b).

HTC also has the potential to be used as fertilizer because its solid phase contains calcium, magnesium, phosphorus, and nitrogen. According to estimates, nutrients incorporated into hydrochar and the liquid phases of landfilled waste might replace 0.96% of nitrogen and 2.30% of phosphorus-based fertilizers in the United States (Le et al., 2022). HTC uses water as a reaction medium, which facilitates the chemical processes and replaces dangerous solvents. Because recovered process water contains high organic acids, recirculating it is a desirable approach that may improve carbon recovery yields and increase the energy density of hydrochar. The HTC of FW has the potential to produce hydrochar that retains about 70% of the fatty acids, which can then be extracted for industrial purposes using ethanol, with an overall recovery rate of 49% (Hoekman et al., 2018). High-quality hydrochar with an energy content of 23 MJ/kg was produced by restaurant FW treated with the HTC process at 200 °C for 6 h [216]. Another study found that hydrochar produced from FW had a higher carbon content and heating value compared to raw biomass (Yan et al., 2024).

5.4 Biofuels from Food Waste

Automobiles can use liquid or gaseous biofuels made from biomass. Energy security and the current reliance on expensive and finite fossil fuels, as well as the production of energy from biofuels, are essential. Power, heat, and cold can all be produced using biofuels. However, sustainable biomass sources need to be developed in order to enhance biofuel uses both economically and environmentally. Because they can be stored, pumped, and handled using current machinery for mineral oil-based fuels, biodiesel and bioethanol are the most widely used biofuels in transportation and are possible substitutes for diesel (Niculescu et al., 2019). Based on where they came from, biofuels are separated into four generations, each of which includes a unique type of biofuel. Edible crops are the source of first-generation biofuels, which puts strain on food supplies and drives up costs. In order to overcome these limitations, second-generation biofuels use inedible food; however, because lignocellulose degrades slowly and poorly, larger areas are needed (Pramanik et al., 2023). In addition to requiring less space and not competing with food, algae are used to create third-generation biofuels. More biomass sources and a new generation of biofuels that are on par with or even better than fossil fuels are required due to the increasing demand for renewable energy (Igwebuike et al., 2024). A sustainable biofuel is offered by the fourth generation of biofuels, which are made from modified algae biomass (Cavelius et al., 2023).

Sustainable biofuel production and an uninterrupted, abundant supply of waste biomass are crucial. In terms of reducing greenhouse gas emissions, second- and third-generation biofuels outperform first-generation biofuels; nevertheless, this should be changed for the fourth generation (Singh et al., 2024). However, less-than-third-generation biofuels have negative effects on the environment and the economy, making them incompatible with fossil fuels. First-generation fuels are affordable, but they are only accessible in regions where production requires a lot of land. A significant amount of soluble and insoluble sugars, including glucose, fructose, and sucrose, are found in fruit waste and are often utilized to make bioethanol (Favaretto et al., 2023). Nevertheless, D-limonene, a strong microbial inhibitor, was present in a lot of fruit wastes, especially orange peels. D-limonene can therefore be removed by pretreatment using physical or thermochemical techniques (milling, steam explosion, and adsorption) as well as by digestion in conjunction with other food waste (Satira, 2022).

To extract bioethanol, food waste is fermented or hydrolyzed enzymatically using a variety of microorganisms. *Trichoderma harzianum*, which generates hydrolytic and cellulolytic enzymes, was to hydrolyze pineapple fruit peel. An increase in enzyme access to pineapple peelings during hydrolysis was demonstrated by the 571.4 g/L carbohydrate content that was obtained after 30 min of sonication (Casabar et al., 2020). Other yeast strains, such as *Hanseniaspora guilliermondii* and *Candida krusei*, *Metchnikowia cibodasensis*, and *Metchnikowia* sp., were employed in addition to the common baker yeast Saccharomyces cerevisiae to ferment food waste and extract ethanol (Dhar et al., 2023). One of the many economic and environmental benefits of burning ethanol is that it produces no net carbon dioxide emissions. Following are biofuels derived from food waste:

5.4.1 Biodiesel

Food waste is transformed into biodiesel through the transesterification of microbial oil. Algae, tallows, and edible and inedible oils are used to make biodiesel (Sharma et al., 2020a, 2020b). Jatropha waste, rapeseed and fish oil waste, tallow from animal fat, and leftover poultry fat are a few of the feedstocks used in the production of biodiesel (Malik et al., 2024). Food waste can be transesterified directly or indirectly to make biodiesel or fatty acid methyl ester. Biodiesel is produced by transesterification of these oils. Barik et al. (2018) produced biodiesel from kitchen garbage. Methanol was utilized as a solvent and the food waste was dried in order to extract 37% of the lipids. More transesterification and extraction were required for the production of biodiesel from mixed food waste, and the results were found to be rather low (Almutairi et al., 2021).

5.4.2 Biohydrogen

Using water photolysis by algae, AD of food containing carbs, organic waste photodecomposition, microbial fermentation, and two-stage fermentation, biohydrogen is a renewable fuel made from fruit waste (Kumari et al., 2024). Additionally, foods high in carbohydrates can produce 20 times as much hydrogen as foods high in fat (Jain et al., 2024). Food waste has a high C:N ratio (9:21), significant moisture content (72–85.2%), and degradable organic components, making it a suitable source for hydrogen generation (Latif, 2021).

Biohydrogen generation is influenced by the type of inoculant, vessel design (e.g., constantly mixed immobilized sludge reactors), and the origin, type, and pH of a substrate. The cultivation of acetogenic bacteria is frequently used for hydrogen production. Food waste is favorable for hydrogen synthesis due to its high carbon content and the presence of native microbial populations. The dark fermentation method is one of them that significantly favorably controls the synthesis of biohydrogen. In fact, the best anaerobes for consuming food waste and producing hydrogen throughout the fermentation process are *Clostridium* and *Enterobacter* (Canto-Robertos et al., 2023). By utilizing all of the metabolites produced during the first stage of fermentation, the combination of photo and anaerobic fermentation for hydrogen production boosts productivity.

5.4.3 Bioethanol

Bioethanol production from food waste involves the fermentation of carbohydrates found in discarded food items, such as fruits, vegetables, and grains. The process begins with the collection and preprocessing of food waste, which is then converted into sugars through hydrolysis (Haleem et al., 2023). These sugars are fermented by yeast or bacteria to produce ethanol, a type of alcohol that can be used as a renewable fuel. The process not only reduces food waste but also contributes to lowering greenhouse gas emissions by providing an alternative to fossil fuels. Bioethanol from food waste offers a sustainable solution to both energy production and waste management. Additionally, this method helps reduce the environmental impact of landfills, as it prevents food waste from decomposing and releasing methane, a potent greenhouse gas (Kiehbadroudinezhad et al., 2024).

5.4.4 Biochar

Through pyrolysis (slow or fast) at 300–800 °C in anoxic conditions, biochar—a carbonaceous, black, porous, and solid organic material is created from organic wastes such as manures and agricultural residues (Bolan et al., 2022). The biochar's energy potential ranged from 15 to 30 kJ/g dry weight, and its carbon content was 45% approximately. Maize (Dietrich et al., 2020), brewery grain (Maqhuzu et al., 2021), distillery residual crumbs, and peanut hulls (Guida & Hannioui, 2016) were among the food wastes that were examined. Because of its high capacity to hold water, biochar is used as a soil enhancer to boost the soil's nutritional content for more effective plant growth (Hegab, 2024).

By decreasing the leaching process in soil, biochar increases phosphorus availability and helps retain nitrogen by reducing gaseous loss and leaching. Under low oxygen circumstances, recycled anaerobically digested food waste as feedstock for the synthesis of biochar at pyrolysis temperatures of 400 and 900 °C for two hours. The maximum amount of biochar was recorded at 400 °C (70.4%), while at 900 °C, the amount of biochar and volatile matter decreased from 40.6% to 4.5% (Maisonnier, 2019). Biochar with a carbon content of 66% and 71%, respectively, using a combination of food waste and rice trash (Jeong et al., 2016).

The overall effectiveness of the process seems to be significantly hampered by the moisture content of both kinds of food waste. Energy recovery through waste incineration is therefore not a practical choice. In one study, food waste was dried by direct burning using thermal losses brought on by the moisture valorization of the trash (Li et al., 2022). Consequently, food waste energy recovery can assist in lowering costs related to trash disposal or conventional fuel usage. To lower the total cost of biochar synthesis, food wastes should be pretreated to remove moisture in order to facilitate gasification and pyrolysis.

5.5 Challenges in Biofuel Production from Food Waste

Without a question, the growing population, together with the corresponding rise in expenses and environmental consequences, is making it extremely difficult for many municipal councils worldwide to manage food waste on a daily basis. The scientific community is well aware of this issue, yet most governments and policymakers are still lagging behind in addressing the difficulties that have been highlighted. The majority of food waste management techniques currently in use rely on uncontrolled combustion and landfills, two traditional and straightforward approaches that have been used for hundreds of years.

Generally speaking, the initial water content of the biomass is one of the key characteristics that determines the conversion process. For instance, biomass with a high moisture content (> 50%) is probably better suited for wet processing methods like hydrothermal conversions and biological processes (such fermentation and anaerobic digestion). However, food waste is a better candidate for thermochemical conversion if it is available dry or can be reduced to low moisture content using economical and technically feasible ways (Kassim et al., 2022). Since the feed needs to be dried first, high moisture biomass would raise the energy needed for the majority of traditional thermal conversion techniques. Wet biomass also presents challenges for material handling and a higher danger of rapid biological breakdown. When the biomass contains a significant amount of organic matter, salt, and oil, biological degradation may become more difficult (Wang et al., 2024). Additionally, there are other issues related to food waste, like the foul odor, annoyance, and several environmental risks brought on by leachate formation and biogas discharge. Methane, the primary component of biogas, has a 20–25 times greater impact on global warming than carbon dioxide (Cho et al., 2024).

Nevertheless, food waste can be a very alluring feedstock for bioenergy if technically and financially viable solutions to these issues are found. An overview of the main social and technological issues, along with suggested fixes, can be seen below.

5.5.1 *Techno-economic Challenges*

- With the exception of a few European nations, the United States, and Canada, there is a dearth of accurate data and statistics regarding the generation of food waste in the region. Such quantitative information is necessary to determine the best ways to reduce losses and to take appropriate corrective action.
- Due to regional differences in food preparation techniques and diet, there are significant differences in the physical and chemical features of food waste.
- In order to prevent biological deterioration, food waste typically has a high moisture content (up to 90% in certain situations), necessitating particular handling and storage procedures.

- Technical and financial analyses of large-scale food waste drying methods (such as solar, open-air, or effluent gas) are scarce.
- To compare the technological and financial elements of each food waste management technique, more research is needed. This is necessary to support decision-making and evaluate the possibilities for scale-up.
- The appropriateness of current technologies, such as facilities for material handling, storage, and heat conversion, is not well understood. This is especially difficult when processing food waste on a big scale because of its complexity and rapid biological deterioration.

5.5.2 Social Challenges

- More work is needed at the community and industrial levels to encourage food waste segregation since, despite global advancements in waste segregation, food waste is still seen as part of the general municipal solid waste.
- Strict guidelines for disposing of food waste must be established by city councils and legislators, who must also distribute educational resources and programs on waste reduction techniques for homes and businesses.

5.6 Conclusion

Food waste (FW) depletes resources and increases carbon footprints, which have a substantial negative influence on society and the environment. Making the switch to circular biorefineries with FW and organic waste valorization will encourage sustainable practices and environmental care in the production of bioenergy. Opportunities for biofuels are presented by the increasing amount of FW, however there are still difficulties because of the variety and type of FW. A potential future solution, however, would be the use of integrated biorefineries to use FW for optimal energy and product recovery by combining biological, chemical, thermal, or electro-chemical processes. While two-stage AD/DF-AD methods show promise for producing biofuels such as biogas, bio-hydrogen, and biochar as well as value-added products, single-stage AD of FW is extensively researched due to its ease of use and economic feasibility. Efficient and economical production techniques are necessary for the incorporation of FW into the bioeconomy. While government laws and actions are essential for a circular economy and sustainable development, effective FW valorisation pathways can control and reduce carbon footprints. In the long run, a cleaner, more sustainable future requires scientific, social, and political breakthroughs in order to implement an integrated bioeconomy for FW management. Due to digesting conditions and the limited commercialization potential of certain technologies, commercial digestate utilization is difficult.

Aside from other inherent environmental benefits, the circular bioeconomy helps us by preserving resources for longer periods of time and pursuing the "waste-to-wealth" idea, which has resulted in the development of new technologies, job types, and livelihoods. In the end, it aims to achieve zero waste and a significant decrease in greenhouse gas emissions. Therefore, the circular bioeconomy's closed-loop system can be employed as a tool to facilitate FW reduction and address related remediation in the future. Furthermore, multi-dimensional examination of carbon footprints, environmental loads, energy sustainability via life cycle assessment studies, and economic rewards is required to design workable biomass valorisation methods that integrate hydrothermal and biochemical approaches.

Acknowledgements Authors are thankful to the Punjab University for providing necessary support.

References

Abubakar, A. M., Silas, K., & Aji, M. M. (2022). An elaborate breakdown of the essentials of biogas production. *Journal of Engineering Research and Sciences, 1*(4), 93–118.

Agarwal, M., Agarwal, S., Ahmad, S., Singh, R., & Jayahari, K. M. (2021). Food loss and waste in india: the knowns and the unknowns. *World Resources Institute, 10.*

Ahmad, A., Rambabu, K., Hasan, S. W., Show, P. L., & Banat, F. (2024). Biohydrogen production through dark fermentation: Recent trends and advances in transition to a circular bioeconomy. *International Journal of Hydrogen Energy, 52*, 335–357.

Almutairi, A. W., Al-Hasawi, Z. M., & Abomohra, A. E. F. (2021). Valorization of lipidic food waste for enhanced biodiesel recovery through two-step conversion: A novel microalgae-integrated approach. *Bioresource Technology, 342*, 125966.

Astuti, R. D. P., & Rauf, A. U. (2024). Outdoor air pollution due to transportation, landfill, and incinerator. In *Health and environmental effects of ambient air pollution* (pp. 257–302). Academic Press.

Aworanti, O. A., Ajani, A. O., Agbede, O. O., Agarry, S. E., Ogunkunle, O., Laseinde, O. T., et al. (2023). Enhancing and upgrading biogas and biomethane production in anaerobic digestion: A comprehensive review. *Frontiers in Energy Research, 11*, 1170133.

Barik, S., Paul, K. K., & Priyadarshi, D. (2018). Utilization of kitchen food waste for biodiesel production. In *IOP conference series: earth and environmental science* (Vol. 167, No. 1, p. 012036). IOP Publishing.

Bhatia, L., Jha, H., Sarkar, T., & Sarangi, P. K. (2023). Food waste utilization for reducing carbon footprints towards sustainable and cleaner environment: A review. *International Journal of Environmental Research and Public Health, 20*(3), 2318.

Bolan, N., Hoang, S. A., Beiyuan, J., Gupta, S., Hou, D., Karakoti, A., et al. (2022). Multifunctional applications of biochar beyond carbon storage. *International Materials Reviews, 67*(2), 150–200.

Burezq, H. A., & Davidson, M. K. (2023). Biochar from date palm (*Phoenix dactylifera* L.) residues—A critical review. *Arabian Journal of Geosciences, 16*(2), 101.

Canto-Robertos, M., Quintal-Franco, C., Ponce-Caballero, C., Vega-De Lille, M., & Moreno-Andrade, I. (2023). Inhibition of hydrogen production by endogenous microorganisms from food waste. *Brazilian Journal of Chemical Engineering, 40*(1), 137–150.

Casabar, J. T., Ramaraj, R., Tipnee, S., & Unpaprom, Y. (2020). Enhancement of hydrolysis with *Trichoderma harzianum* for bioethanol production of sonicated pineapple fruit peel. *Fuel, 279*, 118437.

Cavelius, P., Engelhart-Straub, S., Mehlmer, N., Lercher, J., Awad, D., & Brück, T. (2023). The potential of biofuels from first to fourth generation. *PLoS Biology, 21*(3), e3002063.

Chatterjee, B., & Mazumder, D. (2019). Role of stage-separation in the ubiquitous development of anaerobic digestion of organic fraction of municipal solid waste: A critical review. *Renewable and Sustainable Energy Reviews, 104*, 439–469.

Chisoro, P., Jaja, I. F., & Assan, N. (2023). Incorporation of local novel feed resources in livestock feed for sustainable food security and circular economy in Africa. *Frontiers in Sustainability, 4*, 1251179.

Cho, H. H., Strezov, V., & Evans, T. J. (2024). Life cycle assessment of power-to-methane and renewable methane production technologies. *Renewable and Sustainable Energy Reviews, 206*, 114856.

Dhar, P., Nickhil, C., Pandiselvam, R., & Deka, S. C. (2023). Pineapple waste-based-biorefinery for sustainable generation of value-added products. *Biomass Conversion and Biorefinery*, 1–22.

Dietrich, C. C., Rahaman, M. A., Robles-Aguilar, A. A., Latif, S., Intani, K., Müller, J., & Jablonowski, N. D. (2020). Nutrient loaded biochar doubled biomass production in juvenile maize plants (*Zea mays* L.). *Agronomy, 10*(4), 567.

Favaretto, D. P. C., Rempel, A., Lanzini, J. R., Silva, A. C. M., Lazzari, T., Barbizan, L. D., et al. (2023). Fruit residues as biomass for bioethanol production using enzymatic hydrolysis as pretreatment. *World Journal of Microbiology and Biotechnology, 39*(6), 144.

Food and Agriculture Organization of the United Nations. (2019). The state of food and agriculture 2019: Moving forward on food loss and waste reduction. FAO.

Gbiete, D., Narra, S., Mani Kongnine, D., Narra, M. M., & Nelles, M. (2024). Insights into biohydrogen production through dark fermentation of food waste: Substrate properties, inocula, and pretreatment strategies. *Energies, 17*(24), 6350.

Georganas, A., Giamouri, E., Pappas, A. C., Papadomichelakis, G., Galliou, F., Manios, T., et al. (2020). Bioactive compounds in food waste: A review on the transformation of food waste to animal feed. *Foods, 9*(3), 291.

Guida, M. Y., & Hannioui, A. (2016). A review on thermochemical treatment of biomass: Pyrolysis of olive mill wastes in comparison with other types of biomass. *Progress in Agricultural Engineering Sciences, 12*(1), 1–23.

Gupta, S., Fernandes, A., Lopes, A., Grasa, L., & Salafranca, J. (2024). Photo-fermentative bacteria used for hydrogen production. *Applied Sciences, 14*(3), 1191.

Haleem M. I. A, Ganesh, M. S., Lekha, S. S., Khan, M. S., & Mohamed M. S. N. (2023). Valorization of food waste for optimized production of bio-fuel. *Plant Archives, 23*(2) (09725210).

Hassan, Q., Hafedh, S. A., Mohammed, H. B., Abdulrahman, I. S., Salman, H. M., & Jaszczur, M. (2024). A review of hydrogen production from bio-energy, technologies and assessments. *Energy Harvesting and Systems, 11*(1), 20220117.

Hegab, R. (2024). Impact of Integrating between mineral nitrogen, vermicompost and biochar on nitrogen-use efficiency and productivity of panicum plants grown on sandy soil. *Egyptian Journal of Soil Science, 64*(3), 1273–1284.

Hoekman, S. K., Leland, A., & Felix, L. (2018). Hydrothermal carbonization (HTC) of biomass for energy applications. In *Biomass preprocessing and pretreatments for production of biofuels* (pp. 196–254). CRC Press.

Igwebuike, C. M., Awad, S., & Andrès, Y. (2024). Renewable energy potential: second-generation biomass as feedstock for bioethanol production. *Molecules, 29*(7), 1619.

Jain, R., Panwar, N. L., Jain, S. K., Gupta, T., Agarwal, C., & Meena, S. S. (2024). Bio-hydrogen production through dark fermentation: An overview. *Biomass Conversion and Biorefinery, 14*(12), 12699–12724.

Jeong, C. Y., Dodla, S. K., & Wang, J. J. (2016). Fundamental and molecular composition characteristics of biochars produced from sugarcane and rice crop residues and by-products. *Chemosphere, 142*, 4–13.

Julkipli, J., Babel, S., Bilyaminu, A. M., & Rene, E. R. (2024). Hydrogen and biodiesel production from food waste: A review. *Environmental Chemistry Letters, 22*(2), 585–607.

Kannah, R. Y., Merrylin, J., Devi, T. P., Kavitha, S., Sivashanmugam, P., Kumar, G., & Banu, J. R. (2020). Food waste valorization: Biofuels and value added product recovery. *Bioresource Technology Reports, 11*, 100524.

Karka, P., Johnsson, F., & Papadokonstantakis, S. (2021). Perspectives for greening European fossil-fuel infrastructures through use of biomass: The case of liquid biofuels based on lignocellulosic resources. *Frontiers in Energy Research, 9*, 636782.

Kassim, F. O., Thomas, C. P., & Afolabi, O. O. (2022). Integrated conversion technologies for sustainable agri-food waste valorization: A critical review. *Biomass and Bioenergy, 156*, 106314.

Kaza, S., Yao, L., Bhada-Tata, P., & Van Woerden, F. (2018). *What a waste 2.0: A global snapshot of solid waste management to 2050*. World Bank Publications.

Khan, M. N., Sial, T. A., Ali, A., & Wahid, F. (2024). Impact of agricultural wastes on environment and possible management strategies. *Frontier studies in soil science* (pp. 79–108). Springer International Publishing.

Kiehbadroudinezhad, M., Merabet, A., & Hosseinzadeh-Bandbafha, H. (2024). Landfill source of greenhouse gas emission. In *Advances and technology development in greenhouse gases: Emission, capture and conversion* (pp. 123–145). Elsevier.

Kumari, R., Singh, A., Sharma, R., & Malaviya, P. (2024). Conversion of food waste into energy and value-added products: A review. *Environmental Chemistry Letters*, 1–32.

Lahiri, A., Daniel, S., Kanthapazham, R., Vanaraj, R., Thambidurai, A., & Peter, L. S. (2023). A critical review on food waste management for the production of materials and biofuel. *Journal of Hazardous Materials Advances, 10*, 100266.

Latif, M. B. (2021). *Effect of sludge content on different types of food waste degradation in anaerobic digester*. The University of Texas at Arlington.

Le, H. S., Chen, W. H., Ahmed, S. F., Said, Z., Rafa, N., Le, A. T., et al. (2022). Hydrothermal carbonization of food waste as sustainable energy conversion path. *Bioresource Technology, 363*, 127958.

Li, R., Gong, M., Biney, B. W., Chen, K., Xia, W., Liu, H., & Guo, A. (2022). Three-stage pretreatment of food waste to improve fuel characteristics and incineration performance with recovery of process by-products. *Fuel, 330*, 125655.

Luo, K., Pang, Y., Yang, Q., Wang, D., Li, X., Lei, M., & Huang, Q. (2019). A critical review of volatile fatty acids produced from waste activated sludge: Enhanced strategies and its applications. *Environmental Science and Pollution Research, 26*, 13984–13998.

Magdalena, J. A., Ballesteros, M., & González-Fernández, C. (2020). Acidogenesis and chain elongation for bioproduct development. In *Wastewater treatment residues as resources for biorefinery products and biofuels* (pp. 391–414). Elsevier.

Maisonnier, A. (2019). Comparative study of biochar amendment effects on nutrient cycling in contrasting pedological context of Koumbia, Burkina Faso.

Malik, M. A. I., Zeeshan, S., Khubaib, M., Ikram, A., Hussain, F., Yassin, H., & Qazi, A. (2024). A review of major trends, opportunities, and technical challenges in biodiesel production from waste sources. *Energy Conversion and Management: X*, 100675.

Maqhuzu, A. B., Yoshikawa, K., & Takahashi, F. (2021). Prospective utilization of brewers' spent grains (BSG) for energy and food in Africa and its global warming potential. *Sustainable Production and Consumption, 26*, 146–159.

Mattson, K. R., Pettersen, J. B., & Brattebø, H. (2024). Incineration economy: Waste management policy failing the circular economy transition in Norway. *Resources, Conservation and Recycling, 210*, 107838.

Morison, S. D. (2021). *In situ extraction and recovery of volatile fatty acids from biogas-producing anaerobic digestion* (Doctoral dissertation, Stellenbosch University).

Moya, R., Tenorio, C., Quesada-Kimzey, J., & Másis-Meléndez, F. (2024). Pyrogenic carbonaceous materials production of four tropical wood produced by slow pyrolysis at different temperatures: charcoal and biochar properties. *Energies, 17*(8), 1953.

Munesue, Y., Masui, T., & Fushima, T. (2015). The effects of reducing food losses and food waste on global food insecurity, natural resources, and greenhouse gas emissions. *Environmental Economics and Policy Studies, 17*, 43–77.

Niculescu, R., Clenci, A., & Iorga-Siman, V. (2019). Review on the use of diesel–biodiesel–alcohol blends in compression ignition engines. *Energies, 12*(7), 1194.

Policastro, G., Cesaro, A., & Fabbricino, M. (2024). Valorization of purple phototrophic bacteria biomass resulting from photo fermentation aimed at biohydrogen production. *Molecules, 29*(7), 1679.

Pour, F. H., & Makkawi, Y. T. (2021). A review of post-consumption food waste management and its potentials for biofuel production. *Energy Reports, 7*, 7759–7784.

Pramanik, A., Sinha, A., Chaubey, K. K., Hariharan, S., Dayal, D., Bachheti, R. K., et al. (2023). Second-generation bio-fuels: Strategies for employing degraded land for climate change mitigation meeting United Nation-sustainable development goals. *Sustainability, 15*(9), 7578.

Sahota, S., Kumar, S., & Lombardi, L. (2024). Biohythane, biogas, and biohydrogen production from food waste: Recent advancements, technical bottlenecks, and prospects. *Energies, 17*(3), 666.

Sarker, A., Ahmmed, R., Ahsan, S. M., Rana, J., Ghosh, M. K., & Nandi, R. (2024). A comprehensive review of food waste valorization for the sustainable management of global food waste. *Sustainable Food Technology.*

Satira, A. (2022). Sustainable process for the complete upgrading of orange peel waste into value-added chemicals and bio-carbon materials.

Saxena, A., Parveen, F., Hussain, A., Khubaib, M., & Ashfaque, M. (2024). Second-generation biorefineries: single platform for the conversion of lignocellulosic wastes to environmentally important biofuels. *Environmental Science and Pollution Research*, 1–32.

Scarlat, N., Dallemand, J. F., Monforti-Ferrario, F., & Nita, V. (2015). The role of biomass and bioenergy in a future bioeconomy: Policies and facts. *Environmental Development, 15*, 3–34.

Selvarajoo, A., Wong, Y. L., Khoo, K. S., Chen, W. H., & Show, P. L. (2022). Biochar production via pyrolysis of citrus peel fruit waste as a potential usage as solid biofuel. *Chemosphere, 294*, 133671.

Sharma, H. B., Sarmah, A. K., & Dubey, B. (2020a). Hydrothermal carbonization of renewable waste biomass for solid biofuel production: A discussion on process mechanism, the influence of process parameters, environmental performance and fuel properties of hydrochar. *Renewable and Sustainable Energy Reviews, 123*, 109761.

Sharma, K. D., & Jain, S. (2020). Municipal solid waste generation, composition, and management: The global scenario. *Social Responsibility Journal, 16*(6), 917–948.

Sharma, P., Gaur, V. K., Kim, S. H., & Pandey, A. (2020b). Microbial strategies for bio-transforming food waste into resources. *Bioresource Technology, 299*, 122580.

Shukla, K. A., Sofian, A. D. A. B. A., Singh, A., Chen, W. H., Show, P. L., & Chan, Y. J. (2024). Food waste management and sustainable waste to energy: Current efforts, anaerobic digestion, incinerator and hydrothermal carbonization with a focus in Malaysia. *Journal of Cleaner Production, 448*, 141457.

Singh, P., & Kalamdhad, A. S. (2021). A comprehensive assessment of state-wise biogas potential and its utilization in India. *Biomass Conversion and Biorefinery*, 1–23.

Singh, N., Saluja, R. K., Rao, H. J., Kaushal, R., Gahlot, N. K., Suyambulingam, I., et al. (2024). Progress and facts on biodiesel generations, production methods, influencing factors, and reactors: A comprehensive review from 2000 to 2023. *Energy Conversion and Management, 302*, 118157.

Srisowmeya, G., Chakravarthy, M., & Devi, G. N. (2020). Critical considerations in two-stage anaerobic digestion of food waste—A review. *Renewable and Sustainable Energy Reviews, 119*, 109587.

Subramanian, K., & Suresh, K. (2024). Metabolic and biochemical pathways for anaerobic digestion. In *Anaerobes and waste conversion technologies* (pp. 3–25). Springer Nature Singapore.

Tiang, M. F., Hanipa, M. A. F., Abdul, P. M., Jahim, J. M., Mahmod, S. S., Takriff, M. S., et al. (2020). Recent advanced biotechnological strategies to enhance photo-fermentative biohydrogen production by purple non-sulphur bacteria: An overview. *International Journal of Hydrogen Energy, 45*(24), 13211–13230.

Truong, L., Morash, D., Liu, Y., & King, A. (2019). Food waste in animal feed with a focus on use for broilers. *International Journal of Recycling of Organic Waste in Agriculture, 8*, 417–429.

Varjani, S., Pandey, A., Bhaskar, T., Mohan, S. V., & Tsang, D. C. (Eds.). (2021). *Biomass, biofuels, biochemicals: circular bioeconomy: technologies for biofuels and biochemicals*. Elsevier.

Villanueva-Galindo, E., Vital-Jácome, M., & Moreno-Andrade, I. (2023). Dark fermentation for H_2 production from food waste and novel strategies for its enhancement. *International Journal of Hydrogen Energy, 48*(27), 9957–9970.

Wang, Y., Zhang, Z., Wang, X., Guo, H., Zhu, T., Zhao, Y., et al. (2023). Medium-chain fatty acids production from sewage sludge through anaerobic fermentation: A critical review. *Chemical Engineering Journal*, 147138.

Wang, C., Zhang, W., Qiu, X., & Xu, C. (2024). Hydrothermal treatment of lignocellulosic biomass towards low-carbon development: Production of high-value-added bioproducts. *EnergyChem*, 100133.

Warming, G. (2023). 11 Global environmental issues global warming, biodiversity, resource scarcities Kamal Kumar. *Understanding Globalisation: Challenges and Prospects.*

Wu, J., Zhang, H., Zhao, Y., Yuan, X., & Cui, Z. (2024). Effect of temperature on the inocula preservation, mesophilic anaerobic digestion start-up, and microbial community dynamics. *Agronomy, 14*(12), 2991.

Xue, S., Chen, H., Wang, F., Lv, G., Tan, L., & Liu, G. (2024). The effect of substrate acidification on the biohydrogen production by dark fermentation. *International Journal of Hydrogen Energy, 49*, 177–188.

Yan, M., Wang, H., Liu, Y., Luo, C., Yang, Y., Grisdanurak, N., et al. (2024). Innovative carbonization techniques for food waste: A comparative study of hydrothermal carbonization (HTC) and vapor-thermal carbonization (VTC). *Journal of Analytical and Applied Pyrolysis, 182*, 106685.

Chapter 6
Biofuel Production from Organic Waste Materials: A Sustainable Waste Management Approach

Manish Pant, Nisha Singh, Divya Bisen, and Jabrinder Singh

Abstract Several technologies for converting organic waste to energy have successfully addressed global difficulties including dependence on fossil fuels, production cost optimization, waste management, and emission control. The production of carbon dioxide from fossil fuels has been shown to be one of the primary drivers of global warming, which has many negative consequences. Industrialized and developing countries have increased the proportion of renewable energy in the total energy picture in recent years. Several technologies for converting organic waste to energy have successfully addressed global difficulties including dependence on fossil fuels, production cost optimization, waste management, and emission control. The production of carbon dioxide from fossil fuels has been shown to be one of the primary drivers of global warming, which has many negative consequences. Industrialized and developing countries have increased the proportion of renewable energy in the total energy picture in recent years. Biofuels are seen as a sustainable and clean energy source, unlike fossil fuels. A viable strategy for improving energy security, pollution control, process economics, sustainable production and social advancement is the use of organic waste for the production of biofuels. The most recent developments and perspectives in the field of using organic waste materials for the production of biofuels are presented in this book chapter. Organic waste can support the Sustainable Development Goals (SDGs), including clean energy, sustainable cities and communities, responsible production and consumption, national economic growth and, hence, urban development.

M. Pant (✉)
Centre for Land and Water Resource Management (CLWRM), G.B. Pant National Institute of Himalayan Environment, Kosi-Katarmal, Almora 263643, Uttarakhand, India
e-mail: manishpant66@gmail.com

Department of Chemical Engineering, IISER Bhopal, Bhopal 462066, Madhya Pradesh, India

N. Singh · J. Singh
Department of Chemistry, DIT University, Dehradun 248009, India

D. Bisen
Department of Physical Science and Chemical Engineering, Lovely Professional University, Phagwara, Punjab, India

V. Singh et al. (eds.), *Bioenergy and Value Added Byproducts from Biomass Waste*, Sustainable Environmental Waste Management Strategies,
https://doi.org/10.1007/978-3-032-22849-9_6

Keywords Organic waste · Biofuels · Waste management · Climate action · SDGs

6.1 Introduction

The increasing demand for energy due to uncontrolled population expansion inevitably leads to the overuse of non-renewable energy sources, which in turn promotes the search for alternative sustainable energy sources (Lin et al., 2024). Global environmental issues including air pollution and global warming are caused by the excessive use of fossil fuels (Abdeshahian et al., 2016). As pollution associated with health concerns is actively considered, it is imperative to implement nationwide coordination to reduce climate change. A sustainable clean energy solution is important, and the survival rate of earth's living forms with health risks is constantly monitored (Shariat Panahi et al., 2020). By properly using waste-derived biogas, it is possible to promote the realization of sustainable renewable energy. Biogas production is linked with waste management for sustainability (Pant et al., 2025a). Renewable bioenergy sources can replace non-sustainable energy sources (Rincon et al., 2019).

At the moment, natural gas, coal, and oil are the main sources of fuel, and dependence on these fossil fuels is limited. Alternative energy sources are needed as the growing population uses more fuel, increasing the demand. Finding a suitable, environmentally friendly, renewable energy source is a global problem. According to Sims et al. (2016), bioenergy is produced from wastewater discharged by the food, pulp, and paper sectors. Bioenergy is an important desired goal for an environmentally friendly approach. According to the Energy Information Administration (EIA) of the USA, global energy consumption is expected to increase by 50% to approximately 900 quadrillion BTUs (quad) by 2050. According to the EIA, the growing electricity demand worldwide will be accounted for by developing markets (Angulo-Mosquera et al., 2021). As the world's population continues to grow, it is important to meet their energy needs (Chandrasekar et al., 2020).

According to Durán-Valle and López-Coca (2024) and Sarkar et al. (2024), organic waste undergoes chemical, biological and thermal conversion processes to produce its bioenergy forms. Bioenergy is produced by burning organic waste through a thermal conversion process. Thermal conversion includes both conventional incineration and sophisticated thermal technologies. Different feedstocks generate different amounts of bioenergy. Advanced thermal methods include pyrolysis and gasification (Mekepa et al., 2025). Plasma gasification is a new technology used to promote energy recovery (Nagar and Kaushal, 2024). It has been demonstrated that one of the chemical processes for producing bioenergy is methanation. Biochemical or biological conversion technologies use enzymes to break down the substrate, making them more suitable for highly wet and easily biodegradable biomass than thermochemical processes. Through the use of aerobic and anaerobic bacteria, biochemical processes convert wet biomass waste into biofuels and other products with added value. The two most commonly used biochemical processes to convert such biomass waste into biofuels are anaerobic digestion and fermentation (Awasthi et al., 2023). There are two types of chemical conversion processes: landfill gas recovery and enhanced thermal (Yashika et al., 2022). Landfill gas recovery generates 3.2–106 kW of electricity depending on the technology (Carbon Trust, 2014). Zero-waste discharge and biorefinery ideas aim to effectively create bioenergy while reducing waste and using leftover materials from recycling. To improve the production of various biofuels and charcoal, microalgae are cultivated on various types of organic wastes that are mixed with flue gas, resulting in a sustainable process that produces no waste. Microalgae can be used in biorefineries due to their fast growth rate and biomass output. The biorefinery concept can help overcome tetrameric challenges including sustainability, technical difficulties, environmental problems, and economic constraints. Microalgal feedstock is an effective renewable energy source that can be used to meet the world's energy needs (Peter et al., 2022).

According to Manea et al. (2024), sewage sludge, municipal solid waste (MSW), food waste, kitchen waste, garden waste, agricultural waste and animal waste are all considered solid organic waste. Utilization of organic waste is a practical way to accomplish energy security, environmental control, process economics, sustainable production and development in society. Organic waste conversion scheme is strongly advised because waste management is important and because reducing the problem and hazard of waste is very important. Technological developments in the production of biofuels are important to ensure future energy security and reduce the use of fossil fuels in industrial and infrastructural applications worldwide. To meet the world's energy needs in the future, sustainable biofuel production must advance rapidly. The modern world is constantly searching for innovative solutions to waste management issues, which require the use of additional water and land for waste processing and treatment. By turning waste into energy, these issues can be solved. Bioenergy production from organic waste is associated with reducing the generation and land disposal of bulk waste, such as hazards and negative impacts generated during waste disposal in all industries. Reaction residence time, random feedstock accumulation, high production costs and increased energy consumption are the main obstacles to the commercial use of organic waste. A bioenergy sector that is sustainable and has improved fuel security is also tackling the difficult problem of climate change (Dhanya et al., 2020a, 2020b). Thus, this chapter presents the current trends in processing organic waste for the production of biofuels and bioproducts along with their limitations. A detailed overview of potential organic waste sources and production of bioenergy and other value-added bioproducts through effective conversion processes to enhance circular bioeconomy is given in the present chapter.

6.2 Organic Waste Materials for Biofuel Production

Greenhouse gas emissions and climate change are closely linked. As a result, there is a great deal of attention being paid to reducing greenhouse gas intensity and using organic waste for effective energy recovery. Since the substrate used is more expensive than the cost of production, increasing bioenergy production from sustainable biosources is an issue. Production of bioenergy from organic waste creates secondary waste, which is used as input for subsequent recycling processes (Dhanya et al., 2020a, 2020b; Vieira and Matheus, 2019). Potential sources for biofuel production include organic waste, including sewage sludge, industrial waste, organic municipal solid waste, agricultural residues, and animal/livestock waste (Stephen and Periyasamy, 2018).

6.2.1 *Types of Waste*

The two major categories of waste differ in terms of their content, characteristics, and methods of transformation or disposal. The most prevalent is known as the organic fraction of municipal solid waste (OFMSW), and it originates from food as well as other commercial and household products. The second category is called "green waste", and it originates in both public and privately owned green spaces and vegetated areas (Fig. 6.1). The types of organic waste that fall under this classification are as follows:

Food remains: These include vegetables, fruit, seeds, oils, eggshells and bones. They also include food scraps that have not been sold or eaten (removed from containers), food scraps, spoiled food and waste from food preparation, handling and processing.

Plant remains: Plant residues consist of these (leaves or branches). Issues of collection logistics, treatment, and generation time (frequency and seasonality) require special management of this waste.

Material and nature remains: This includes cotton, cloth and netting made from it, clothing made from animal skins, and the remains of dead animals.

Paper waste: This includes napkins, paper towels, cardboard, and paper. Figure 6.1 illustrates the various types of organic waste.

6.3 Waste to Energy Technologies

The following technologies are used for producing energy from waste.

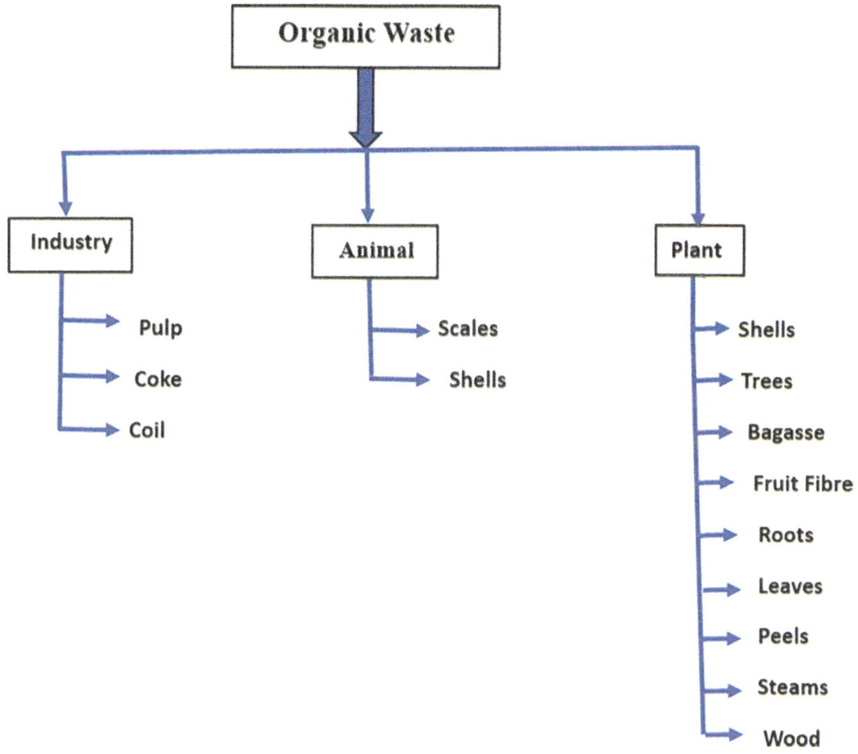

Fig. 6.1 Various types of organic waste

6.3.1 Thermo-chemical Conversion

Conversion by thermochemical technology involves a high temperature chemical reforming process where bond disruption and organic matter solution are required to produce biochar (solid), synthesis gas and highly oxygenated bio-oil (liquid). Gasification, pyrolysis and liquefaction are the three main process options accessible in thermochemical conversion. The detailed thermochemical conversion flowchart is displayed in Fig. 6.2. Carbonaceous substances are converted into synthetic gas when the right catalysts and reagents are present. With the gasification method, biomass is heated to high temperatures (500–1400 °C), atmospheric pressure up to 33 bar, and little or no oxygen content to create a combination of flammable gases. This technique is considered effective in producing H2 gas from organic waste (Ahmed et al., 2016). It has also been found that the use of metal-based catalysts speeds up the reverse process, ultimately leading to the synthesis of more hydrogen and methane.

The two processes of biomass pyrolysis and liquefaction yield products such as bio-oil or bio-crude. The process of thermochemical liquefaction, which can be carried out with or without a catalyst, produces bio-oil at high pressure and low temperature in the presence of hydrogen. Hydrothermal liquefaction (HTL), often

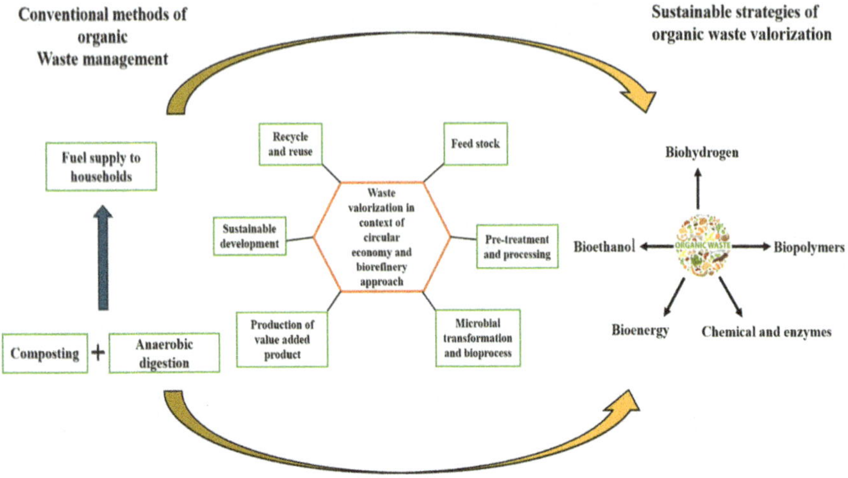

Fig. 6.2 Sustainable strategies of organic waste valorization

referred to as hydrous pyrolysis, is a well-known process for converting biomass to bio-oil. It uses subcritical water (SCW) at moderate temperatures between 250 and 374 °C and working pressures between 40 and 220 bar. Biomass with high moisture content can reduce the cost of drying or water removal processes.

Therefore, bio-oil can be produced from biomass that has different moisture content, such as biomass obtained from wood waste and biomass obtained from algae. Among the thermochemical biomass conversion routes, pyrolysis and gasification are two processes that are being investigated. The process of heat-based disintegration, known as pyrolysis, occurs between 3500 and 5500 °C and up to 7000°. Due to this tremendous heat, all organic wastes turned into a mixture of gases. Gasification produces fuel gas that can be ignited to produce heat, while pyrolysis produces liquid fuel called pyrolysis oil (py-oil) or bio-oil. Since the biofuel generated in the liquid phase can be easily stored and transported, the pyrolysis method was better than the gasification process (Dhyani & Bhaskar, 2018).

6.3.2 Bio-chemical Conversion

Using microorganisms, biochemical (or biological) conversion methods transform biomass into commercially viable gaseous products with various applications (Leonardo et al., 2020). The two main techniques used are fermentation and anaerobic digestion. A commercial process called fermentation uses biomass that is high in sugar and starch, such as corn and sugarcane, to make bioethanol. Saccharides must be broken down, the sugars must be converted into ethanol, and the byproducts must be purified. Sugarcane's great productivity and energy potential make it an ideal feedstock (Muritala et al., 2020; Srivastava et al., 2019). Anaerobic Digestion Methane-rich biogas is produced during the anaerobic digestion process, in which bacteria are used to break down organic matter without oxygen. Both fuel manufacturing and electricity generation use this environmentally beneficial and energy-efficient technology. According to Aldayat et al. (2021), biosolids are used for soil conditioning, digestate is used as fertilizer, and biogas is used as fuel. The waste is converted into organic waste into compost and used either commercially or in an environmentally friendly way using bioproducts. Figure 6.2 depicts the sustainable strategies of organic waste valorization.

6.3.3 Chemical Conversion

Transesterification (alcoholization), the most popular process for producing biodiesel, occurs when alcohols, ideally short-chain (methanol and ethanol) and triglycerides (a lipid supply), are combined with a homogeneous, heterogeneous, or biocatalyst (Pidimalla et al. 2023). Temperature, stirring, reaction time, catalyst type, loading, concentration, and alcohol-to-oil molar ratio are often the variables that have the greatest impact on the profitability of a transesterification process (Guné et al. 2019). According to Gunay et al. (2019), the (reversible) transesterification method promotes the production of biodiesel and glycerol, both of which are beneficial byproducts. The stoichiometric alcohol-to-oil molar ratio must be at least 3/1 for the reaction to be complete. According to Ahmed et al. (2023), more alcohol is often needed to speed up the reaction and increase the yield of products. Currently, one of the most popular and commonly used methods is catalytic transesterification, which produces high conversion and biodiesel production under moderate reaction conditions. This is due to its convenience of use and the fact that a range of lipid sources and catalysts can be used (Pandit et al., 2023). Some of the innovative and unique methods for synthesis for the production of biofuels include reactive distillation, transesterification using supercritical fluid technology (a non-catalytic process), transesterification using membrane technology, transesterification using microwave technology, transesterification using in situ technology, transesterification using ultrasound-assisted transesterification, and transesterification using membrane technology (Quah et al., 2019).

6.4 Applications

6.4.1 *Biogas*

Biogas is the gas that forms when organic matter breaks down without oxygen. It is derived from biomass, which includes plant and animal remains, which is why it is called biogas (Hendriks and Zeeman, 2009). It is formed when organic matter is broken down by anaerobic microorganisms. With the help of enzymes, they metabolize organic matter, breaking down large molecules such as proteins, lipids and carbohydrates into smaller molecules such as H_2 from CH_4 and CO_2. This process cannot occur unless oxygen is present at all. Carbon (IV) oxide, hydrogen sulfide and methane are among the combinations of gases that are formed (Okonkwo et al., 2018). Digestate, which contains a variety of macro- and micronutrients, is the material left over from the production of biogas from a former organic feedstock. It is suitable for use as a soil fertilizer due to its favorable C/N ratio, homogeneity, nutrient availability, and significantly reduced odor. When it comes to biogas conversion, anaerobic digestion (AD) has clear advantages over many other biowaste treatment methods (Pant et al., 2025b). The hydrolytic products are processed into low molecular weight chemicals such as propanoic acid, butanoic acid, ethanoic acid, aliphatic acid, alcohol and carbon dioxide by Streptococcus, Lactobacillus, Bacillus, Escherichia and Salmonella. The technique of acetic acidification develops acetic acid from all the low molecular weight molecules of the previous step. The result depends on the growth conditions and type of bacteria. *Methanomicrobiales*, *Methanobacteriales* and *Methanococcales* are among the methane forming bacteria that form acetic acid, hydrogen and carbon dioxide in the second and third steps (Kothari et al., 2014).

6.4.2 *Bioethanol*

Using conventional techniques, bioalcohol has also been used as a fuel source for several decades. Bioalcohol is now seen as a non-fossil alternative mode of transportation. Currently, plant materials with high starch and sugar content—such as cereal crops and sugar cane—are the primary source of bioalcohol; but, in recent years, attention has focused on perennial grasses such as switchgrass and miscanthus. Although fermentation and distillation processes are necessary to convert cellulosic biomass into sugars beforehand, these crops do not compete with food consumption. While biopropanol and biobutanol are less popular, ethanol is the most commonly used bioalcohol. These alcohols are produced when microorganisms ferment carbohydrate-derived feedstock (Shah and Sen, 2011; Singh et al., 2022). The H058 strain of Saccharomyces cerevisiae was used to hydrolyze food waste and produce ethanol (Yan et al., 2013). Through simultaneous saccharification and fermentation processes, S. cerevisiae K35 was used to convert starch residues from instant noodle waste into bioethanol. According to Yang et al. (2014), the fermentation was optimized to achieve 96.8% ethanol conversion in 36 h.

6.4.3 Biodiesel

Catalysts are responsible for initiating the reaction and making the alcohols in the oil more soluble during the transesterification process. Also, pyrolysis and other thermochemical processes, where the use of waste biomass reduces the cost and increases the supply of such biodiesel, require catalysts (Bisen et al., 2025). Using calcium oxide (CaO) produced by calcination of leftover chicken bones at 800 °C, Suwannasom et al. (2016) studied the transesterification of discarded cooking oil and obtained a 96.31% yield. The biodiesel concentration was lower for catalysts made at lower calcination temperatures. Using cow bones and soybean oil, Ayodeji et al. (2018) made a calcium oxide catalyst with a 93% yield at 60 °C and a reaction time of 3 h.

6.4.4 Biofuel from Algae

Although algae do not have roots, leaves, and stems, they are still able to perform photosynthesis using chlorophyll. It can be unicellular or multicellular. According to Narayan (2024), microalgae can fix CO_2 10 times more effectively than terrestrial plants. According to several studies, the fatty acids found in algae vary depending on the species. Despite difficult conditions and low nutrition levels, microalgae were able to grow. To produce biofuel from microalgae, the sample must be handled carefully. In the presence of inorganic nitrate, carbon, etc., the formation of CO_2-tolerant algae Chlorococcum littorale at 295 K and 170 μmol-photon/m2/s light intensity was reported (Burman et al., 2022). Algae at 5% CO_2 concentration produced a maximum of 34 weight fatty acids, which is almost similar to the seed oil of various terrestrial plants. The concentration of CO_2, nitrate and salts affected the lipid content and concentration of the biomass. About 56% lipids and 1.75 g/l biomass content were obtained (Burman et al., 2022).

6.5 Environmental Impacts

The desire for more sustainable energy sources stems from the environmental damage caused by the use of fossil fuels and the ever-increasing need for energy (Alzaiza et al., 2023). Biodiesel as an alternative fuel source has generated great interest because it can reduce the negative environmental impacts of conventional fossil fuels. According to Ramachandran et al. (2023) and Shahbaz et al. (2013), biodiesel is promoted as an environmentally benign and renewable alternative that can reduce emissions of hazardous pollutants such as particulate matter and greenhouse gases that deteriorate air quality. It aims to provide an in-depth analysis of the environmental impacts of biodiesel, emphasizing both its potential benefits and ecological problems regarding sustainable energy. With varying environmental impacts and varying alignments with the Sustainable Development Goals (SDGs), different biodiesel resources offer different contributions to the field of environmentally friendly energy options. When biodiesel is used instead of petroleum fuels, CO_2 emissions can be greatly reduced. This is because, to some extent, the cycle is carbon-neutral, in that the carbon dioxide absorbed by plants during their growth phase compensates for the carbon dioxide emitted during burning (Gur, 2022).

Biodiesel based on waste oils, especially fish oil, and using waste cooking oil reduces the reliance on renewable energy sources, which helps meet SDGs 7 (affordable and clean energy) and 13 (climate action). This innovative application is a great example of SDG 9 (industry, innovation and infrastructure), which emphasizes the creation of environmentally friendly technologies and resource efficiency. However, when making biodiesel from bones, careful attention to the environment is required to minimize any potential harm to both terrestrial and aquatic ecosystems. According to SDGs 14 (life below water) and 15 (life on land), biodiesel production processes must take into account biological diversity and habitat conservation. SDG 8 (decent work and economic growth) requires strict environmental controls for biodiesel made from waste animal fats, regardless of its potential economic benefits, without compromising food supply chains or social justice (Malik et al., 2024a, 2024b).

6.6 Challenges

Since waste streams vary in both physical and chemical characteristics, the valorization of organic waste can present some problems. This can lead to low conversion yields, costly processing, and extremely complex processes. More studies are needed worldwide to solve these problems and improve the efficiency, competitiveness, and sustainability of biorefineries. According to Ma and Liu (2019), the process of valorization of organic waste is usually not sustainable in the long term. Future biorefinery technologies must be both environmentally friendly and highly productive to facilitate the shift to a bioeconomy (Lam et al., 2012). The sustainability of organic waste assessment can be examined using life cycle assessment techniques. To fully utilize the potential of high-value resource recovery from organic waste, research and innovation are needed, and the bioeconomy and waste systems must be prepared to adapt to new technologies in the future (Schot & Steinmueller, 2018). But when analyzing bioeconomic value chains, it is important to include the context of production and commerce. Research on bioeconomic processes has often also taken a sectoral approach, including forestry, chemistry, agriculture, pharmacy, aquaculture, agri-food, paper, energy, etc. Nevertheless, collaboration across different elements and biorefineries is important for all industries and supply chain components. For example, the integration of multiple waste streams may be necessary to make a biorefinery more economically efficient. Building cross-sectoral partnerships and collaborations can be challenging in reality and may require significant adjustments. Examples of such changes include new infrastructure, services and technologies, as well as updated production methods, new habits and new behaviors. An important point to keep in mind is that production, consumption and innovation regimes are all based on existing value chains, organizations and geographies. Businesses looking to gain a competitive edge through cost reduction or specific products can help enhance the value of organic waste. Because systems can serve as a barrier to more long-term change and because repetitive behaviors become embedded and accepted over time, it can be difficult to change an established system (Bauer et al., 2018). Therefore, large expenditures in this sector for technology and infrastructure can hinder constructive change. There is no motivation to develop waste avoidance systems because, for example, an industrial recycling system uses the generated waste as input. According to Ashokkumar et al. (2022), a bioeconomy based on organic waste has the potential to provide new market possibilities and help achieve environmental goals.

6.7 Future Perspectives

The availability of organic waste materials has fluctuated in recent years, as has the manufacturing cost. As a result, it is challenging to select the most efficient technology for the route to produce biofuels through thermochemical, chemical and biochemical processes. The different steps required for research, development and innovation can be combined into different processes for large-scale production of biofuels depending on the size of the market value. Third-generation biofuels are still being researched and developed, while second-generation biofuel production has begun to reach the market. However, at the moment, the production of first-generation biofuels has been established at commercial or large scale scales. However, it is expected that by the next period, they will all be transferred to industrial or commercial scale. Many countries are now developing, evaluating or promoting biofuel-based initiatives; yet, only a few of these initiatives have shown promising results when the right productivity and production system are employed (Le et al., 2013; Pant & Singh, 2025). However, additional work needs to be done to create and demonstrate the technology in the future. Additionally, the best conversion technologies—which are state-of-the-art, economical and feasible—can be used to produce biofuels from a variety of waste feedstocks. To reduce costs and increase production efficiency, new routes and processes should be used in the production of new generation biofuels. Biofuel development should be economically viable if various biomass is properly utilized, not only for fuel production but also for the production of energy and other important commodities. One of the special properties of organic waste is that it contains renewable carbon sources that can be converted into non-toxic, environmentally acceptable solid, liquid, and gaseous fuels (Malik et al., 2024a, 2024b). As a result, biofuels made from various organic wastes are important in this scenario as they help replace conventional fuels. Future studies should focus more on developing cost-effective conversion methods for producing biofuels, rather than on using new feedstocks, reducing the cost of enzymes, or improving downstream manufacturing processes. The future of biofuels is still unknown. However, new technologies may make biofuels more efficient in the future.

Commercial biofuel production is currently difficult, and other types with commercial potential need to be investigated. Combining multiple genetic engineering techniques to maximize biofuel production would undoubtedly be beneficial. Global cooperation is therefore crucial to integrate biofuels into our current energy system and overcome the energy problem. This will also help to enhance farming techniques and advance the technology of biofuel production (Malik et al., 2024a, 2024b; Pant & Kumari, 2026).

6.8 Conclusion

Organic waste has the potential to produce biofuels and other valuable commodities, it can also have very negative impacts on the environment and human health. Thus, waste management combined with various bioproducts has led to a significant increase in interest in biorefinery strategies from organic waste. To strengthen the circular bioeconomy, the present assessment emphasized the potential and existing possibilities of using organic waste to produce several value-added bioproducts. Organic waste often contains a significant amount of proteins, carbohydrates and lignocellulosic components, mainly cellulose and hemicellulose. However, when this waste is dumped in landfills without being properly treated, it causes serious health and environmental problems. This study clearly demonstrates the potential of converting organic waste into value-added biofuels (bioethanol, bioelectricity, biochar and biodiesel) and other bioproducts (bioplastics, organic acids, etc.). Commercial use of this technology can reduce waste management costs and increase government and private sector revenues.

Acknowledgements Authors are thankful to the IISER Bhopal and DIT University Dehradun for providing research support.

References

Abdeshahian, P., Lim, J. S., Ho, W. S., Hashim, H., & Lee, C. T. (2016). Potential of biogas production from farm animal waste in Malaysia. *Renewable and Sustainable Energy Reviews, 60*, 714–723. https://doi.org/10.1016/j.rser.2016.01.117

Ahmad, A. A., Zawawi, N. A., Kasim, F. H., Inayat, A., & Khasri, A. (2016). Assessing the gasification performance of biomass: A review on biomass gasification process conditions, optimization and economic evaluation. *Renewable and Sustainable Energy Reviews, 53*, 1333–1347. https://doi.org/10.1016/j.rser.2015.09.030

Ahmed, M., Ahmad, K. A., Vo, D. V. N., Yusuf, M., Haq, A., Abdullah, A., Aslam, M., Patle, D. S., Ahmad, Z., Ahmad, E., & Athar, M. (2023). Recent trends in sustainable biodiesel production using heterogeneous nanocatalysts: Function of supports, promoters, synthesis techniques, reaction mechanism, and kinetics and thermodynamic studies. *Energy Conversion and Management, 280*, 116821. https://doi.org/10.1016/j.enconman.2023.116821

Alazaiza, M. Y., Albahnasawi, A., Al Maskari, T., Abujazar, M. S. S., Bashir, M. J., Nassani, D. E., & Abu Amr, S. S. (2023). Biofuel production using cultivated algae: Technologies, economics, and its environmental impacts. *Energies, 16*(3), 1316.

AlDayyat, E. A., Saidan, M. N., Al-Hamamre, Z., Al-Addous, M., & Alkasrawi, M. (2021). Pyrolysis of solid waste for bio-oil and char production in refugees' camp: A case study. *Energies, 14*(13), 3861. https://doi.org/10.3390/en14133861

Angulo-Mosquera, L. S., Alvarado-Alvarado, A. A., Rivas-Arrieta, M. J., Cattaneo, C. R., Rene, E. R., & García-Depraect, O. (2021). Production of solid biofuels from organic waste in developing countries: A review from sustainability and economic feasibility perspectives. *Science of the Total Environment, 795*, 148816.

Ashokkumar, V., Flora, G., Venkatkarthick, R., SenthilKannan, K., Kuppam, C., Stephy, G. M., Kamyab, H., Chen, W. H., Thomas, J., & Ngamcharussrivichai, C. (2022). Advanced technologies on the sustainable approaches for conversion of organic waste to valuable bioproducts: Emerging circular bioeconomy perspective. *Fuel, 324*, 124313. https://doi.org/10.1016/j.fuel.2022.124313

Awasthi, M. K., Sar, T., Gowd, S. C., Rajendran, K., Kumar, V., Sarsaiya, S., Li, Y., Sindhu, R., Binod, P., Zhang, Z., & Pandey, A. (2023). A comprehensive review on thermochemical, and biochemical conversion methods of lignocellulosic biomass into valuable end product. *Fuel, 342*, 127790. https://doi.org/10.1016/j.fuel.2023.127790

Ayodeji, A. A., Blessing, I. E., & Sunday, F. O. (2018). Data on calcium oxide and cow bone catalysts used for soybean biodiesel production. *Data in Brief, 18*, 512–517. https://doi.org/10.1016/j.dib.2018.03.057

Barman, A., Bhattacharjee, S., Dutta, T., Pal, S., Chatterjee, S., Karmakar, P. and Mondal, S., 2022. Biofuel from organic waste-a smart solution to conserve nonrenewable resources—A review. In *Journal of physics: Conference series* (Vol. 2286, No. 1, p. 012028). IOP Publishing. https://doi.org/10.1088/1742-6596/2286/1/012028.

Bauer, F., Hansen, T., & Hellsmark, H. (2018). Innovation in the bioeconomy—Dynamics of biorefinery innovation networks. *Technology Analysis & Strategic Management, 30*(8), 935–947.

Bisen, D., Chouhan, A. P. S., & Pant, M. (2025). Bio-refinery approach for a sustainable ecosystem and circular bio-economy. In *Sustainable biotechnological remedial frameworks for the rejuvenation of heavily polluted environments* (pp. 304–324). CRC Press.

Carbon Trust, 2014. Waste to energy in Indonesia, assesing opportunities and barriers using insights from the UK and beyond. https://www.carbontrust.com/media/512147/ctc831-waste-to-energy-in-indonesia.pdf

Chandrasekhar, K., Kumar, S., Lee, B. D., & Kim, S. H. (2020). Waste based hydrogen production for circular bioeconomy: Current status and future directions. *Bioresource Technology, 302*, 122920.

Dhanya, B. S., Mishra, A., Chandel, A. K., & Verma, M. L. (2020a). Development of sustainable approaches for converting the organic waste to bioenergy. *Science of the Total Environment, 723*, 138109.

Dhanya, B. S., Mishra, A., Chandel, A. K., & Verma, M. L. (2020b). Development of sustainable approaches for converting the organic waste to bioenergy. *Science of the Total Environment, 723*, 138109. https://doi.org/10.1016/j.scitotenv.2020.138109

Dhyani, V., & Bhaskar, T. (2018). A comprehensive review on the pyrolysis of lignocellulosic biomass. *Renewable Energy, 129*, 695–716. https://doi.org/10.1016/j.renene.2017.04.035

Durán-Valle, C. J., & López-Coca, I. M. (2024). Biochemical and thermochemical conversion technologies for agriculture waste transformation. In *Transforming agriculture residues for sustainable development: From waste to wealth* (pp. 47–84). Cham.

Günay, M. E., Türker, L., & Tapan, N. A. (2019). Significant parameters and technological advancements in biodiesel production systems. *Fuel, 250*, 27–41. https://doi.org/10.1016/j.fuel.2019.03.147

Gür, T. M. (2022). Carbon dioxide emissions, capture, storage and utilization: Review of materials, processes and technologies. *Progress in Energy and Combustion Science, 89*, 100965. https://doi.org/10.1016/j.pecs.2021.100965

Hendriks, A. T. W. M., & Zeeman, G. (2009). Pretreatments to enhance the digestibility of lignocellulosic biomass. *Bioresource Technology, 100*(1), 10–18. https://doi.org/10.1016/j.biortech.2008.05.027

Kothari, R., Pandey, A. K., Kumar, S., Tyagi, V. V., & Tyagi, S. K. (2014). Different aspects of dry anaerobic digestion for bio-energy: An overview. *Renewable and Sustainable Energy Reviews, 39*, 174–195. https://doi.org/10.1016/j.rser.2014.07.011

Lam, M. K., Lee, K. T., & Mohamed, A. R. (2012). Current status and challenges on microalgae based carbon capture. *International Journal of Greenhouse Gas Control, 10*, 456–469. https://doi.org/10.1016/j.ijggc.2012.07.010

Le, L. T., van Ierland, E. C., Zhu, X., & Wesseler, J. (2013). Energy and greenhouse gas balances of cassava-based ethanol. *Biomass and Bioenergy, 51*, 125–135. https://doi.org/10.1016/j.biombioe.2013.01.011

Leonard, M. D., Michaelides, E. E., & Michaelides, D. N. (2020). Energy storage needs for the substitution of fossil fuel power plants with renewables. *Renewable Energy, 145*, 951–962. https://doi.org/10.1016/j.renene.2019.06.066

Lin, J., Zhang, Y., Ibrahim, R. L., & Abrorov, S. (2024). Paving the way for sustainable environment in BRICS through technological innovation, sustainable energy, and financial development: Heterogenous analyses. *International Journal of Sustainable Development & World Ecology*, 1–17. https://doi.org/10.1080/13504509.2024.2335270

Ma, Y., & Liu, Y. (2019). Turning food waste to energy and resources towards a great environmental and economic sustainability: An innovative integrated biological approach. *Biotechnology Advances, 37*, 107414.

Makepa, D. C., Chihobo, C. H., & Musademba, D. (2025). Technological advancements in biofuel, bioproducts, and bioenergy production from fast pyrolysis of lignocellulosic biomass. In *Biofuels and bioenergy* (pp. 85–119). Elsevier Science Ltd.. https://doi.org/10.1016/B978-0-443-29254-5.00005-9

Malik, K., Capareda, S. C., Kamboj, B. R., Malik, S., Singh, K., Arya, S., & Bishnoi, D. K. (2024a). Biofuels production: A review on sustainable alternatives to traditional fuels and energy sources. *Fuels, 5*(2), 157–175. https://doi.org/10.3390/fuels5020010

Malik, M. A. I., Zeeshan, S., Khubaib, M., Ikram, A., Hussain, F., Yassin, H., & Qazi, A. (2024b). A review of major trends, opportunities, and technical challenges in biodiesel production from waste sources. *Energy Conversion and Management, X*, 100675. https://doi.org/10.1016/j.ecmx.2024.100675

Manea, E. E., Bumbac, C., Dinu, L. R., Bumbac, M., & Nicolescu, C. M. (2024). Composting as a sustainable solution for organic solid waste management: Current practices and potential improvements. *Sustainability, 16*(15), 6329. https://doi.org/10.3390/su16156329

Muritala, I. K., Guban, D., Roeb, M., & Sattler, C. (2020). High temperature production of hydrogen: Assessment of non-renewable resources technologies and emerging trends. *International Journal of Hydrogen Energy, 45*(49), 26022–26035. https://doi.org/10.1016/j.ijhydene.2019.08.154

Nagar, V., & Kaushal, R. (2024). A review of recent advancement in plasma gasification: A promising solution for waste management and energy production. *International Journal of Hydrogen Energy, 77*, 405–419. https://doi.org/10.1016/j.ijhydene.2024.06.180

Narayanan, M. (2024). Marine algae biomass: A viable and renewable resource for biofuel production: A review. *Algal Research*, 103687. https://doi.org/10.1016/j.algal.2024.103687

Okonkwo, U. C., Onokpite, E., & Onokwai, A. O. (2018). Comparative study of the optimal ratio of biogas production from various organic wastes and weeds for digester/restarted digester. *Journal of King Saud University-Engineering Sciences, 30*(2), 123–129. https://doi.org/10.1016/j.jksues.2016.02.002

Pandit, C., Banerjee, S., Pandit, S., Lahiri, D., Kumar, V., Chaubey, K. K., Al-Balushi, R., Al-Bahry, S., & Joshi, S. J. (2023). Recent advances and challenges in the utilization of nanomaterials in transesterification for biodiesel production. *Heliyon, 9*(4).

Pant, M., Bisen, D., Kewlani, P., Srivastav, A. L., Bhatt, I. D., & Chakma, S. (2025a). Review of food waste valorization technologies: A sustainable approach to resource recovery and utilization. *Biomass Futures*, 100001.

Pant, M., Bisen, D., & Singh, J. (2025b). Biomass residue utilization: Harnessing unconventional resources. In *Bioenergy: Sustainable solutions for a greener planet* (pp. 67–86). American Chemical Society.

Pant, M., & Kumari, A. (2026). Evolution of agriculture technology: Assimilation of biotechnology and food technology for smart and sustainable agriculture. In *Digital twin technology for sustainable agriculture: applications, implementation and future trends* (pp. 195–217). Springer Nature Singapore.

Pant, M., & Singh, J. (2025). Integrating AI, machine learning, and nanotechnology: Shaping the future of water quality management. In *Computational automation for water security* (pp. 111–134). Elsevier.

Peter, A. P., Koyande, A. K., Chew, K. W., Ho, S. H., Chen, W. H., Chang, J. S., Krishnamoorthy, R., Banat, F., & Show, P. L. (2022). Continuous cultivation of microalgae in photobioreactors as a source of renewable energy: Current status and future challenges. *Renewable and Sustainable Energy Reviews, 154*, 111852. https://doi.org/10.1016/j.rser.2021.111852

Pydimalla, M., Husaini, S., Kadire, A., & Verma, R. K. (2023). Sustainable biodiesel: A comprehensive review on feedstock, production methods, applications, challenges and opportunities. *Materials Today: Proceedings, 92*, 458–464. https://doi.org/10.1016/j.matpr.2023.03.593

Quah, R. V., Tan, Y. H., Mubarak, N. M., Khalid, M., Abdullah, E. C., & Nolasco-Hipolito, C. (2019). An overview of biodiesel production using recyclable biomass and non-biomass derived magnetic catalysts. *Journal of Environmental Chemical Engineering, 7*(4), 103219. https://doi.org/10.1016/j.jece.2019.103219

Ramachandran, E., Krishnaiah, R., Venkatesan, E. P., Shaik, S., Saleel, C. A., & Hussain, F. (2023). Investigation into the ideal concoction for performance and emissions enhancement of jatropha biodiesel-diesel with CuO nanoparticles using response surface methodology. *ACS Omega, 8*(42), 39067–39079. https://doi.org/10.1021/acsomega.3c03890

Rincon, L., Puri, M., Kojakovic, A., & Maltsoglou, I. (2019). The contribution of sustainable bioenergy to renewable electricity generation in Turkey: Evidence based policy from an integrated energy and agriculture approach. *Energy Policy, 130*, 69–88.

Sarker, T. R., Ethen, D. Z., Asha, H. H., Islam, S., & Ali, M. R. (2024). Transformation of municipal solid waste to biofuel and bio-chemicals–a review. *International journal of Environmental Science and Technology*, 1–22. https://doi.org/10.1007/s13762-024-05975-0

Schot, J., & Steinmueller, W. E. (2018). Three frames for innovation policy: R&D systems of innovation and transformative change. *Research Policy, 47*(9), 1554–1567.

Shah, Y. R., & Sen, D. J. (2011). Bioalcohol as green energy-a review. *International Journal of Current Science Research, 1*(2), 57–62.

Shahbaz, M., Rasheed, M. A., Tahir, M. S., Ahmed, M. A., & Ikram, A. (2013). A parametric study on the production of bio fuels from rapeseed oil. *International Journal of Environmental Science and Development, 4*(6), 613–616.

Shariat Panahi, H. K., Dehhaghi, M., Aghbashlo, M., Karimi, K., & Tabatabaei, M. (2020). Conversion of residues from agro-food industry into bioethanol in Iran: An undervalued biofuel additive to phase out MTBE in gasoline. *Renewable Energy, 145*, 699–710.

Sims, R. C., Bedingfield, S. K., Thompson, R., & Sims, J. L. (2016). Bioenergy from wastewater-based biomass. *AIMS Bioengineering, 3*(1), 103.

Singh, A., Singhania, R. R., Soam, S., Chen, C. W., Haldar, D., Varjani, S., Chang, J. S., Dong, C. D., & Patel, A. K. (2022). Production of bioethanol from food waste: Status and perspectives. *Bioresource Technology, 360*, 127651. https://doi.org/10.1016/j.biortech.2022.127651

Srivastava, N., Kharwar, R. K., & Mishra, P. K. (2019). Cost economy analysis of biomass-based biofuel production. In *New and future developments in microbial biotechnology and bioengineering* (pp. 1–10). Elsevier. https://doi.org/10.1016/B978-0-444-64223-3.00001-1

Stephen, J. L., & Periyasamy, B. (2018). Innovative developments in biofuels production from organic waste materials: A review. *Fuel, 214*, 623–633. https://doi.org/10.1016/j.fuel.2017.11.042

Suwannasom, P., Tansupo, P., & Ruangviriyachai, C. (2016). A bone-based catalyst for biodiesel production from waste cooking oil. *Energy Sources, Part A: Recovery, Utilization, and Environmental Effects, 38*(21), 3167–3173. https://doi.org/10.1080/15567036.2015.1137998

Vieira, V. H. A. D. M., & Matheus, D. R. (2019). Environmental assessments of biological treatments of biowaste in life cycle perspective: A critical review. *Waste Management & Research, 37*(12), 1183–1198. https://doi.org/10.1177/0734242X19879222

Yaashikaa, P. R., Kumar, P. S., Nhung, T. C., Hemavathy, R. V., Jawahar, M. J., Neshaanthini, J. P., & Rangasamy, G. (2022). A review on landfill system for municipal solid wastes: Insight into leachate, gas emissions, environmental and economic analysis. *Chemosphere, 309*, 136627. https://doi.org/10.1016/j.chemosphere.2022.136627

Yan, S., Chen, X., Wu, J., & Wang, P. (2013). Pilot-scale production of fuel ethanol from concentrated food waste hydrolysates using Saccharomyces cerevisiae H058. *Bioprocess and Biosystems Engineering, 36*, 937–946. https://doi.org/10.1007/s00449-012-0827-9

Yang, X., Lee, S. J., Yoo, H. Y., Choi, H. S., Park, C., & Kim, S. W. (2014). Biorefinery of instant noodle waste to biofuels. *Bioresource Technology, 159*, 17–23. https://doi.org/10.1016/j.biortech.2014.02.068

Chapter 7
Green Synthesized Nanomaterials and Its Role in the Biofuel Production

Ayush Saxena, Akhtar Hussain, Fouziya Parveen, Nafisa Shaheen, Zeeshan Quraishi, Mariyam and Mohammad Ashfaque

Abstract The continuously increasing demand for fossil fuels leads to a negative impact on the environment causing the increase in greenhouse gases and depletion of the ozone layer. To solve this problem, various countries are developing green fuels. Green fuels or biofuels are characterized into four generations depending upon the source used for their production. Developing biofuels such as bioethanol, biodiesel, biogas, etc. requires multiple steps. These steps are based on different techniques, and out of all steps some steps require high energy and chemicals for the production of desired products which ultimately cause negative impacts on the environment as well as on the yield of the product. To combat this major hurdle concerning biofuel production researchers around the globe nowadays involve the use of green nanotechnology in their work. Green nanotechnology mainly involves bioactive agents such as plant extracts, lignocellulose wastes, microbes, and other biowastes for synthesizing green nanomaterials that leverage the sustainable development of various biofuels. The application of green nanomaterials is vast due to their, nanoscale size, surface area, stability, catalytic efficiency, no toxicity, low environmental footprints, and reduced energy consumption. This chapter showcases the type, versatility, applications, and challenges of green nanomaterials used in biofuel production.

Keywords Green nanotechnology · Biofuels · Renewable energy · Nanocatalyst · Zero waste

A. Saxena · A. Hussain · F. Parveen · Z. Quraishi · Mariyam · M. Ashfaque (✉)
Natural Products Research and Biofuel Research Laboratory, Department of Biosciences, Integral University, Lucknow, India
e-mail: mohdashfaquekhan@rediffmail.com

N. Shaheen
Department of Environmental Science, Integral University, Lucknow, India

V. Singh et al. (eds.), *Bioenergy and Value Added Byproducts from Biomass Waste*, Sustainable Environmental Waste Management Strategies,
https://doi.org/10.1007/978-3-032-22849-9_7

7.1 Introduction

It is widely recognized that the need for petroleum-based fuels is increasing daily. The improper use of fossil fuels negatively impacts the earth's health, and their exhaustion is imminent, leading to price hikes. Therefore, considering the environment and other risks, lowering the use of fossil fuels is necessary, and for this, significant research has been focused on developing biofuels (Arya et al., 2021). Biofuels are green fuels developed through various processes utilizing different substrates (Saravanan et al., 2022). Based on substrates used for their production, biofuels are categorized into four generations, i.e., first-generation biofuels, which are developed using feedstocks; second-generation biofuels use lignocellulose biomass; third-generation biofuels involve algae, and lastly, fourth-generation biofuels use a combination of recombinant microbes and plants (Saxena et al., 2024). The global biofuel production for the years 2023–2028 is estimated to expand by 30% i.e. 38 billion liters in comparison to the last five-year period (International Energy Agency, 2020). Moreover, India has become the third largest consumer and producer with three times the production of bioethanol in the last five years (International Energy Agency, 2024). Furthermore, increasing the yield of biofuel production takes several steps depending on the substrate used (Boro et al., 2022). Various pretreatment methods are used in second-generation biofuel production to loosen plant biomass's cell wall (Sinitsyn & Sinitsyna, 2021) so that more sugar can be utilized in fermentation for bioethanol production. Furthermore, to enhance the production of biofuels, the involvement of nanotechnology has become an advanced approach (Assad et al., 2022), especially the use of green nanoparticles and/or nanomaterials is gaining interest due to their reusability, cost-effectiveness, and sustainability (Ahmed et al., 2023; Khan et al., 2022).

Nanotechnology has become the fastest-growing field for research due to its multifaceted approach (Murtaza et al., 2024). Multiple nanoparticles and nanomaterials are used at different steps of biofuel production (Devi et al., 2021). These nanoscale materials with less than 100 nm in size at least in one dimension give a large surface area to volume ratio, provide various active sites, and faster reaction rate (Khoo et al., 2020). Nanomaterials act as promising candidates in different biofuel systems, not only due to their large surface area but also some other characteristics like crystallinity, catalytic activity, storage efficiency, durability, adsorption capacity, and stability making them integral to advancing renewable energy technologies (Nizami & Rehan, 2018). Green synthesis of nanomaterials replaced chemical and physical methods due to several issues, such as the generation of heat and toxic substances, and multiple instruments and chemicals are required to prepare nanomaterials. Whereas the green synthesis approach involves eco-friendly reducing materials (proteins, polyphenols, pigments, etc.) that can also be used as dispersants and end-capping agents to help reduce energy consumption, and cost (Fig. 7.1) (Somya et al., 2025; Ying et al., 2022), moreover, the synthesis of green nanoparticles takes place at room temperature (Abuzeid et al., 2023) (Fig. 7.2).

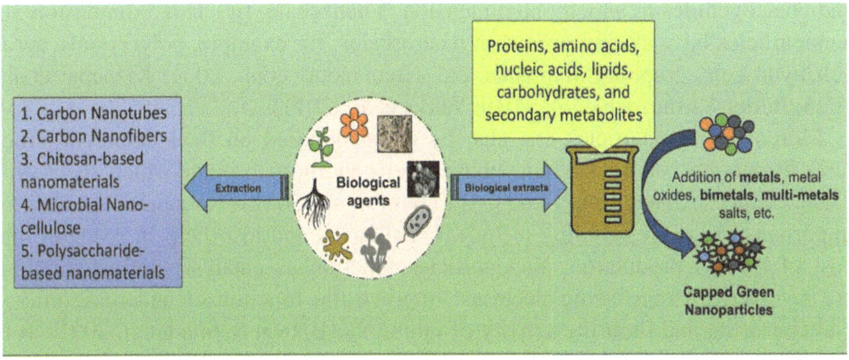

Fig. 7.1 Simple representation of green synthesis of nanomaterials

Fig. 7.2 Various dimensions of nanomaterials and their examples

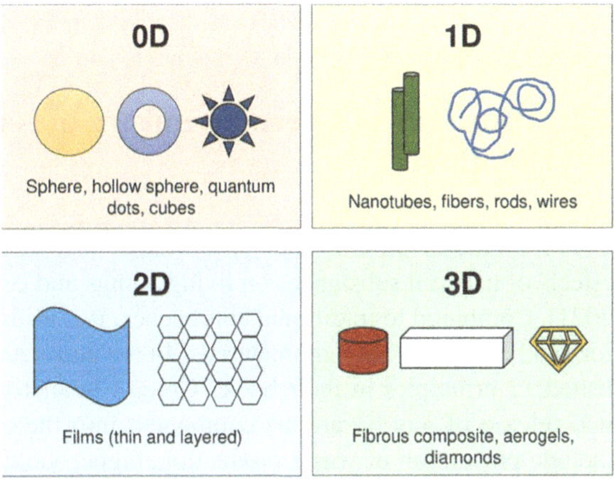

7.1.1 Nanotechnology: A Science of Minions

Nanotechnology is one of the growing fields of science involved in various fields. Compared to large-size materials, nanoparticles have excellent physicochemical, mechanical, magnetic, electrical, and optical properties (Biswas et al., 2012) such as nanoscale size and multi-dimension shapes, high surface-to-volume ratio and porosity with small pore size, highly crystalline, good catalytic activity, and many other make nanoparticles a valuable resource for biofuel production (Sharma et al., 2020; Biswal & Shadangi, 2021). Furthermore, based on dimensions nanoparticles are classified into four categories (Fig. 7.2). (i) Zero-dimension/0D, (hollow spheres, spheres, cubes, metals, quantum dots, etc. i.e. sized under 100 nm) (ii) One-dimension/1D nanoparticles have two dimensions in the nanoscale range and another one is not, this category involves nanofibers, metals, rods, ceramic, wires, and tubes) (iii) Two-dimension/2D nanoparticles include graphene family, oxides, and chalcogenides in which one dimension is nanoscale. (iv) Three-dimension/3D nanoparticles have all dimensions in macroscales, for example, polycrystals, aerogels/hydrogels, graphite, diamonds, etc. (Betancourt et al., 2021; Kizhepat et al., 2023; Rafiei-Sarmazdeh et al., 2019; Yahya et al., 2011).

Moreover, nanoparticles can also be grouped based on their chemical structure, dispersion, shapes, origin, phase, and condition through which their synthesis occurs (Ahmed et al., 2023; Barhoum et al., 2022). Nanoparticles play an important role in overcoming technical challenges and also help in reducing the cost of biofuel production. Nanoparticles function as catalysts during reaction, are involved in transferring electrons, improve the function of anaerobic microbial consortia, and limit the activity of inhibitors (Biswal & Shadangi, 2021). It is noticed that the size and geometry of nanoparticles are important which regulates their catalytic activity and stability. The catalytic activity of nanoparticles is sensitive to their structure which depends upon the presence of low-energy electronic fluctuation sites (Kumar et al., 2021).

7.2 Principles of Green Chemistry in Nanomaterial Synthesis

Green synthesis differs from biosynthesis and targets minimizing the negative effects of harmful substances on living beings and ecosystems (Sharma & Sharma, 2021). Compared to traditional approaches, it is a sustainable and effective method for synthesizing green nanomaterials. In the early 2000s, Anastas and Warner published 12 principles in their book "Green Chemistry" to be followed for the limited release of any hazardous component into the environment. These principles include prevention of waste production, higher yield, lower production of dangerous material, use of non-toxic chemicals/solvents and renewable feedstocks, the process must be energy efficient, use of stoichiometric catalysis agents, the end product must be easily degradable into non-toxic components, reduce the use of derivatives, real-time monitoring of the process to detect the production of toxic compound, and agents used must be safe and not responsible for any hazardous accident (Anastas & Warner, 2000). Further (Gałuszka et al., 2013) these principles were reviewed and expanded for better and easier understanding.

7.3 Synthesis of Nanomaterials

Due to different synthesis processes nanoparticles are differentiated from each other on several parameters. Multiple methods have been used to synthesize nanoparticles according to the requirements of the process and product. The synthesis of nanoparticles is based on two approaches, first, the top-down approach (using bulky materials and breaking them down to form nano-size particles), and second,

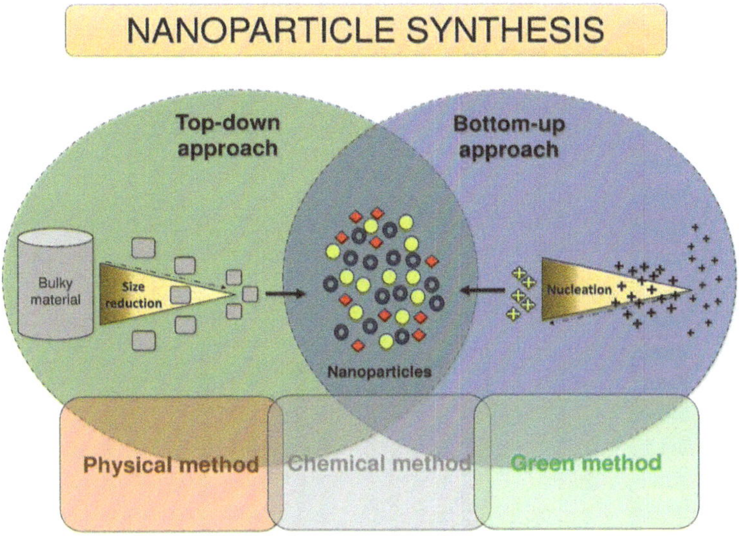

Fig. 7.3 Simple representation of methodology for nanoparticle synthesis

the bottom-up approach (using tiny components and combining them to form nanoparticles/nanomaterials (Zhang et al., 2022) (Fig. 7.3; Table 7.1). Green synthesis of nanoparticles uses a bottom-up approach, avoiding harsh conditions and toxic chemicals (Arum et al., 2024).

Some traditional methods include co-precipitation (nanomaterial production based on precipitation), laser ablation, and sol–gel. Furthermore, a study mentioned the synthesis of the nanotube with carbon sources using electricity, and heat through the arc discharge method or chemical vapor deposition (Huston et al., 2021; Jume et al., 2020). Moreover, a study mentioned the synthesis of nanofibers from an electro-spinning technique that requires polymer, collecting plates, high-volt electricity, and spinneret (Ahmadian-Fard-Fini et al., 2020).

7.4 Role of Capping and Reducing Agents

Capping agents are binding molecules that are used in the synthesis of nanoparticles to regulate their size, and morphology, provide stability by forming an interface between nanoparticle and medium and prevent agglomeration (Gulati et al., 2018; Javed et al., 2020). Whereas, reducing agents reduce the metal ions into metal nanoparticles. The size of nanoparticles depends upon the strength of the reducing agent. A stronger agent produces smaller-size nanoparticles (Aragaw et al., 2022; Villaverde-Cantizano et al., 2021). Green agents like, amino acids, proteins, carbohydrates, lipids, nucleic acids, and biological extracts from plants, microbes, and biowaste are used to synthesize capping and reducing agents

Table 7.1 Various methods for the synthesis of nanoparticles

Type	Method		Advantages and disadvantages	References
Top-down	Ball milling	High energy collision/grinding of powdered materials to form nanoparticles	Advantages—cost-effective, eco-friendly, and simple method Disadvantages—requires high energy and long time to operate	Al-Harbi and Abd-Elrahman (2024), Joy et al. (2022) and Xiao et al. (2022)
	Laser ablation	Continuous intense waves are used to irradiate atoms of a solid surface	Advantages—fast process, requires less energy, and simple method Disadvantages—agglomeration of particles	Balachandran et al. (2022) and Wang et al. (2023)
	Electrical wire explosion	High-density current pulses pass through the wire to synthesize nanoparticles and nanopowders	Advantages—energy efficient, simple method, and gives stabilized nanoparticles Disadvantages—low productivity	Kotov (2003), Kryzhevich et al. (2017) and Lerner et al. (2021)
	Metal alloying	It is a vapor deposition technique used for preparing nanoparticles for films and coating surfaces	Advantages—this method provides high catalytic and photocatalytic properties with high stability Disadvantages—crystal structure and atomic size of precursor can affect the property of nanoparticles	Huynh et al. (2020), Lemoisson and Froyen (2005), Nguyen et al. (2024) and Williams et al. (2020)
	Etching	Use various chemicals or etchant solutions for preparing nanomaterials	Advantages—controlled size and shape of nanoparticles, plasma etching can be used to create nanoscale-level patterns, etc Disadvantages—gives rough surface, produces a large amount of waste, uses various chemicals including toxic chemicals	Harish et al. (2022), He et al. (2013), Cao et al. (2021), Nowicki and Wandelt (2018) and Zhang et al. (2016)
	Lithography	It involves the patterning of bulky material into small nanoparticles	Advantages—can be used in the preparation of hybrid nanoparticles Disadvantages—expensive, requires longer duration	Ahmed (2020) and Ebad-Sichani et al. (2023)
Bottom-up	Co-precipitation	It is also known as the solvent displacement method and precipitates more than one compound from the solution. Metal precipitates as hydroxide from salt in the presence of a base in a solvent. Nanoparticles found at the interfacial phase during the reaction	Advantages—high productivity, high yield, fast method, etc. Disadvantages—downstream processing requires to get pure phase, increases water absorption of nanoparticles	Nawaz et al. (2019), Patil et al. (2021), Prakash et al. (2023), Pudovkin et al. (2018) and Srivastava and Katiyar (2022)
	Sol–gel	This is a diphasic solution having both liquid and solid phases. A Series of irreversible reactions take place to form the final product	Advantages—eco-friendly, require low temperature, produce homogenous structures Disadvantages—a costly and slow process, requires toxic chemicals, etc.	Bokov et al. (2021), Baig et al. (2021) and Nathawat et al. (2023)

(continued)

Table 7.1 (Continued)

Type	Method		Advantages and disadvantages	References
	Chemical vapor deposition	This technique is used for the development of nanomaterials. In this technique, the deposition of a thin film of material on the substrate occurs due to the reaction of vapor or gaseous substances at the interface	Advantages—Co-deposition of materials with a high-deposition rate Disadvantages—unsuitable for substrates that are sensitive to higher temperatures (> 600 °C)	Hamedani et al. (2016), Li et al. (2018) and Xie et al. (2020)
	Flame spraying pyrolysis	It is a single-step technique for nanoparticle preparation utilizing liquid precursor and burns in a flame. In this process, combustion of hydrocarbons is achieved in the presence of an oxidizer	Advantages—use low-cost precursors for the synthesis of nanoparticles, easily scalable process, etc. Disadvantages—use of low-cost metal precursor can lead to contamination in nanopowder formation, moreover water condensation can cause flame extinction affecting the quality of the nanomaterial	Rosebrock et al. (2016), Seah et al. (2011), Serrano-Bayona et al. (2023), Solero (2017) and Teoh et al. (2010)
	Microemulsion	In this process, self-aggregation of particles occurs. Water and oil are mixed with surfactants acting as nanoreactors. Further precipitating agents are added and energy in the form of a laser or pulse is given in the reaction which triggers the nucleation to form nanoparticles	Advantages—production of controlled size, shape, and stable nanoparticles Disadvantages—the problem arises due to the production of varied shape nanoparticles, high deposition parameter	Cid (2018), Malik et al. (2012), Salvador et al. (2021), Arredondo-Ochoa and Silva-Martínez (2022) and Modan and Plăiașu (2020)
	Green synthesis	It is an eco-friendly approach that utilizes biodegradable sources for the preparation of nanoparticles. Extracts from these sources are mixed with metal precursor to synthesize the nanoparticle	Advantages—eco-friendly, cost-effective, non-toxic Disadvantages—some precursors used in synthesis may be low soluble, and toxic. Another problem with this method is the lack of understanding of the synthesis pathway	Singh et al. (2018, 2023), Gupta et al. (2023), Pal et al. (2022) and Kakakhel et al. (2021)

(Khodashenas et al., 2020; Raveendran et al., 2003). During the synthesis of nanoparticles, the capping agents interact with metal salts present in the precursor and enhance the reduction kinetics of the nanoparticles (Sidhu et al., 2022). The synthetic and traditional methods use hetero-atom-functionalized long-chain hydrocarbons which harm the environment (Duan et al., 2015). Moreover, due to their biodegradability into simple non-toxic forms, the green synthesized capping agents have fetched the attention of nanotechnologists (Priya et al., 2021). Capping and reducing agents are also used to increase activity and stability (Bahari et al., 2023) and play an important role in the stabilization and functionalization of nanoparticles. Thus, selecting a capping agent is critical in achieving the targeted work (Javed et al., 2020; Marişca & Leopold, 2019).

7.5 Factors Required for Nanoparticles in the Biofuel Production System

The nanoscale size and properties of nanoparticles have drawn the attention of researchers and industrialists for biofuel production focus in enhancing the yield. Nanoparticles have fast reaction rates and many active sites compared to any other oversized molecules (Contreras et al., 2017). Functionalization of nanoparticles with different groups increases their capability of binding various class enzymes (Pavlidis et al., 2010). Properties like high surface area help in enhancing the catalytic efficiency and reaction rates (Anekwe et al., 2024), highly stable nanoparticles can be used multiple times to achieve the reaction (Panahi et al., 2024), and diverge surface chemistry helps in interaction with catalyst (Sharif et al., 2024). Yet, with so many advantageous properties the role of nanoparticles in the production of biofuels is still challenging due to the factors that hinder the activity, stability, and functionality of these particles. These factors include the use of precursors for developing these nanoparticles, synthesis approaches, reaction temperature and pH, pressure requirement, size, and toxicity against the environment, etc. (Hossain et al., 2019; Ahmed et al., 2023).

7.6 Biological Precursors for Green Synthesis of Nanoparticles

The naturally present biological sources have been used for the green synthesis of several metallic nanoparticles (El-Seedi et al., 2019). Green synthesis uses no hazardous secondary sources or any toxic chemicals (Banjara et al., 2024). In past years scientists have developed nanoparticles from green synthesis methods using various biological sources (Ahmed et al., 2017) (Table 7.2). Plants have been frequently used as biological sources rather than microbes and animals due to the alluring mixture of phytochemicals that act as stabilizing and capping agents

Table 7.2 Different nanomaterials and their applications

Nanoparticles type	Size and shape	Biological sources	Part	Biological activity	References
Silver nanoparticles	47 nm	*Cymbopogon citratus*	Leaves	Antimicrobial	Rakib-Uz-Zaman et al. (2022)
	30–45 nm (spherical)	*Brassica oleracea*	Leaves	Antifungal (*Aspergillus* and *Pneumocystis*)	Kuppusamy et al. (2015)
	35 nm	*Cassia roxburghii*	Leaves	Antifungal (*A. fumigates, A. niger, A. flavus, Candida albicans, Penicillium* sp.,*Curvularia* sp., *Rhizoctonia solani,* and*F. oxysporum)*	Balashanmugam et al. (2016)
	10 nm	*W. coagulant*	Fruit	Anti-fungal and anti-bacterial	Hussain et al. (2023)
	80 nm (spherical)	*Pinus roxburghii*	Pine needles	Anticancer (prostate cancer cells)	Kumari et al. (2020a, 2020b)
	80 nm (spherical)	*Moringa oleifera*	Leaves	Antibacterial and antioxidant activities	Din et al. (2024)
	18–100 nm (spherical crystalline)	*Bacillus subtilis*	–	Antibacterial activity (pathogenic bacteria)	Mathivanan et al. (2019)
Gold nanoparticles	90 nm	*Cucumis prophetarum*	Leaf	Antibacterial and antiproliferative activity	Meena et al. (2020)
	38.8 nm (spherical)	*Jatropha integerrima* Jacq	Flower	Antibacterial activity	Suriyakala et al. (2022)
	53.29 nm (spherical)	*Panchagavya* (PG)	PG extract	Antibacterial activity	Sathiyaraj et al. (2021)
	8–37 nm (spherical)	Jasminum auriculatum	Leaf	Antibacterial, antifungal, and anticancer activity	Balasubramanian et al. (2020)
	25.5 nm (spherical mono-dispersed)	*Annona muricata*	Leaf	Antibacterial and antifungal activity	Folorunso et al. (2019)
Carbon nanotubes (CNTs)	1.5–1.9 nm	Sugarcane	Bagasse	Antibacterial activity and biosensor potential	Pandiyan et al. (2020)
	71.2 nm	*Moringa oleifera*	Leaf	Antibacterial activity	Banerjee and Rajeswari (2024)

(continued)

Table 7.2 (Continued)

Nanoparticles type	Size and shape	Biological sources	Part	Biological activity	References
Iron nanoparticles	6092 d nm	*Phoenix dactylifera*	Leaf	Antibacterial activity	Batool et al. (2021)
	51 nm (spherical)	*Hibiscus rosa sinensis*	Flower	Antibacterial activity	Buarki et al. (2022)
	50 nm (spheroidal)	*Plumeria obtusa*	Leaf	Antimicrobial, antifungal, antioxidant, and biocompatibility activity	Perveen et al. (2022)
	52.78 nm (hexagonal nanorods with agglomeration)	*Ruellia tuberosa*	Leaf	Antibacterial and photocatalytic ability activity	Vasantharaj et al. (2019)
Copper nanoparticles	10–20 nm	*Kigelia africana*	Fruit	Antibacterial activity	Alao et al. (2022)
	11.7 nm	Rosa 'Andeli' double delight fresh petals (RAFE) or Gardenia jasminoides Ellis	Leaf	Antibacterial activity	Nieto-Maldonado et al. (2022)
	~ 80 nm (cubic, hexagonal, and rectangular)	*Ageratum houstonianum* Mill	Leaf	Antibacterial and Photocatalytic activity	Chandraker et al. (2020)
	44.9 nm (crystalline shape)	*Azadirachta indica* (Neem)	Flower	Antifungal and antibacterial activities	Gopalakrishnan and Muniraj (2021)
	7 nm (spherical)	*Hyptis suaveolens* (L.)	Leaf	Antifungal, cytotoxicity, and antibacterial activities	Shubhashree et al. (2022)
Carbon nanodots	3–7 nm	*Mahonia oiwakensis*	Stem and root	Cytotoxic activity	Chen et al. (2021)
	<10 nm (quasi-spherical amorphous particles)	*Hemigraphis alternata*	Leaf	Cytotoxic activity	Manisha et al. (2023)
	5 nm (spherical)	*Plectranthus amboinicus* and *Leucas lavandulifolia*	Leaf	Antibacterial activity	Raveendran and Renuka (2022)
	25 nm	*Malva Sylvestris*	Leaf	Cytotoxic activity	Moaddab and Ghasemi (2021)

(continued)

Table 7.2 (Continued)

Nanoparticles type	Size and shape	Biological sources	Part	Biological activity	References
Cellulose nanocrystal (CNC)	221 nm	*Gossypium* sp.	Raw material	Nanocomposite, sensor, electronic device	Pandi et al. (2021)
CNC incorporated with 5-fluorouracil	69.53 ± 1.14 nm (length) and 8.13 ± 0.72 nm (width)	*Oryza* sp.	Straw	Anticancer	Yusefi et al. (2022)
Magnetite-functionalized cellulose nanocrystals-hydrogel	CNC- 15–25 nm (diameter) and 40–150 nm (length)	*Beta vulgaris* spp.	Pulp	Cationic dye removal	Moharrami and Motamedi (2020)
Cellulose nanofibers (CNF)	2–30 nm	*Bambusa vulgaris*	Bamboo	Biomaterials	Xie et al. (2016)
	240 ± 12 nm and 10 μm (median size), ropelike shape	*Saccharum officinarum*	Bagasse	–	Sofla et al. (2016)
Bacterial Nanocellulose	5 mm (sphere)	*Gluconacetobacter xylinus*	–	Antimicrobial	Cesca et al. (2020)
	–	*Beta vulgaris* spp. and tea fungus	Beetroot peels	–	Beluhan et al. (2022)
AgNPs/Bacterial Nanocellulose	*Komagataeibacter hansenii* NBRC 14820	*Malus domestica, Cucumis melo var. reticulatus,Musa acuminata*	–	Antimicrobial wound healing dressing	Shaheen et al. (2021)

(Ruddaraju et al., 2020). Moreover, algae are also used in the synthesis of nanoparticles due to their extraordinary capability to convert metals into usable forms (Chaudhary et al., 2020). Furthermore, carbon and its allotropes have also been used for the synthesis of various nanomaterials. The exceptional properties of these nanomaterials make them a promising area of research for various applications including biofuels (Rao et al., 2021).

7.6.1 Plant-Based Precursors for Green Nanoparticle Synthesis

In addition to being highly costly, physical and chemical methods lead to the generation of toxic by-products. The biological methods are cost-effective, reduce the chemical load on the environment, are easy to synthesize, and eliminate unnecessary processes during synthesis (Kumar et al., 2011). Various types of nanoparticles are made from leaves, seeds, fruits, flowers, stems, roots, rhizomes, bark, and fruit peel extracts that contain phytochemicals that act like reducing and stabilizing agents in nanoparticle synthesis. Therefore, incorporating green synthesis into mass production and scientific research offers a potential solution to the drawbacks of conventional synthesis methods (Jadoun et al., 2021).

As mentioned above, various parts of the plant and their sources are being used to synthesize the nanoparticles. Plant extract consists of phenols, flavonoids, alkaloids, antioxidants, etc. which act as good capping and reducing agents in the fabrication of the nanoparticles (Nguyen et al., 2022). These phytochemicals enhance the catalytic activity and absorptivity by enriching the surface of the nanoparticles with new functional groups leading to more physicochemical interactions (Wu et al., 2020). Synthesis of metal and metal oxide nanoparticles utilizing plants is simple, gives a higher yield, and is a one-pot synthesis that works on the principles of Green Chemistry (Sabouri et al., 2020). Plant-extracted phytochemicals donate electrons to metal ions and convert them into nanoparticles. Carboxylic groups, polyphenols, flavonoids, carbonyl groups, hydroxy groups of polysaccharides, and amine groups of proteins chelate Cu^{2+}, Fe^{2+}, Fe^{3+}, Zn^{2+}, and Pb^{2+} and form stable metal cations (Marslin et al., 2018).

Furthermore, cellulose a major component of the plant cell walls is also used to synthesize various structured types of nanomaterials (Sheikhi, 2019). Cellulose has attracted much interest in a variety of applications, including dielectric layers, derived carbon materials, and flexible electronic devices because of its biodegradability, cost-effectiveness, exceptional mechanical properties, and abundant reserves (Cui et al., 2020). The most significant cellulose source is plants composed of oxygen, carbon, and hydrogen. Meanwhile, microbes and other sea animals are used to synthesize nanocellulose (Barhoum et al., 2020). Naturally present cellulose is fibrous with infrequently present amorphous and crystalline sections. These cellulose fibers are nanoscale cellulose which shows different morphologies such as crystals, whiskers, fibers, etc. (Huang et al., 2019).

7.6.2 Microbial-Based Precursors for Green Nanoparticle Synthesis

Plants as well as microbes are used for the synthesis of metal nanoparticles. The phytochemical of plants is composed of various important chemicals and thus used preferably in nanoparticle synthesis (Shahid et al., 2017). Microbes considered as "nano-factories" bring microbiology, nanotechnology, and biotechnology together (Alloosh et al., 2021). Synthesis of nanoparticles via microbes is another cost-effective and sustainable approach. Microbes like bacteria, fungi, and algae are easy to grow and easily adapt to environmental conditions (Li et al., 2017). Metal ions can be reduced or adsorbed by microorganisms to synthesize nanoparticles (Osman et al., 2024). Bacteria such as *Pseudomonas* sp., *Bacillus* sp., *Pyrococcus* sp., *Pyrobaculum* sp., etc. are used in Au, Ag, Zn, Fe, Ni nanoparticle synthesis (Sanaeimehr et al., 2018). Moreover, fungi such as *Verticillium luteolbum*, *Aspergillus oryza*, *Fusarium oxysporum*, *Trichothecium* sp., *Collototrichum* sp., etc. have also been seen in the production of nanoparticles (Jeevanandam et al., 2016). Yeasts also induce biomineralization and can synthesize nanoparticles. An advantage of using yeast is that yeast cells can encapsulate and do not require any stabilizing agent (Roychoudhury, 2020). Several studies also used algal extracts for the preparation of metal nanoparticles. These extracts are easy to handle and non-toxic. Extracellular and intracellular pathways are used for the synthesis of nanoparticles from algae. The extracellular pathway acts on the surface of the cell and involves the biological reduction of metal ions into nanoparticles, while the intracellular pathway occurs inside the cell wall (Negi & Singh, 2018). Moreover, microbial-synthesized enzymes have enormous potential in biotechnology research and applications and may be used in complex synthesis and catalytic reactions. The close interaction of the active site's naturally screened and ordered reaction components is responsible for remarkable catalytic efficiency. Furthermore, studies have shown that catalysis can be improved by attaching enzymes or their substrates to the surface of the nanoparticles (Jin et al., 2023). Moreover, besides bacteria, fungi, and algae, actinomycetes are also used for the synthesis of nanoparticles (Skalickova et al., 2017). Members of the actinobacteria phylum are mostly responsible for producing antimicrobial compounds (Kumari et al., 2020a, 2020b). Many studies showed that researchers have used actinomycetes in the production of metal nanoparticles and used them in anticancer, anti-biofouling, anti-parasitic, antimicrobial, anti-malarial, and antioxidant activity (Manivasagan et al., 2016). Also, nanoparticles synthesized from various actinomycetes have shown good polydispersity and stability (Mabrouk et al., 2021).

Bacterial nanocellulose (BNC) is another type of naturally synthesized green nanomaterial composed of β-D glucopyranose. *Gluconacetobacter xylinus*/*Acetobacter xylinum* is effectively used in the production of BNC. Some other microbes such as *Agrobacterium*, *Pseudomonas*, *Acanthamoeba*, *Achromobacter*, *Acetobacter*, *Sarcina*, *Rhizobium*, and *Alcaligenes* as well as the algae *Cladophora*, *Microdiction*, *Rhizoclonium*, and *Chaetomorpha* have used to develop BNC (Randhawa et al., 2022). Majorly BNC is composed of water yet has exceptional mechanical properties. The water-holding capacity and nanostructure of BNC are similar to collagen which makes it suitable for adhesion and cellular immobilization (El-Hoseny et al., 2015; Gama et al., 2012; Numata et al., 2015). BNC is categorized under generally recognized as safe (GRAS) products and can be used in multiple applications (Sharma & Bhardwaj, 2019). Synthesis of BNC is a complex and multistep process that involves enzymes and both regulatory and catalytic proteins. Uridine diphosphoglucose which is a precursor of cellulose is synthesized in this process. Glucose units polymerize into β-1,4-glucan chains, forming ribbon-like structures by hundreds or even thousands of individual cellulose chains (Foresti et al., 2015). A study showed the preparation of bacterial nanocellulose from banana peels using *Gluconacetobacter xylinus*. Synthesized BNC was used as a water filtration membrane (Sijabat et al., 2020). Another study used a cell-free supernatant of *Lactobacillus acidophilus* for the preparation of bacterial nanocellulose with carbon dots through the hydrothermal method. The antimicrobial activity of prepared carbon dots was tested against various pathogenic strains. Furthermore, a nanopaper with antimicrobial and UV resistance properties was developed by incorporating carbon dots into nanocellulose (Kousheh et al., 2020).

7.6.3 Carbon-Based Nanomaterials

7.6.3.1 Graphene Family Nanomaterials (GFNs)

Graphene is a second-dimension, single-layered trigonal (sp^2) hybridized hexagonal network of carbon atoms that is bound together through covalent bonding. Due to its exceptional chemical and physical properties, graphene has been used as a nanomaterial in various scientific research and industries. Graphene is a basic unit for all carbon nanomaterials including 0D-bucky ball to 3D-graphite. It is the tiniest inorganic nanomaterial extremely tough, and 100 × stronger than the steel (Wang et al., 2011, 2019a, 2019b). Several other graphene nanomaterials that come under the graphene family nanomaterials (GFNs) are developed by a range of layers that are modified into sheets and chemically built forms. GFNs such as reduced graphene oxides (rGOs), multi-layered graphene oxides, bi-layered, and single or mono-layered differ from each other on several properties like horizontal dimensions, layers, purity, composition, and surface chemistry and thus have been used in multiple applications (Arfin & Athar, 2018). Furthermore, graphene oxide (GO), reduced graphene oxide (rGO), and graphene are the most used nanomaterials for catalysis research (Zhu et al., 2015).

7.6.3.2 Carbon Nanotubes (CNTs)

One of the most important nano allotropes of carbon is CNT (Nasir et al., 2018). CNT is a cylindrical-shaped 1D nanocarbon tube that was first discovered by Sumio Iijima in 1991 (Devi et al., 2023). Depending upon the number of rolled overlapping layers, CNT can be classified into single-walled carbon nanotube (SWCNT), double-walled carbon nanotube, and multi-walled nanotube (MWCNT). The cylindrical structure and C–C bonding of CNT give high strength to it (Kausar, 2022). CNTs have properties like mechanical, high tensile, and flexural strengths, thermal, electrical, high elastic modulus, and high aspect ratio. CNTs have attracted the attention of many researchers since their discovery. Due to these properties, CNTs have been used in various fields providing novel strategies and replacing conventional methods (Guo, 2020). The surface activity of the CNTs can be enhanced by functionalizing the sides by immobilizing various chemical groups and enzymes on the outer and inner portions (Kushwaha et al., 2018). Moreover, due to their electrical capabilities, CNTs are the most utilized carbon-based nanomaterial for immobilizing enzymes, mainly for biofuel production. Enzyme immobilization on CNTs is performed by the covalent and non-covalent interaction. The non-covalent method is preferably used as it does not denature the enzymes. According to the requirement of the method, SWCNTs and MWCNTs are used in various areas, and MWCNTs are frequently used in the immobilization of cellulase enzymes as they are cost-effective, multi-layered graphite structures with good dispersibility.

7.6.3.3 Carbon Nanofibers (CNFs)

Carbon nanofibers (CNFs) are also 1D nanofillers made up of multiple graphite planes arranged longitudinally and tilted at a 25° angle. The diameter of CNFs is greater than CNTs, this makes more exposure of active carbons at the outer surface of CNFs leading to its increased chemical reactivity. CNFs are also used in various industrial applications due to their cost-effectiveness. CNFs have inferior thermal, mechanical, and electrical properties than CNFs. Methods like heat treatment, salinization, in-situ, solvent casting, ultrasonication, mechanical, stirring, surfactants, and plasma can be used in obtaining the uniform dispersion of CNFs (Nasajpour-Esfahani et al., 2023). Furthermore, except for diameter, the structure of CNFs differs from conventional carbon fibers (CCFs). CNFs are mainly synthesized using two approaches, (i) catalytic vapor deposition growth (CVDG) and electro-spinning (ES) (Feng et al., 2014)., CNFs have properties like electrochemical catalysis, selective adsorption, hydrogen storage, polymer reinforcement, etc. They are sp^2-based fibrous nanostructures with non-continuous linear filaments, and aspect ratios higher than 100:1 (Yadav et al., 2020). Because of their graphitic structure, CNFs have special characteristics like high mechanical strength, resistance to corrosion, and thermal and electrical conductivities (Sharma & Kar, 2020).

7.6.3.4 Nanoflowers (NFs)

Nanoflowers are another example of carbon-based nanoparticles which are flower-shaped, composed of various layers arranged in petal-like structure increasing surface area with the size range between 100 and 500 nm. Nanoflower is highly stable and easier to synthesize using both organic and inorganic materials alone or in combination (Shende et al., 2018). In a study, researchers used cellulosic material and transformed it into bioproducts with the help of recombinant cellobiohydrolase enzymes i.e. endo-glucanase, and β-glucosidase immobilized over nanoflowers and integrated with ELP and His-tag. These nanoflowers act as biocatalysts for the immobilized enzymes and help in catalyzing the hydrolysis of cellulose to monomer forms. The developed nanoflower has shown excellent catalytic efficiency and stability in comparison to the unbounded enzymes. Furthermore, the study also mentioned that these enzyme-bound nanoflowers showed their activity up to the 8 cycles. The Km (Michaelis constant) and catalytic efficiency of enzyme-bound nanoflowers were 9.33 g L^{-1} and 0.0051 L min–^{1} g^{-1}, respectively, which indicates a strong affinity of the enzyme for substrate than unbound enzymes (Han et al., 2021). Another study showed biofuel production from sunflower oil using lipase inorganic hybrid nanoflowers. The study mentioned that these hybrid nanoflowers showed a 96.5% conversion rate even after five continuous cycles, with 72.5% retention (Jiang et al., 2018). Furthermore, in another study researchers (Zhong et al., 2021) used magnetic-activated lipase-inorganic hybrid nanoflowers for the production of biodiesel from soybean oil. In comparison to unbound lipase, the modified nanoflowers showed tolerance against methanol and activity recovery of up to 190% along with improved stability and storage. Due to the magnetic property of the nanoparticles, 84% of the initial activity was maintained even after reusing for 10 cycles with the yield of biodiesel after the 1st cycle was 88% and in the 6th cycle, it was 76%.

7.7 Role of Green Nanomaterials in Biofuel Production: Case Studies

As mentioned above, biologically active agents are used to prepare green nanomaterials. Mostly, plant extracts are used in the synthesis of nanomaterials due to the presence of phytochemicals which act as reducing and stabilizing agents (Jadoun et al., 2021). Two methods, top-down and bottom-up approaches are used to synthesize the nanomaterials (Khan, 2020). The synthesis of green nanomaterials comes under the bottom-up approach (Salem & Fouda, 2021). Green nanomaterials have shown a crucial role in the production of biofuels due to their catalytic properties help alter the reaction mechanism (Srivastava et al., 2021). Moreover, nanotechnology research is also used in the modification of feedstocks, immobilizing enzymes over nanomaterials, acts as catalyst support for the accumulation of lipids, extraction, and transesterification reaction, etc. (Rodríguez-Couto, 2019; Panahi et al., 2024).

In a study, Stegarescu et al. developed an MnO_2 nanoparticle using different extracts from plants *Rosmarinus officinalis*, *Origanum vulgare*, and *Artemisia dracunculus* utilizing both chemical and biological synthesis routes. The prepared MnO_2 nanoparticles were used as a catalyst in biofuel production from grape residue and seeds oil. The study revealed that the biochemically synthesized nanoparticles have four times higher surface area than chemically synthesized nanoparticles. However, the nanoparticles prepared from *Origanum vulgare* showed the highest Mn^{2+} content in low oxidation state, surface area, pore volume, and largest crystalline core. Due to these properties, the MnO_2-*Origanum vulgare* nanoparticles were used for microwave-assisted transesterification of grape residue and seeds oil. These nanoparticles showed the best catalytic properties and a 3.5 times increased transesterification rate for biodiesel production compared to the control i.e. yeast catalyst (Stegarescu et al., 2020). In another study, scientists have used extracts of pods and seeds of *Acacia furnesiana* fruit to synthesize $NiFe_2O_4$ nanoparticles as well as a substrate to produce biofuel. The study resulted in 100% recovery of nanoparticles after the transesterification process and reusability up to four cycles with more than 80% *A. furnesiana* oil methyl ester yield (Kattimani et al., 2023). Ariaeenejad et al. developed biogenic mesoporous silica nanoparticles from rice husk biomass in a study. These nanoparticles were used as a nanocarrier to improve recyclability and recombinant xylanase enzyme stability. Further, these nanoparticles were modified with bis-(2-aminoethyl) ether to immobilize the recombinant xylanase. The study revealed that after immobilizing the enzyme on the nanomaterial the optimal temperature for enzyme activity increased 10 °C compared to non-immobilized enzyme. It was also found that immobilized enzyme maintains its 90% activity at temperature ranges 30–60 °C with enhanced biomass degradation and reducing sugar production. Moreover, the immobilized enzyme showed its activity up to nine cycles of use (Ariaeenejad et al., 2020). Another study by Ariaeenejad et al. showed the synthesis of cellulose nanocrystals functionalized with dopamine and magnetite nanoparticles. The study aims at the biological conversion of sugars to fermentable sugars. Developed nanomaterial is immobilized with enzyme cascade, resulting in an increased yield (20–76%) of fermentable sugars compared to the free enzyme cocktail. The initial activity of the prepared nanocatalyst was found to be > 50% up to 10 cycles of use (Ariaeenejad et al., 2021). Zhang et al. used water hyacinth extract to prepare the magnetite nanoparticles to increase the yield of biohydrogen by utilizing lignocellulose biomass as a substrate. The hydrogen produced was found 23.49% more with $Y(H_2/S)$ of 83.20 $mL/g_{substrate}$ at 20 mg/L of nanoparticle concentration. Furthermore, the activity of hydrogenase in the presence of nanoparticles was found to be enhanced 2.1 times in comparison to the control which was analyzed through the RT-PCR technique (Q. Zhang et al., 2021). In a study, Sawaira et al. synthesized a stable bismuth oxide nanocatalyst from the leaf extract of Royle's spurge to produce biodiesel from the seed oil of *Cannabis sativa* using a membrane reactor. The highest biodiesel yield of 92% was achieved under the optimal conditions provided (Alsaiari et al., 2023). In a study, Yatish et al. prepared copper oxide nanoparticles from leaves and seeds extract of *Terminalia chebula* for heterogenous use in biodiesel production from the oil of the same plant. The maximum biodiesel yield was achieved 97.1% at 3 wt% of nanocatalyst (Yatish et al., 2021). In another study, Yatish et al. synthesized the calcium titanate nanoparticles from the leaf extract of *Ochrocarpus longifolius* and used them as a heterogeneous catalyst in the production of biodiesel from dairy waste scum oil. Study results in 97.7% yield of biodiesel and nanoparticles showed catalytic stability up to 5 cycles (Yatish et al., 2020). Furthermore, Aziz et al. used coconut shell extract and iron salt for the production of magnetic iron nanoparticles as a micronutrient for biohydrogen production. The addition of nanoparticles in 200 mg Fe/L achieved maximum hydrogen production (101.33 mL), glucose utilization (9.12 g/L), and hydrogen yield (0.79 mol H_2/mol glucose). The kinetic study also revealed that the addition of nanoparticles in optimal concentration increases the hydrogen production potential (50.69 mL) and hydrogen production (3.30 mL/h) rate with reduced lag phase time in comparison to the control (Aziz et al., 2022).

In a study, Kandasamy et al. used biochar synthesized from *Spirulina platensis* which was impregnated into CeO_2 and used as a nanocomposite for hydrothermal liquefaction of microalgae to produce bio-oil. It was found at temperature 250 °C the highest production of bio-oil was achieved. Also, utilizing biochar nanocomposites decreases nitrogen and oxygen content and increases heating value. Furthermore, in situ inclusion of nanocomposite showed increased energy recovery up to 65.34% (Kandasamy et al., 2021).

Furthermore, several studies also used microbial-synthesized nanoparticles and nanomaterials to increase biofuel yield. The nanomaterials able to biologically stimulate microbes enable them to enhance biofuel production (Salah Eldin et al., 2024). A study by Gupta et al., developed zinc oxide nanoparticles from a high zinc metal tolerant *Fusarium oxysporum* strain. The study used rice straw as a source for bioethanol production where the substrate is pretreated with ammonia for delignification and *F. oxysporum* itself is used to produce enzymes to achieve monomeric sugars. The nanoparticles synthesized were used in increasing the concentration of bioethanol which was a maximum of 0.0359 g/g at 200 mg/L of the nanoparticles (Gupta & Chundawat, 2020). Aarti et al. synthesized the gold nanoparticles using partially purified cellulase as a reducing agent from *Glutamicibacter arilaitensis* strain ALA4 for increasing the yield of reducing sugars from alkali-pretreated *Brachiaria mutica* and *Alternanthera philoxeroides*. Both biomass on treatment with nanoparticles showed increased total reducing sugar percentage after 72 h with increased saccharification degree up to 96 h. Further, bioethanol was produced using yeast immobilized in calcium alginate beads (Aarti et al., 2022). In a study, Kalpana et al., synthesized carbon quantum dots from the exopolysaccharide of *Exiguobacterium* sp. VK2 through hydrothermal reaction. The prepared nanodots are used as a catalyst for biodiesel production from waste cooking oil. The study produced 90% fatty acid methyl ester using nanodots (Kalpana et al., 2023).

7.8 Challenges and Prospects for Biofuel Production in Association with Green Nanomaterials

Despite exemplary properties and an environment-friendly approach, the synthesis of green nanomaterials has several challenges and limitations (Ahmed et al., 2022). These challenges are rooted in balancing environmental safety and economic feasibility (Wahab et al., 2024). Stability and reusability are the primary issues for the synthesis processes (Wang et al., 2019a, 2019b). Moreover, the major hurdle is scaling up the synthesis of nanoparticles at the industrial level. Identifying stabilizing and reducing agents involved in nanoparticle synthesis is also challenging. Other challenges such as maintenance of optimum conditions at a large-scale level, non-uniform size, and shape are problems in achieving the targeted results (Aswathi et al., 2023). It is necessary to synthesize a nanoparticle according to the step, biomass, and type of biofuel to be produced. For example, nanomaterials used during thermochemical conversion can face morphological changes, sintering, and even deactivation due to the generation of high temperatures (Ramesh et al., 2020). Furthermore, the improper and careless use of nanomaterials or nanoparticles can also cause serious damage to the health of human beings (Asmatulu et al., 2024). The different attributes of nanomaterials can also be a possible cause directly or indirectly to human health, for instance, many metal-oxide nanoparticles are found to be extremely harmful if entered into the human body. Therefore, concerning not only the use and advantages of such nanoparticles in large-scale industries but also proposing safety measures on their use and disposal can reduce the chances of getting impact (Pandit et al., 2023). On the other hand, microbes also have the potential to synthesize nanoparticles at a large scale, but their extraction is tough and needs to be optimized to conserve their activity (Sarkar et al., 2021). Microbially synthesized metal nanoparticles have shown promising and eco-friendly approaches in various applications in the agriculture sector. Yet their role against soil-crop ecosystems is not completely known. Thus, their potential must be evaluated, especially toxicity and long-term use, before utilizing them with living systems (Vecchio et al., 2012).

Despite such challenges, the possibility of green nanomaterials in recasting biofuel production remains important. Continuous research and interdisciplinary collaboration are required to overcome these problems for the proper implementation of nanotechnology in the sustainable production of biofuels.

7.9 Conclusion

Green-synthesized nanomaterials are emerging as promising candidates in biofuel production, offering eco-friendly and sustainable solutions worldwide to meet the growing energy needs. Green nanomaterials synthesized using biological agents hold immense potential due to their availability, cost-effectiveness, less toxic or non-toxic, and biodegradability. These nanomaterials have unique properties like catalytic activity, high surface area, physicochemical properties, and stability making them efficient candidates in enhancing the yield of biofuels. Furthermore, green nanomaterials support sustainable development goals, zero-waste, and circular economy strategy by converting organic wastes into valuable bioenergy sources. Integrating green-synthesized nanomaterials with various cutting-edge technologies can significantly help standardize biofuel production leading to a safer environment by fulfilling the day-to-day human needs. However, various challenges such as long-term stability, toxicity, scalability, and optimization of synthesis methods need to be revised and addressed to unlock their full potential for large-scale roles. Moreover, the use of green nanomaterials for biofuel production is one of the important steps in minimizing the use of fossil fuels and achieving environment-friendly energy solutions.

Acknowledgements Authors are thankful to the Integral University for providing necessary research support.

References

Aarti, C., Khusro, A., Agastian, P., Kuppusamy, P., & Al Farraj, D. A. (2022). Synthesis of gold nanoparticles using bacterial cellulase and its role in saccharification and bioethanol production from aquatic weeds. *Journal of King Saud University-Science, 34*(4), 101974.

Abuzeid, H. M., Julien, C. M., Zhu, L., & Hashem, A. M. (2023). Green synthesis of nanoparticles and their energy storage, environmental, and biomedical applications. *Crystals, 13*(11), 1576.

Ahmadian-Fard-Fini, S., Ghanbari, D., Amiri, O., & Salavati-Niasari, M. (2020). Electrospinning of cellulose acetate nanofibers/Fe/carbon dot as photoluminescence sensor for mercury (II) and lead (II) ions. *Carbohydrate Polymers, 229*, 115428.

Ahmed, M. (2020). Nanomaterial synthesis. In *Polymer science and nanotechnology* (pp. 361–399). Elsevier.

Ahmed, S., Chaudhry, S. A., & Ikram, S. (2017). A review on biogenic synthesis of ZnO nanoparticles using plant extracts and microbes: A prospect towards green chemistry. *Journal of Photochemistry and Photobiology B: Biology, 166*, 272–284.

Ahmed, S. F., Debnath, J. C., Mehejabin, F., Islam, N., Tripura, R., Mofijur, M., Hoang, A. T., Rasul, M. G., & Vo, D.-V.N. (2023). Utilization of nanomaterials in accelerating the production process of sustainable biofuels. *Sustainable Energy Technologies and Assessments, 55*, 102894.

Ahmed, S. F., Mofijur, M., Rafa, N., Chowdhury, A. T., Chowdhury, S., Nahrin, M., Islam, A. B. M. S., & Ong, H. C. (2022). Green approaches in synthesising nanomaterials for environmental nanobioremediation: Technological advancements, applications, benefits and challenges. *Environmental Research, 204*, 111967.

Alao, I. I., Oyekunle, I. P., Iwuozor, K. O., & Emenike, E. C. (2022). Green synthesis of copper nanoparticles and investigation of its anti-microbial properties. *Advanced Journal of Chemistry-Section B, 4*(1), 39–52.

Al-Harbi, N., & Abd-Elrahman, N. K. (2024). Physical methods for preparation of nanomaterials, their characterization and applications: A review. *Journal of Umm Al-Qura University for Applied Sciences*, 1–22.

Alloosh, M. T., Khaddam, W. I., & Almuhammady, A. K. (2021). Biosynthesis of metal nanoparticles using microorganisms and its medicinal applications. *Novel Research in Microbiology Journal*, *5*(1), 1077–1090.

Alsaiari, M., Ahmad, M., Munir, M., Zafar, M., Sultana, S., Dawood, S., Almohana, A. I., MH, A.-M. H., Alharbi, A. F., & Ahmad, Z. (2023). Efficient application of newly synthesized green Bi2O3 nanoparticles for sustainable biodiesel production via membrane reactor. *Chemosphere*, *310*, 136838.

Anastas, P. T., & Warner, J. C. (2000). *Green chemistry: Theory and practice*. Oxford University Press.

Anekwe, I. M. S., Tetteh, E. K., Akpasi, S. O., & Isa, Y. M. (2024). Current Advances and Potentials of Nanotechnology for Biofuel Production. *Sustainable Engineering: Concepts and Practices*, 379–394.

Aragaw, B. A., Alula, M. T., Majoni, S., & King'ondu, C. K. (2022). Chemical synthesis of silver nanoparticles. In *Green Synthesis of Silver Nanomaterials* (pp. 21–53). Elsevier.

Arfin, T., & Athar, S. (2018). Graphene for advanced organic photovoltaics. *Nanomaterials: Biomedical, Environmental, and Engineering Applications*, 93–103.

Ariaeenejad, S., Jokar, F., Hadian, P., Ma'mani, L., Gharaghani, S., Fereidoonnezhad, M., & Salekdeh, G. H. (2020). An efficient nano-biocatalyst for lignocellulosic biomass hydrolysis: Xylanase immobilization on organically modified biogenic mesoporous silica nanoparticles. *International Journal of Biological Macromolecules*, *164*, 3462–3473.

Ariaeenejad, S., Motamedi, E., & Salekdeh, G. H. (2021). Immobilization of enzyme cocktails on dopamine functionalized magnetic cellulose nanocrystals to enhance sugar bioconversion: A biomass reusing loop. *Carbohydrate Polymers*, *256*, 117511.

Arredondo-Ochoa, T., & Silva-Martínez, G. A. (2022). Microemulsion based nanostructures for drug delivery. *Frontiers in Nanotechnology*, *3*, 753947.

Arum, C. T., Bbumba, S., Kigozi, M., Karume, I., Ssekatawa, J., & Kiganda, I. (2024). Synthesis of metallic nanomaterials using green chemistry approaches. *Asian Journal of Chemical Sciences*, *14*(5).

Arya, I., Poona, A., Dikshit, P. K., Pandit, S., Kumar, J., Singh, H. N., Jha, N. K., Rudayni, H. A., Chaudhary, A. A., & Kumar, S. (2021). Current trends and future prospects of nanotechnology in biofuel production. *Catalysts*, *11*(11), 1308.

Asmatulu, R., Khan, W. S., & Asmatulu, E. (2024). *Nanotechnology safety*. Elsevier.

Assad, H., Kaya, S., Kumar, P. S., Vo, D.-V.N., Sharma, A., & Kumar, A. (2022). Insights into the role of nanotechnology on the performance of biofuel cells and the production of viable biofuels: A review. *Fuel*, *323*, 124277.

Aswathi, V. P., Meera, S., Maria, C. G. A., & Nidhin, M. (2023). Green synthesis of nanoparticles from biodegradable waste extracts and their applications: A critical review. *Nanotechnology for Environmental Engineering*, *8*(2), 377–397.

Aziz, A. H. A., Engliman, N. S., Mansor, M. F., Abdul, P. M., Arisht, S. N., Jamali, N. S., & Tiang, M. F. (2022). Synergistic enhancement of biohydrogen production by supplementing with green synthesized magnetic iron nanoparticles using thermophilic mixed bacteria culture. *International Journal of Hydrogen Energy*, *47*(96), 40683–40695.

Bahari, N., Hashim, N., Abdan, K., Md Akim, A., Maringgal, B., & Al-Shdifat, L. (2023). Role of honey as a bifunctional reducing and capping/stabilizing agent: Application for silver and zinc oxide nanoparticles. *Nanomaterials*, *13*(7), 1244.

Baig, N., Kammakakam, I., & Falath, W. (2021). Nanomaterials: A review of synthesis methods, properties, recent progress, and challenges. *Materials Advances*, *2*(6), 1821–1871.

Balachandran, A., Sreenilayam, S. P., Madanan, K., Thomas, S., & Brabazon, D. (2022). Nanoparticle production via laser ablation synthesis in solution method and printed electronic application-A brief review. *Results in Engineering*, *16*, 100646.

Balashanmugam, P., Balakumaran, M. D., Murugan, R., Dhanapal, K., & Kalaichelvan, P. T. (2016). Phytogenic synthesis of silver nanoparticles, optimization and evaluation of in vitro antifungal activity against human and plant pathogens. *Microbiological Research*, *192*, 52–64.

Balasubramanian, S., Kala, S. M. J., & Pushparaj, T. L. (2020). Biogenic synthesis of gold nanoparticles using Jasminum auriculatum leaf extract and their catalytic, antimicrobial and anticancer activities. *Journal of Drug Delivery Science and Technology, 57*, 101620.

Banerjee, M., & Rajeswari, V. D. (2024). Green synthesis of selenium nanoparticles using leaf extract of Moringa oleifera, their biological applications, and effects on the growth of *Phaseolus vulgaris*: Agricultural synthetic biotechnology for sustainable nutrition. *Biocatalysis and Agricultural Biotechnology, 55*, 102978.

Banjara, R. A., Kumar, A., Aneshwari, R., Satnami, M. L., & Sinha, S. K. (2024). A comparative analysis of chemical vs green synthesis of nanoparticles and their various applications. *Environmental Nanotechnology, Monitoring & Management*, 100988.

Barhoum, A., García-Betancourt, M. L., Jeevanandam, J., Hussien, E. A., Mekkawy, S. A., Mostafa, M., Omran, M. M., S. Abdalla, M., & Bechelany, M. (2022). Review on natural, incidental, bioinspired, and engineered nanomaterials: history, definitions, classifications, synthesis, properties, market, toxicities, risks, and regulations. *Nanomaterials, 12*(2), 177.

Barhoum, A., Jeevanandam, J., Rastogi, A., Samyn, P., Boluk, Y., Dufresne, A., Danquah, M. K., & Bechelany, M. (2020). Plant celluloses, hemicelluloses, lignins, and volatile oils for the synthesis of nanoparticles and nanostructured materials. *Nanoscale, 12*(45), 22845–22890.

Batool, F., Iqbal, M. S., Khan, S.-U.-D., Khan, J., Ahmed, B., & Qadir, M. I. (2021). Biologically synthesized iron nanoparticles (FeNPs) from *Phoenix dactylifera* have anti-bacterial activities. *Scientific Reports, 11*(1), 22132.

Beluhan, S., Herceg, F., Leboš Pavunc, A., & Djaković, S. (2022). Preparation and structural properties of bacterial nanocellulose obtained from beetroot peel medium. *Energies, 15*(24), 9374.

Betancourt, M. L. G., Ramirez-Jimenez, S., Gonzalez-Hodges, A. N., Nunez-Salazar, Z., Garcia, I. L. E., & Aparicio, J. R. (2021). Low dimensional nanostructures: Measurement and remediation technologies applied to trace heavy metals in water. *Chapters*.

Biswal, T., & Shadangi, K. P. (2021). Application of nanotechnology in the production of biofuel. *Liquid Biofuels: Fundamentals, Characterization, and Applications*, 487–515.

Biswas, A., Bayer, I. S., Biris, A. S., Wang, T., Dervishi, E., & Faupel, F. (2012). Advances in top–down and bottom–up surface nanofabrication: Techniques, applications & future prospects. *Advances in Colloid and Interface Science, 170*(1–2), 2–27.

Bokov, D., Turki Jalil, A., Chupradit, S., Suksatan, W., Javed Ansari, M., Shewael, I. H., Valiev, G. H., & Kianfar, E. (2021). Nanomaterial by sol-gel method: Synthesis and application. *Advances in Materials Science and Engineering, 2021*(1), 5102014.

Boro, M., Verma, A. K., Chettri, D., Yata, V. K., & Verma, A. K. (2022). Strategies involved in biofuel production from agro-based lignocellulose biomass. *Environmental Technology & Innovation, 28*, 102679.

Buarki, F., AbuHassan, H., Al Hannan, F., & Henari, F. Z. (2022). Green synthesis of iron oxide nanoparticles using Hibiscus rosa sinensis flowers and their antibacterial activity. *Journal of Nanotechnology, 2022*(1), 5474645.

Cao, Y., Liu, T., Chen, T., Zhang, B., Jiang, D., & Xie, J. (2021). Revealing the etching process of water-soluble Au25 nanoclusters at the molecular level. *Nature Communications, 12*(1), 3212.

Cesca, K., Netto, M. S., Ely, V. L., Dotto, G. L., Foletto, E. L., & Hotza, D. (2020). Synthesis of spherical bacterial nanocellulose as a potential silver adsorption agent for antimicrobial purposes. *Cellulose Chemistry and Technology, 54*, 285–290.

Chandraker, S. K., Lal, M., Ghosh, M. K., Tiwari, V., Ghorai, T. K., & Shukla, R. (2020). Green synthesis of copper nanoparticles using leaf extract of *Ageratum houstonianum* Mill. and study of their photocatalytic and antibacterial activities. *Nano Express, 1*(1), 10033.

Chaudhary, R., Nawaz, K., Khan, A. K., Hano, C., Abbasi, B. H., & Anjum, S. (2020). An overview of the algae-mediated biosynthesis of nanoparticles and their biomedical applications. *Biomolecules, 10*(11), 1498.

Chen, Y.-Y., Jiang, W.-P., Chen, H.-L., Huang, H.-C., Huang, G.-J., Chiang, H.-M., Chang, C.-C., Huang, C.-L., & Juang, T.-Y. (2021). Cytotoxicity and cell imaging of six types of carbon nanodots prepared through carbonization and hydrothermal processing of natural plant materials. *RSC Advances, 11*(27), 16661–16674.

Cid, A. (2018). Synthesis of NPs by microemulsion method. *Microemulsion-A Chemical Nanoreactor.*

Contreras, J. E., Rodriguez, E. A., & Taha-Tijerina, J. (2017). Nanotechnology applications for electrical transformers—A review. *Electric Power Systems Research, 143*, 573–584.

Cui, S., Song, N., Shi, L., & Ding, P. (2020). Enhanced thermal conductivity of bioinspired nanofibrillated cellulose hybrid films based on graphene sheets and nanodiamonds. *ACS Sustainable Chemistry & Engineering, 8*(16), 6363–6370.

Devi, N., Kumar, R., Chen, Y.-S., & Singh, R. K. (2023). Carbon-based nanomaterials: carbon nanotube, fullerene, and carbon dots. In *Nanomaterials: Advances and Applications* (pp. 27–57). Springer.

Devi, A., Singh, A., Bajar, S., & Owamah, H. I. (2021). Nanomaterial in liquid biofuel production: Applications and current status. *Environmental Sustainability, 4*(2), 343–353.

Din, M. M. U., Batool, A., Ashraf, R. S., Yaqub, A., Rashid, A., & Din, N. M. U. (2024). Green synthesis and characterization of biologically synthesized and antibiotic-conjugated silver nanoparticles followed by post-synthesis assessment for antibacterial and antioxidant applications. *ACS Omega, 9*(17), 18909–18921.

Duan, H., Wang, D., & Li, Y. (2015). Green chemistry for nanoparticle synthesis. *Chemical Society Reviews, 44*(16), 5778–5792.

Ebad-Sichani, S., Sharafi-Badr, P., Hayati, P., & Soltanian-Fard, M. J. (2023). Nano-fabrication methods. In *Nanofabrication techniques-principles, processes and applications*. IntechOpen.

El-Hoseny, S. M., Basmaji, P., de Olyveira, G. M., Costa, L. M. M., Alwahedi, A. M., da Costa Oliveira, J. D., & Francozo, G. B. (2015). Natural ECM-bacterial cellulose wound healing—Dubai study. *Journal of Biomaterials and Nanobiotechnology, 6*(4), 237–246.

El-Seedi, H. R., El-Shabasy, R. M., Khalifa, S. A. M., Saeed, A., Shah, A., Shah, R., Iftikhar, F. J., Abdel-Daim, M. M., Omri, A., & Hajrahand, N. H. (2019). Metal nanoparticles fabricated by green chemistry using natural extracts: Biosynthesis, mechanisms, and applications. *RSC Advances, 9*(42), 24539–24559.

Feng, L., Xie, N., & Zhong, J. (2014). Carbon nanofibers and their composites: A review of synthesizing, properties and applications. *Materials, 7*(5), 3919–3945.

Folorunso, A., Akintelu, S., Oyebamiji, A. K., Ajayi, S., Abiola, B., Abdusalam, I., & Morakinyo, A. (2019). Biosynthesis, characterization and antimicrobial activity of gold nanoparticles from leaf extracts of Annona muricata. *Journal of Nanostructure in Chemistry, 9*, 111–117.

Foresti, M. L., Cerrutti, P., & Vazquez, A. (2015). Bacterial nanocellulose: Synthesis, properties and applications. *Polymer Nanocomposites Based on Inorganic and Organic Nanomaterials*, 39–61.

Gałuszka, A., Migaszewski, Z., & Namieśnik, J. (2013). The 12 principles of green analytical chemistry and the significance mnemonic of green analytical practices. *TrAC Trends in Analytical Chemistry, 50*, 78–84.

Gama, M., Gatenholm, P., & Klemm, D. (2012). *Bacterial nanocellulose: A sophisticated multifunctional material*. CRC Press.

Gopalakrishnan, V., & Muniraj, S. (2021). Neem flower extract assisted green synthesis of copper nanoparticles–optimisation, characterisation and anti-bacterial study. *Materials Today: Proceedings, 36*, 832–836.

Gulati, S., Sachdeva, M., & Bhasin, K. K. (2018). Capping agents in nanoparticle synthesis: Surfactant and solvent system. *AIP Conference Proceedings, 1953*(1).

Guo, K. W. (2020). Carbon nanotubes synthesized by green/eco-friendly technique potential for bioenergy applications. *Nanomaterials in Biofuels Research*, 251–274.

Gupta, D., Boora, A., Thakur, A., & Gupta, T. K. (2023). Green and sustainable synthesis of nanomaterials: Recent advancements and limitations. *Environmental Research, 231*, 116316.

Gupta, K., & Chundawat, T. S. (2020). Zinc oxide nanoparticles synthesized using *Fusarium oxysporum* to enhance bioethanol production from rice-straw. *Biomass and Bioenergy, 143*, 105840.

Hamedani, Y., Macha, P., Bunning, T. J., Naik, R. R., & Vasudev, M. C. (2016). Plasma-enhanced chemical vapor deposition: Where we are and the outlook for the future. *Chemical Vapor Deposition-Recent Advances and Applications in Optical, Solar Cells and Solid State Devices, 4*, 243–280.

Han, J., Feng, H., Wu, J., Li, Y., Zhou, Y., Wang, L., Luo, P., & Wang, Y. (2021). Construction of multienzyme co-immobilized hybrid nanoflowers for an efficient conversion of cellulose into glucose in a cascade reaction. *Journal of Agricultural and Food Chemistry, 69*(28), 7910–7921.

Harish, V., Ansari, M. M., Tewari, D., Gaur, M., Yadav, A. B., García-Betancourt, M.-L., Abdel-Haleem, F. M., Bechelany, M., & Barhoum, A. (2022). Nanoparticle and nanostructure synthesis and controlled growth methods. *Nanomaterials, 12*(18), 3226.

He, B., Yang, Y., Yuen, M. F., Chen, X. F., Lee, C. S., & Zhang, W. J. (2013). Vertical nanostructure arrays by plasma etching for applications in biology, energy, and electronics. *Nano Today, 8*(3), 265–289.

Hossain, N., Mahlia, T. M. I., & Saidur, R. (2019). Latest development in microalgae-biofuel production with nano-additives. *Biotechnology for Biofuels, 12*, 1–16.

Huang, J., Ma, X., Yang, G., & Alain, D. (2019). Introduction to nanocellulose. *Nanocellulose: From Fundamentals to Advanced Materials*, 1–20.

Hussain, R., Hasan, M., Iqbal, K. J., Zafar, A., Tariq, T., Saif, M. S., Hassan, S. G., Shu, X., Caprioli, G., & Anjum, S. I. (2023). Nano-managing silver and zinc as bio-conservational approach against pathogens of the honey bee. *Journal of Biotechnology, 365*, 1–10.

Huston, M., DeBella, M., DiBella, M., & Gupta, A. (2021). Green synthesis of nanomaterials. *Nanomaterials, 11*(8), 2130.

Huynh, K.-H., Pham, X.-H., Kim, J., Lee, S. H., Chang, H., Rho, W.-Y., & Jun, B.-H. (2020). Synthesis, properties, and biological applications of metallic alloy nanoparticles. *International Journal of Molecular Sciences, 21*(14), 5174.

International Energy Agency. (2024). *India could triple its biofuel use and accelerate global deployment.*

International Energy Agency. (2020). Renewables 2020. *Paris*. https://doi.org/10.1002/peng.20026

Jadoun, S., Arif, R., Jangid, N. K., & Meena, R. K. (2021). Green synthesis of nanoparticles using plant extracts: A review. *Environmental Chemistry Letters, 19*(1), 355–374.

Javed, R., Zia, M., Naz, S., Aisida, S. O., Ain, N. ul, & Ao, Q. (2020). Role of capping agents in the application of nanoparticles in biomedicine and environmental remediation: recent trends and future prospects. *Journal of Nanobiotechnology, 18*, 1–15.

Jeevanandam, J., Chan, Y. S., & Danquah, M. K. (2016). Biosynthesis of metal and metal oxide nanoparticles. *ChemBioEng Reviews, 3*(2), 55–67.

Jiang, W., Wang, X., Yang, J., Han, H., Li, Q., & Tang, J. (2018). Lipase-inorganic hybrid nanoflower constructed through biomimetic mineralization: A new support for biodiesel synthesis. *Journal of Colloid and Interface Science, 514*, 102–107.

Jin, Z., Dridi, N., Palui, G., Palomo, V., Jokerst, J. V., Dawson, P. E., Amy Sang, Q.-X., & Mattoussi, H. (2023). Evaluating the catalytic efficiency of the human membrane-type 1 matrix metalloproteinase (MMP-14) using AuNP–peptide conjugates. *Journal of the American Chemical Society, 145*(8), 4570–4582.

Joy, J., Krishnamoorthy, A., Tanna, A., Kamathe, V., Nagar, R., & Srinivasan, S. (2022). Recent developments on the synthesis of nanocomposite materials via ball milling approach for energy storage applications. *Applied Sciences, 12*(18), 9312.

Jume, B. H., Gabris, M. A., Nodeh, H. R., Rezania, S., & Cho, J. (2020). Biodiesel production from waste cooking oil using a novel heterogeneous catalyst based on graphene oxide doped metal oxide nanoparticles. *Renewable Energy, 162*, 2182–2189.

Kakakhel, M. A., Sajjad, W., Wu, F., Bibi, N., Shah, K., Yali, Z., & Wang, W. (2021). Green synthesis of silver nanoparticles and their shortcomings, animal blood a potential source for silver nanoparticles: A review. *Journal of Hazardous Materials Advances, 1*, 100005.

Kalpana, R., Vignesh, N. S., Vinothini, K., Rajan, M., Ashokkumar, B., Brindhadevi, K., Chi, N. T. L., Pugazhendhi, A., & Varalakshmi, P. (2023). Carbon quantum dots (CQD) fabricated from *Exiguobacterium* sp. VK2 exopolysaccharide (EPS) using hydrothermal reaction and its biodiesel applications. *Fuel, 333*, 126426.

Kandasamy, S., Devarayan, K., Bhuvanendran, N., Zhang, B., He, Z., Narayanan, M., Mathimani, T., Ravichandran, S., & Pugazhendhi, A. (2021). Accelerating the production of bio-oil from hydrothermal liquefaction of microalgae via recycled biochar-supported catalysts. *Journal of Environmental Chemical Engineering, 9*(4), 105321.

Kattimani, V. R., Yatish, K. V., Pramoda, K., Sakar, M., & Balakrishna, R. G. (2023). Acacia furnesiana plant as a novel green source for the synthesis of $NiFe_2O_4$ magnetic nanocatalyst and as feedstock for sustainable high quality biofuel production. *Fuel, 348*, 128549.

Kausar, A. (2022). Carbonaceous nanofillers in polymer matrix. *Polymeric Nanocomposites with Carbonaceous Nanofillers for Aerospace Applications*, 23.

Khan, F. A. (2020). Synthesis of nanomaterials: Methods & technology. *Applications of Nanomaterials in Human Health*, 15–21.

Khan, F., Shariq, M., Asif, M., Siddiqui, M. A., Malan, P., & Ahmad, F. (2022). Green nanotechnology: Plant-mediated nanoparticle synthesis and application. *Nanomaterials, 12*(4), 673.

Khodashenas, B., Ardjmand, M., Sharifzadeh Baei, M., Shokuhi Rad, A., & Akbarzadeh Khiyavi, A. (2020). Bovine serum albumin/gold nanoparticles as a drug delivery system for Curcumin: Experimental and computational studies. *Journal of Biomolecular Structure and Dynamics, 38*(15), 4644–4654.

Khoo, K. S., Chia, W. Y., Tang, D. Y. Y., Show, P. L., Chew, K. W., & Chen, W.-H. (2020). Nanomaterials utilization in biomass for biofuel and bioenergy production. *Energies, 13*(4), 892.

Kizhepat, S., Rasal, A. S., Chang, J.-Y., & Wu, H.-F. (2023). Development of two-dimensional functional nanomaterials for biosensor applications: Opportunities, challenges, and future prospects. *Nanomaterials, 13*(9), 1520.

Kotov, Y. A. (2003). Electric explosion of wires as a method for preparation of nanopowders. *Journal of Nanoparticle Research, 5*, 539–550.

Kousheh, S. A., Moradi, M., Tajik, H., & Molaei, R. (2020). Preparation of antimicrobial/ultraviolet protective bacterial nanocellulose film with carbon dots synthesized from lactic acid bacteria. *International Journal of Biological Macromolecules, 155*, 216–225.

Kryzhevich, D. S., Zolnikov, K. P., Korchuganov, A. V, & Psakhie, S. G. (2017). Nanopowder synthesis based on electric explosion technology. *AIP Conference Proceedings, 1893*(1).

Kumar, K. P., Paul, W., & Sharma, C. P. (2011). Green synthesis of gold nanoparticles with Zingiber officinale extract: Characterization and blood compatibility. *Process Biochemistry, 46*(10), 2007–2013.

Kumar, Y., Yogeshwar, P., Bajpai, S., Jaiswal, P., Yadav, S., Pathak, D. P., Sonker, M., & Tiwary, S. K. (2021). Nanomaterials: Stimulants for biofuels and renewables, yield and energy optimization. *Materials Advances, 2*(16), 5318–5343.

Kumari, R., Saini, A. K., Kumar, A., & Saini, R. V. (2020a). Apoptosis induction in lung and prostate cancer cells through silver nanoparticles synthesized from Pinus roxburghii bioactive fraction. *JBIC Journal of Biological Inorganic Chemistry, 25*, 23–37.

Kumari, S., Tehri, N., Gahlaut, A., & Hooda, V. (2020b). Actinomycetes mediated synthesis, characterization, and applications of metallic nanoparticles. *Inorganic and Nano-Metal Chemistry, 51*(10), 1386–1395.

Kuppusamy, P., Ichwan, S. J. A., Parine, N. R., Yusoff, M. M., Maniam, G. P., & Govindan, N. (2015). Intracellular biosynthesis of Au and Ag nanoparticles using ethanolic extract of *Brassica oleracea* L. and studies on their physicochemical and biological properties. *Journal of Environmental Sciences, 29*, 151–157.

Kushwaha, D., Upadhyay, S. N., & Mishra, P. K. (2018). Nanotechnology in bioethanol/biobutanol production. *Green Nanotechnology for Biofuel Production*, 115–127.

Lemoisson, F., & Froyen, L. (2005). Understanding and improving powder metallurgical processes. In *Fundamentals of metallurgy* (pp. 471–502). Elsevier.

Lerner, M., Pervikov, A., Glazkova, E., Rodkevich, N., & Toropkov, N. (2021). Electrical explosion synthesis, oxidation and sintering behavior of Ti-Al intermetallide powders. *Metals, 11*(5), 760.

Li, C., Zhou, L., Yang, H., Lv, R., Tian, P., Li, X., Zhang, Y., Chen, Z., & Lin, F. (2017). Self-assembled exopolysaccharide nanoparticles for bioremediation and green synthesis of noble metal nanoparticles. *ACS Applied Materials & Interfaces, 9*(27), 22808–22818.

Li, M., Xiong, P., Yan, F., Li, S., Ren, C., Yin, Z., Li, A., Li, H., Ji, X., & Zheng, Y. (2018). An overview of graphene-based hydroxyapatite composites for orthopedic applications. *Bioactive Materials, 3*(1), 1–18.

Mabrouk, M., Elkhooly, T. A., & Amer, S. K. (2021). Actinomycete strain type determines the monodispersity and antibacterial properties of biogenically synthesized silver nanoparticles. *Journal of Genetic Engineering and Biotechnology, 19*, 1–12.

Malik, M. A., Wani, M. Y., & Hashim, M. A. (2012). Microemulsion method: A novel route to synthesize organic and inorganic nanomaterials: 1st Nano Update. *Arabian Journal of Chemistry, 5*(4), 397–417.

Manisha, H., Velayudham, M., Kumara, B. N., Naveen, M. H., Shim, Y.-B., & Prasad, K. S. (2023). Revelation of fluorophore impurities among biocompatible blue fluorescent carbon nanodots derived from Hemigraphis alternata plant and bioimaging. *Carbon Letters, 33*(3), 931–946.

Manivasagan, P., Venkatesan, J., Sivakumar, K., & Kim, S.-K. (2016). Actinobacteria mediated synthesis of nanoparticles and their biological properties: A review. *Critical Reviews in Microbiology, 42*(2), 209–221.

Marişca, O. T., & Leopold, N. (2019). Anisotropic gold nanoparticle-cell interactions mediated by collagen. *Materials, 12*(7), 1131.

Marslin, G., Siram, K., Maqbool, Q., Selvakesavan, R. K., Kruszka, D., Kachlicki, P., & Franklin, G. (2018). Secondary metabolites in the green synthesis of metallic nanoparticles. *Materials, 11*(6), 940.

Mathivanan, K., Selva, R., Chandirika, J. U., Govindarajan, R. K., Srinivasan, R., Annadurai, G., & Duc, P. A. (2019). Biologically synthesized silver nanoparticles against pathogenic bacteria: Synthesis, calcination and characterization. *Biocatalysis and Agricultural Biotechnology, 22*, 101373.

Meena, P. R., Singh, A. P., & Tejavath, K. K. (2020). Biosynthesis of silver nanoparticles using *Cucumis prophetarum* aqueous leaf extract and their antibacterial and antiproliferative activity against cancer cell lines. *ACS Omega, 5*(10), 5520.

Moaddab, A., & Ghasemi, S. (2021). Green synthesis of silver/carbon dot nanoparticles from Malva Sylvestris for fluorescence determination of tetracycline. *Microchemical Journal, 166*, 106222.

Modan, E. M., & Plăiaşu, A. G. (2020). Advantages and disadvantages of chemical methods in the elaboration of nanomaterials. *The Annals of "Dunarea de Jos" University of Galati. Fascicle IX, Metallurgy and Materials Science, 43*(1), 53–60.

Moharrami, P., & Motamedi, E. (2020). Application of cellulose nanocrystals prepared from agricultural wastes for synthesis of starch-based hydrogel nanocomposites: Efficient and selective nanoadsorbent for removal of cationic dyes from water. *Bioresource Technology, 313*, 123661.

Murtaza, G., Ahmad, A., Ullah, M. R., Binyameen, M., Kausar, R., Fakhar, M., Amin, M., & Batool, S. (2024). *Innovations in synthetic techniques, characterization, and pharmaceutical applications of nanoparticles*.

Nasajpour-Esfahani, N., Dastan, D., Alizadeh, A. ad, Shirvanisamani, P., Rozati, M., Ricciardi, E., Lewis, B., Aphale, A., & Toghraie, D. (2023). A critical review on intrinsic conducting polymers and their applications. *Journal of Industrial and Engineering Chemistry, 125*, 14–37.

Nasir, S., Hussein, M. Z., Zainal, Z., & Yusof, N. A. (2018). Carbon-based nanomaterials/allotropes: A glimpse of their synthesis, properties and some applications. *Materials, 11*(2), 295.

Nathawat, R., Rathore, S. S., Kharangarh, P. R., Devi, R., & Kumari, A. (2023). Synthesis and application of carbon-based nanocomposite. In *Carbon nanomaterials and their nanocomposite-based chemiresistive gas sensors* (pp. 169–203). Elsevier.

Nawaz, M., Sliman, Y., Ercan, I., Lima-Tenório, M. K., Tenório-Neto, E. T., Kaewsaneha, C., & Elaissari, A. (2019). Magnetic and pH-responsive magnetic nanocarriers. In *Stimuli responsive polymeric nanocarriers for drug delivery applications* (pp. 37–85). Elsevier.

Negi, S., & Singh, V. (2018). Algae: A potential source for nanoparticle synthesis. *Journal of Applied and Natural Science, 10*(4), 1134–1140.

Nguyen, M. T., Pattanasattayavong, P., & Yonezawa, T. (2024). Detailed discussion on the structure of alloy nanoparticles synthesized via magnetron sputter deposition onto liquid poly (ethylene glycol). *Nanoscale Advances, 6*(7), 1822–1836.

Nguyen, N. T. T., Nguyen, L. M., Nguyen, T. T. T., Nguyen, T. T., Nguyen, D. T. C., & Tran, T. V. (2022). Formation, antimicrobial activity, and biomedical performance of plant-based nanoparticles: A review. *Environmental Chemistry Letters, 20*(4), 2531–2571.

Nieto-Maldonado, A., Bustos-Guadarrama, S., Espinoza-Gomez, H., Flores-López, L. Z., Ramirez-Acosta, K., Alonso-Nuñez, G., & Cadena-Nava, R. D. (2022). Green synthesis of copper nanoparticles using different plant extracts and their antibacterial activity. *Journal of Environmental Chemical Engineering, 10*(2), 107130.

Nizami, A.-S., & Rehan, M. (2018). Towards nanotechnology-based biofuel industry. *Biofuel Research Journal, 5*(2), 798.

Nowicki, M., & Wandelt, K. (2018). *Electrochemical scanning tunneling microscopy.*

Numata, Y., Sakata, T., Furukawa, H., & Tajima, K. (2015). Bacterial cellulose gels with high mechanical strength. *Materials Science and Engineering: C, 47*, 57–62.

Osman, A. I., Zhang, Y., Farghali, M., Rashwan, A. K., Eltaweil, A. S., Abd El-Monaem, E. M., Mohamed, I. M. A., Badr, M. M., Ihara, I., & Rooney, D. W. (2024). Synthesis of green nanoparticles for energy, biomedical, environmental, agricultural, and food applications: A review. *Environmental Chemistry Letters, 22*(2), 841–887.

Pal, K., Chakroborty, S., & Nath, N. (2022). Limitations of nanomaterials insights in green chemistry sustainable route: Review on novel applications. *Green Processing and Synthesis, 11*(1), 951–964.

Panahi, H. K. S., Hosseinzadeh-Bandbafha, H., Dehhaghi, M., Orooji, Y., Mahian, O., Shahbeik, H., Kiehbadroudinezhad, M., Kalam, M. A., Karimi-Maleh, H., & Jouzani, G. S. (2024). Nanotechnology applications in biodiesel processing and production: A comprehensive review. *Renewable and Sustainable Energy Reviews, 192*, 114219.

Pandi, N., Sonawane, S. H., & Kishore, K. A. (2021). Synthesis of cellulose nanocrystals (CNCs) from cotton using ultrasound-assisted acid hydrolysis. *Ultrasonics Sonochemistry, 70*, 105353.

Pandit, C., Banerjee, S., Pandit, S., Lahiri, D., Kumar, V., Chaubey, K. K., Al-Balushi, R., Al-Bahry, S., & Joshi, S. J. (2023). Recent advances and challenges in the utilization of nanomaterials in transesterification for biodiesel production. *Heliyon, 9*(4).

Pandiyan, S., Arumugam, L., Srirengan, S. P., Pitchan, R., Sevugan, P., Kannan, K., Pitchan, G., Hegde, T. A., & Gandhirajan, V. (2020). Biocompatible carbon quantum dots derived from sugarcane industrial wastes for effective nonlinear optical behavior and antimicrobial activity applications. *ACS Omega, 5*(47), 30363–30372.

Patil, N., Bhaskar, R., Vyavhare, V., Dhadge, R., & Patil, Y. (2021). Overview on methods of synthesis of nanoparticles. *International Journal of Current Pharmaceutical Research*, 11–16.

Pavlidis, I. V., Tsoufis, T., Enotiadis, A., Gournis, D., & Stamatis, H. (2010). Functionalized multi-wall carbon nanotubes for lipase immobilization. *Advanced Engineering Materials, 12*(5), B179–B183.

Perveen, S., Nadeem, R., ur Rehman, S., Afzal, N., Anjum, S., Noreen, S., Saeed, R., Amami, M., Al-Mijalli, S. H., & Iqbal, M. (2022). Green synthesis of iron (Fe) nanoparticles using *Plumeria obtusa* extract as a reducing and stabilizing agent: Antimicrobial, antioxidant and biocompatibility studies. *Arabian Journal of Chemistry, 15*(5), 103764.

Prakash, J., Kumari, N., Sharma, P., Godara, S. K., & Kumar, G. (2023). Potential applications of transition and rare-earth metal substituted magnesium nanoferrites. In *Magnetic nanoferrites and their composites* (pp. 141–164). Elsevier.

Priya, N., Kaur, K., & Sidhu, A. K. (2021). Green synthesis: An eco-friendly route for the synthesis of iron oxide nanoparticles. *Frontiers in Nanotechnology, 3*, 655062.

Pudovkin, M. S., Zelenikhin, P. V., Shtyreva, V., Morozov, O. A., Koryakovtseva, D. A., Pavlov, V. V., Osin, Y. N., Evtugyn, V. G., Akhmadeev, A. A., & Nizamutdinov, A. S. (2018). Coprecipitation method of synthesis, characterization, and cytotoxicity of Pr^{3+}: LaF_3 (CPr= 3, 7, 12, 20, 30%) nanoparticles. *Journal of Nanotechnology, 2018*(1), 8516498.

Rafiei-Sarmazdeh, Z., Zahedi-Dizaji, S. M., & Kang, A. K. (2019). *Two-dimensional nanomaterials*. IntechOpen London, UK.

Rakib-Uz-Zaman, S. M., Hoque Apu, E., Muntasir, M. N., Mowna, S. A., Khanom, M. G., Jahan, S. S., Akter, N. R., Khan, M. A., Shuborna, N. S., & Shams, S. M. (2022). Biosynthesis of silver nanoparticles from *Cymbopogon citratus* leaf extract and evaluation of their antimicrobial properties. *Challenges, 13*(1), 18.

Ramesh, D., Kiruthika, T., Prabha, B., Djanaguiraman, M., & Karthikeyan, S. (2020). Nanocatalysts and biofuels: Applications and future challenges. *Green Synthesis of Nanomaterials for Bioenergy Applications*, 1–22.

Randhawa, A., Dutta, S. D., Ganguly, K., Patil, T. V., Patel, D. K., & Lim, K.-T. (2022). A review of properties of nanocellulose, its synthesis, and potential in biomedical applications. *Applied Sciences, 12*(14), 7090.

Rao, N., Singh, R., & Bashambu, L. (2021). Carbon-based nanomaterials: Synthesis and prospective applications. *Materials Today: Proceedings, 44*, 608–614.

Raveendran, P., Fu, J., & Wallen, S. L. (2003). Completely "green" synthesis and stabilization of metal nanoparticles. *Journal of the American Chemical Society, 125*(46), 13940–13941.

Raveendran, P. T. V., & Renuka, N. K. (2022). Carbon dots as a sustainable alternative to plant extracts for the green synthesis of noble metal nanoparticles. *Environmental Nanotechnology, Monitoring & Management, 18*, 100676.

Rodríguez-Couto, S. (2019). Green nanotechnology for biofuel production. In *Sustainable approaches for biofuels production technologies: From current status to practical implementation* (pp. 73–82).

Rosebrock, C. D., Wriedt, T., Mädler, L., & Wegner, K. (2016). The role of microexplosions in flame spray synthesis for homogeneous nanopowders from low-cost metal precursors. *AIChE Journal, 62*(2), 381–391.

Roychoudhury, A. (2020). Yeast-mediated green synthesis of nanoparticles for biological applications. *Indian J Pharm Biol Res, 8*(03), 26–31.

Ruddaraju, L. K., Pammi, S. V. N., Sankar Guntuku, G., Padavala, V. S., & Kolapalli, V. R. M. (2020). A review on anti-bacterials to combat resistance: From ancient era of plants and metals to present and future perspectives of green nano technological combinations. *Asian Journal of Pharmaceutical Sciences, 15*(1), 42–59.

Sabouri, Z., Akbari, A., Hosseini, H. A., Khatami, M., & Darroudi, M. (2020). Egg white-mediated green synthesis of NiO nanoparticles and study of their cytotoxicity and photocatalytic activity. *Polyhedron, 178*, 114351.

Salah Eldin, A., Ahmed, D., Khalifa, D., Ahmed, M., Gamil, M., Mohamed, M. S., Alaa, N., & Kamal, R. (2024). Conversion of biomass wastes into biofuels: a sustainable technology. البحوث التطبيقية في العلوم والانسانيات, *1*(1), 27–40.

Salem, S. S., & Fouda, A. (2021). Green synthesis of metallic nanoparticles and their prospective biotechnological applications: An overview. *Biological Trace Element Research, 199*(1), 344–370.

Salvador, M., Gutiérrez, G., Noriega, S., Moyano, A., Blanco-López, M. C., & Matos, M. (2021). Microemulsion synthesis of superparamagnetic nanoparticles for bioapplications. *International Journal of Molecular Sciences, 22*(1), 427.

Sanaeimehr, Z., Javadi, I., & Namvar, F. (2018). Antiangiogenic and antiapoptotic effects of green-synthesized zinc oxide nanoparticles using Sargassum muticum algae extraction. *Cancer Nanotechnology, 9*, 1–16.
Saravanan, A., Kumar, P. S., Jeevanantham, S., Karishma, S., & Vo, D.-V.N. (2022). Recent advances and sustainable development of biofuels production from lignocellulosic biomass. *Bioresource Technology, 344*, 126203.
Sarkar, R. D., Singh, H. B., & Kalita, M. C. (2021). Enhanced lipid accumulation in microalgae through nanoparticle-mediated approach, for biodiesel production: A mini-review. *Heliyon, 7*(9).
Sathiyaraj, S., Suriyakala, G., Gandhi, A. D., Babujanarthanam, R., Almaary, K. S., Chen, T.-W., & Kaviyarasu, K. (2021). Biosynthesis, characterization, and antibacterial activity of gold nanoparticles. *Journal of Infection and Public Health, 14*(12), 1842–1847.
Saxena, A., Parveen, F., Hussain, A., Khubaib, M., & Ashfaque, M. (2024). Second-generation biorefineries: single platform for the conversion of lignocellulosic wastes to environmentally important biofuels. *Environmental Science and Pollution Research*, 1–32.
Seah, C.-M., Chai, S.-P., & Mohamed, A. R. (2011). Synthesis of aligned carbon nanotubes. *Carbon, 49*(14), 4613–4635.
Serrano-Bayona, R., Chu, C., Liu, P., & Roberts, W. L. (2023). Flame synthesis of carbon and metal-oxide nanoparticles: flame types, effects of combustion parameters on properties and measurement methods. *Materials, 16*(3), 1192.
Shaheen, T. I., El-Gamal, M. S., Desouky, S. E., Hassan, S. E. D., & Alemam, A. M. (2021). Benign production of AgNPs/bacterial nanocellulose for wound healing dress: Antioxidant, cytotoxicity and in vitro studies. *Journal of Cluster Science*, 1–17.
Shahid, M., Dumat, C., Khalid, S., Schreck, E., Xiong, T., & Niazi, N. K. (2017). Foliar heavy metal uptake, toxicity and detoxification in plants: A comparison of foliar and root metal uptake. *Journal of Hazardous Materials, 325*, 36–58.
Sharif, A., Khan, S. M., Gull, N., Rizwan, K., Munir, S., Shakeel, M., & Islam, A. (2024). Introduction to nanomaterials: A strategic tool for production of biofuel and bioenergy from biomass. In *Nanomaterials in biomass conversion* (pp. 57–84). Elsevier.
Sharma, S. K., Kumari, A., Dwivedi, V., Rai, P. K., & Gupta, M. (2020). Nanomaterials and its application in biofuel production. *Substrate Analysis for Effective Biofuels Production*, 263–272.
Sharma, R., & Kar, K. K. (2020). Characteristics of carbon nanofibers. *Handbook of Nanocomposite Supercapacitor Materials I: Characteristics*, 215–245.
Sharma, M., & Sharma, M. (2021). Fundamental and principles of green synthesis. In *Advances in green synthesis: Avenues and sustainability* (pp. 257–265). Springer.
Sharma, C., & Bhardwaj, N. K. (2019). Bacterial nanocellulose: Present status, biomedical applications and future perspectives. *Materials Science and Engineering: C, 104*, 109963.
Sheikhi, A. (2019). Emerging cellulose-based nanomaterials and nanocomposites. In *Nanomaterials and polymer nanocomposites* (pp. 307–351). Elsevier.
Shende, P., Kasture, P., & Gaud, R. S. (2018). Nanoflowers: The future trend of nanotechnology for multi-applications. *Artificial Cells, Nanomedicine, and Biotechnology, 46*(sup1), 413–422.
Shubhashree, K. R., Reddy, R., Gangula, A. K., Nagananda, G. S., Badiya, P. K., Ramamurthy, S. S., Aramwit, P., & Reddy, N. (2022). Green synthesis of copper nanoparticles using aqueous extracts from *Hyptis suaveolens* (L.). *Materials Chemistry and Physics, 280*, 125795.
Sidhu, A. K., Verma, N., & Kaushal, P. (2022). Role of biogenic capping agents in the synthesis of metallic nanoparticles and evaluation of their therapeutic potential. *Frontiers in Nanotechnology, 3*, 801620.
Sijabat, E. K., Nuruddin, A., Aditiawati, P., & Purwasasmita, B. S. (2020). Optimization on the synthesis of bacterial nano cellulose (BNC) from banana peel waste for water filter membrane applications. *Materials Research Express, 7*(5), 55010.

Singh, H., Desimone, M. F., Pandya, S., Jasani, S., George, N., Adnan, M., Aldarhami, A., Bazaid, A. S., & Alderhami, S. A. (2023). Revisiting the green synthesis of nanoparticles: uncovering influences of plant extracts as reducing agents for enhanced synthesis efficiency and its biomedical applications. *International Journal of Nanomedicine*, 4727–4750.

Singh, J., Dutta, T., Kim, K.-H., Rawat, M., Samddar, P., & Kumar, P. (2018). 'Green'synthesis of metals and their oxide nanoparticles: Applications for environmental remediation. *Journal of Nanobiotechnology, 16*, 1–24.

Sinitsyn, A. P., & Sinitsyna, O. A. (2021). Bioconversion of renewable plant biomass. Second-generation biofuels: Raw materials, biomass pretreatment, enzymes, processes, and cost analysis. *Biochemistry (Moscow), 86*, S166–S195.

Skalickova, S., Baron, M., & Sochor, J. (2017). Nanoparticles biosynthesized by yeast: A review of their application. *Kvasny Prumysl, 63*(6), 290–292.

Sofla, M. R. K., Brown, R. J., Tsuzuki, T., & Rainey, T. J. (2016). A comparison of cellulose nanocrystals and cellulose nanofibres extracted from bagasse using acid and ball milling methods. *Advances in Natural Sciences: Nanoscience and Nanotechnology, 7*(3), 35004.

Solero, G. A. G. (2017). Synthesis of nanoparticles through flame spray pyrolysis: Experimental apparatus and preliminary results. *Nanoscience and Nanotechnology, 7*(1), 21–25.

Somya, A., Varshney, A. P., & Khan, A. (2025). Green nanoparticles: Innovative solutions for sustainable wastewater treatment. In *Synthesizing and characterizing plant-mediated biocompatible metal nanoparticles* (pp. 91–112). IGI Global.

Srivastava, A., & Katiyar, A. (2022). Zinc oxide nanostructures. In *Ceramic science and engineering* (pp. 235–262). Elsevier.

Srivastava, M., Srivastava, N., Saeed, M., Mishra, P. K., Saeed, A., Gupta, V. K., & Malhotra, B. D. (2021). Bioinspired synthesis of iron-based nanomaterials for application in biofuels production: A new in-sight. *Renewable and Sustainable Energy Reviews, 147*, 111206.

Stegarescu, A., Lung, I., Leoştean, C., Kacso, I., Opriş, O., Lazăr, M. D., Copolovici, L., Guţoiu, S., Stan, M., & Popa, A. (2020). Green synthesis, characterization and test of MnO_2 nanoparticles as catalyst in biofuel production from grape residue and seeds oil. *Waste and Biomass Valorization, 11*, 5003–5013.

Suriyakala, G., Sathiyaraj, S., Babujanarthanam, R., Alarjani, K. M., Hussein, D. S., Rasheed, R. A., & Kanimozhi, K. (2022). Green synthesis of gold nanoparticles using *Jatropha integerrima* Jacq. flower extract and their antibacterial activity. *Journal of King Saud University-Science, 34*(3), 101830.

Teoh, W. Y., Amal, R., & Mädler, L. (2010). Flame spray pyrolysis: An enabling technology for nanoparticles design and fabrication. *Nanoscale, 2*(8), 1324–1347.

Vasantharaj, S., Sathiyavimal, S., Senthilkumar, P., LewisOscar, F., & Pugazhendhi, A. (2019). Biosynthesis of iron oxide nanoparticles using leaf extract of *Ruellia tuberosa*: Antimicrobial properties and their applications in photocatalytic degradation. *Journal of Photochemistry and Photobiology b: Biology, 192*, 74–82.

Vecchio, G., Galeone, A., Brunetti, V., Maiorano, G., Rizzello, L., Sabella, S., Cingolani, R., & Pompa, P. P. (2012). Mutagenic effects of gold nanoparticles induce aberrant phenotypes in *Drosophila melanogaster*. *Nanomedicine: Nanotechnology, Biology and Medicine, 8*(1), 1–7.

Villaverde-Cantizano, G., Laurenti, M., Rubio-Retama, J., & Contreras-Cáceres, R. (2021). *Reducing agents in colloidal nanoparticle synthesis—An introduction.*

Wahab, A., Muhammad, M., Ullah, S., Abdi, G., Shah, G. M., Zaman, W., & Ayaz, A. (2024). Agriculture and environmental management through nanotechnology: Eco-friendly nanomaterial synthesis for soil-plant systems, food safety, and sustainability. *Science of the Total Environment*, 171862.

Wang, S., Chen, D., Hong, Q., Gui, Y., Cao, Y., Ren, G., & Liang, Z. (2023). Surface functionalization of metal and metal oxide nanoparticles for dispersion and tribological applications-a review. *Journal of Molecular Liquids*, 122821.

Wang, J., Mu, X., & Sun, M. (2019a). The thermal, electrical and thermoelectric properties of graphene nanomaterials. *Nanomaterials, 9*(2), 218.

Wang, K., Ruan, J., Song, H., Zhang, J., Wo, Y., Guo, S., & Cui, D. (2011). Biocompatibility of graphene oxide. *Nanoscale Research Letters, 6*, 1–8.

Wang, Y., O'connor, D., Shen, Z., Lo, I. M. C., Tsang, D. C. W., Pehkonen, S., Pu, S., & Hou, D. (2019b). Green synthesis of nanoparticles for the remediation of contaminated waters and soils: Constituents, synthesizing methods, and influencing factors. *Journal of Cleaner Production, 226*, 540–549.

Williams, B. P., Qi, Z., Huang, W., & Tsung, C.-K. (2020). The impact of synthetic method on the catalytic application of intermetallic nanoparticles. *Nanoscale, 12*(36), 18545–18562.

Wu, J., Lin, Z., Weng, X., Owens, G., & Chen, Z. (2020). Removal mechanism of mitoxantrone by a green synthesized hybrid reduced graphene oxide@ iron nanoparticles. *Chemosphere, 246*, 125700.

Xiao, F., Yang, R., & Liu, Z. (2022). Active aluminum composites and their hydrogen generation via hydrolysis reaction: A review. *International Journal of Hydrogen Energy, 47*(1), 365–386.

Xie, J., Hse, C.-Y., Cornelis, F., Hu, T., Qi, J., & Shupe, T. F. (2016). Isolation and characterization of cellulose nanofibers from bamboo using microwave liquefaction combined with chemical treatment and ultrasonication. *Carbohydrate Polymers, 151*, 725–734.

Xie, T., Kao, W., Sun, L., Wang, J., Dai, G., & Li, Z. (2020). Preparation and characterization of self-matting waterborne polymer–An overview. *Progress in Organic Coatings, 142*, 105569.

Yadav, D., Amini, F., & Ehrmann, A. (2020). Recent advances in carbon nanofibers and their applications–a review. *European Polymer Journal, 138*, 109963.

Yahya, N., Lobnik, A., Turel, M., Urek, Š. K., & Košak, A. (2011). Nanostructured materials use in sensors: Their benefits and drawbacks. *Carbon and Oxide Nanostructures: Synthesis, Characterisation and Applications*, 307–354.

Yatish, K. V., Lalithamba, H. S., Suresh, R., & Latha, H. K. E. (2020). Ochrocarpus longifolius assisted green synthesis of $CaTiO_3$ nanoparticle for biodiesel production and its kinetic study. *Renewable Energy, 147*, 310–321.

Yatish, K. V., Prakash, R. M., Ningaraju, C., Sakar, M., GeethaBalakrishna, R., & Lalithamba, H. S. (2021). Terminalia chebula as a novel green source for the synthesis of copper oxide nanoparticles and as feedstock for biodiesel production and its application on diesel engine. *Energy, 215*, 119165.

Ying, S., Guan, Z., Ofoegbu, P. C., Clubb, P., Rico, C., He, F., & Hong, J. (2022). Green synthesis of nanoparticles: Current developments and limitations. *Environmental Technology & Innovation, 26*, 102336.

Yusefi, M., Soon, M.L.-K., Teow, S.-Y., Monchouguy, E. I., Neerooa, B. N. H. M., Izadiyan, Z., Jahangirian, H., Rafiee-Moghaddam, R., Webster, T. J., & Shameli, K. (2022). Fabrication of cellulose nanocrystals as potential anticancer drug delivery systems for colorectal cancer treatment. *International Journal of Biological Macromolecules, 199*, 372–385.

Zhang, L., Yang, X., Yin, Z., & Sun, L. (2022). A review on carbon quantum dots: Synthesis, photoluminescence mechanisms and applications. *Luminescence, 37*(10), 1612–1638.

Zhang, Q., Zhang, Y., Li, Y., Ding, P., Xu, S., & Cao, J. (2021). Green synthesis of magnetite nanoparticle and its regulatory effect on fermentative hydrogen production from lignocellulosic hydrolysate by Klebsiella sp. *International Journal of Hydrogen Energy, 46*(39), 20413–20424.

Zhang, S., Zeng, X., Matthews, D. T. A., Igartua, A., Rodriguez-Vidal, E., Contreras Fortes, J., Saenz de Viteri, V., Pagano, F., Wadman, B., & Wiklund, E. D. (2016). Selection of micro-fabrication techniques on stainless steel sheet for skin friction. *Friction, 4*, 89–104.

Zhong, L., Jiao, X., Hu, H., Shen, X., Zhao, J., Feng, Y., Li, C., Du, Y., Cui, J., & Jia, S. (2021). Activated magnetic lipase-inorganic hybrid nanoflowers: A highly active and recyclable nanobiocatalyst for biodiesel production. *Renewable Energy, 171*, 825–832.

Zhu, S., Wang, J., & Fan, W. (2015). Graphene-based catalysis for biomass conversion. *Catalysis Science & Technology, 5*(8), 3845–3858.

Chapter 8
Emerging Significance of Life Cycle Analysis in Biofuels Towards Paving the Road to Sustainability

Radhika Jaithaliya, Ayushi Soni, and Kakoli Dutt

Abstract The task of replacing conventional fossil fuels with greener, renewable and sustainable fuels is an uphill battle. In the past four decades, several biofuels like biogas, biodiesel, bioethanol, biooil and recently biohydrogen are being considered as potential replacements with evident technological, performance and market proofs. This chapter attempts to identify the suitability of biofuels with respect to both performance and environmental sustainability goals. Despite the developing market, there is an urgent need for other techno-economically feasible biofuel production, particularly in growing economies with increasing fuel or energy dependence. But to meet the goals of sustainable development, studies on technical innovations or economic and market suitability are not enough. This life cycle assessment (LCA) identifies all relevant parameters including substrate availability, process economics, energy consumption and release and environmental impacts. Thus, it is gaining momentum across the globe and being developed into policy documents. Identifying any form of biofuel without the LCA analysis leaves a significant lacuna with respect to its actual performance. In this chapter, the different aspects of LCA analysis for various biofuels is presented along with a study on the factors influencing it and its significance.

Keywords Life cycle assessment · Energy balance · Life cycle inventory · Greenhouse gas · Balance · Biofuel · Systemic boundaries

---

R. Jaithaliya
Department of Bioscience and Biotechnology, Banasthali Vidyapith, P.O. Banasthali Vidyapith 304022, Rajasthan, India

A. Soni · K. Dutt (✉)
Department of Life Sciences, J.C. Bose University of Science and Technology, YMCA, Faridabad 121006, Haryana, India
e-mail: kakolidutt@jcboseust.ac.in

V. Singh et al. (eds.), *Bioenergy and Value Added Byproducts from Biomass Waste*, Sustainable Environmental Waste Management Strategies,
https://doi.org/10.1007/978-3-032-22849-9_8

8.1 Introduction

Fossil fuels share in the global energy market has declined from 82% to 80% between 2013 and 2023 which shows promise for the steady shift to alternative fuels. In the past half-decade, COVID-19, geopolitical instability and raising concerns over climate change along with disruptions in fuel supply has exacerbated energy security risks leading to further emphasis on alternative energy sources. Additionally, with a 15% rise in overall energy demand, such energy sources, like renewables, nuclear, and low-emission fuels have contributed to 40% of the overall requirement. The liquid biofuel production has expanded to 2.3 million barrels of oil equivalent per day while biomethane supply reached 10 billion cubic meters in 2022 (Muazu et al., 2021). Biofuels constitute a diverse category of liquid, solid, and gaseous fuels derived from biological sources, such as biodiesel, Fischer-Tropsch fuels, alcohols (methanol, ethanol, butanol), and biogas etc. (Muazu et al., 2021). To meet the goals of sustainability it is imperative to increase the share of the alternative energy sources and their output. In this, enhancing the biofuels is highly desirable as it will serve in valorization of waste, utilize non arable land and waste water with potential of serving as carbon sinks (Hassan et al., 2021).

Despite the potential of biofuels, the first-generation biofuels faced criticism concerning food security and low energy yield issues while second-generation biofuels overcame these hurdles by avoiding food competition and reinforcing energy efficiency. The third-generation biofuels, emphasizing algae cultivation contribute to enhance sustainability, high yields, and the possibility to use non-arable land and waste water. Since the biofuels essentially trap the solar energy and convert it to different precursors of fuel, they are inherently linked with the different parts of environmental biogeochemical cycles. Additionally, these biofuels upon usage have the capacity to affect the environment to some extent.

In the current scenario, ability of biofuels to garner the major energy market is dependent not only on the economic sustainability but their intrinsic environmental sustainability. These biofuels act as CO_2 traps and the biomass generated is then converted into energy which upon combustion will release CO_2 and other gases. Therefore, in context of global warming, a major concern is the effect of biofuel production and combustion on current green GHG values. With respect to environmental impact of biofuel production, three conversion modes have been proposed. Firstly, conventional production mode (CPM) which focuses on economic benefit in conversion process and environment benefit is neglected. Second mode is cogeneration mode (CGM) in which coproduct formed in conversion of biofuel is utilized for production of electricity and heat. For example, in Japan, biogas formed in bioethanol production is used for electricity (24%) and heat (50%) generation (Zhang et al., 2017). Next is circular economy mode (CEM), waste product are recycled aiming for reduction of emission and saving energy as, biogas produced is used for biofertilizers production and CO_2 is compressed and liquified for further consumption (Zhang et al., 2017).

Life cycle assessment (LCA) is a methodology (Fig. 8.1) to evaluate environmental impacts, energy consumption, resource depletion, and other impacts for an entire product system (Sharma et al., 2023). Standards to guide the conduct of LCA have been developed and published by the International Organization for Standardization (ISO 14040-14,049). Generally, LCA allows decision makers in industry and government to take informed decisions for the best product or technology alternative to satisfy a particular customer or societal need in an environmentally sound manner. This now been extended to biofuels where researchers have employed numerous tools for the sustainability assessment of biofuels like energy analysis, renewability and exergy analysis, and life cycle assessment (LCA). The exergy analysis integrates the second law of thermodynamics and provides a more sophisticated evaluation by assessing energy quality and determining thermodynamic inefficiencies. The LCA technique, is a standardized environmental tool analyzing the entire life cycle stages of biofuel production, including all the inputs and outputs, from cradle to grave. This LCA analysis is significant as it impacts biofuel development from its production, to scale up while extending to sales, procurement and supply chain along with implementation strategies. The LCA study serves as an essential tool for addressing a completely sustainable scenario for economic, environmental, and social factors. The combination of exergy analysis with LCA facilitates the exergo economic (identifies thermodynamic limitation of a system) and exergo environmental (approaches towards understanding the formation in chemical or energy conversion system) analyses (Singh et al., 2013). The ISO 14040 and EU describes four LCA phases (Tahir et al., 2021):

1. Objectives and scope: In this the goals for the study are defined along with the necessity of the study
2. Life-cycle inventory (LCI) analysis: This focusses towards analysis of input and output parameters and the production pathway for the biofuel in study
3. Life cycle impact assessment: This is the evaluation of the following key environmental impacts

   (i) Abiotic depletion potential (ADP)
   (ii) Quantifying resource depletion (QRD)
   (iii) Eutrophication potential (ETP) assessing the possibility of nutrient pollution resulting in excessive plant growth in the aquatic ecosystem
   (iv) Human toxicity potential (HTP) calculating potential hazard to human health
   (v) Global warming potential (GWP) addressing greenhouse gas (GHG) emissions
   (vi) Acidification potential (ACP) estimating the contribution to acid rain.

4. Interpretation of results: This is the final step where the entire data set is analysed for the possible environmental effects including GHG emission.

The LCA is gained relevance in biofuel research for establishment of environmental performance against traditional fuels (gasoline, diesel), while giving insights on the influence of global warming through government-developed GHG calculators aiding biofuel policies (Rocha et al., 2014). The LCA frameworks are of two basic

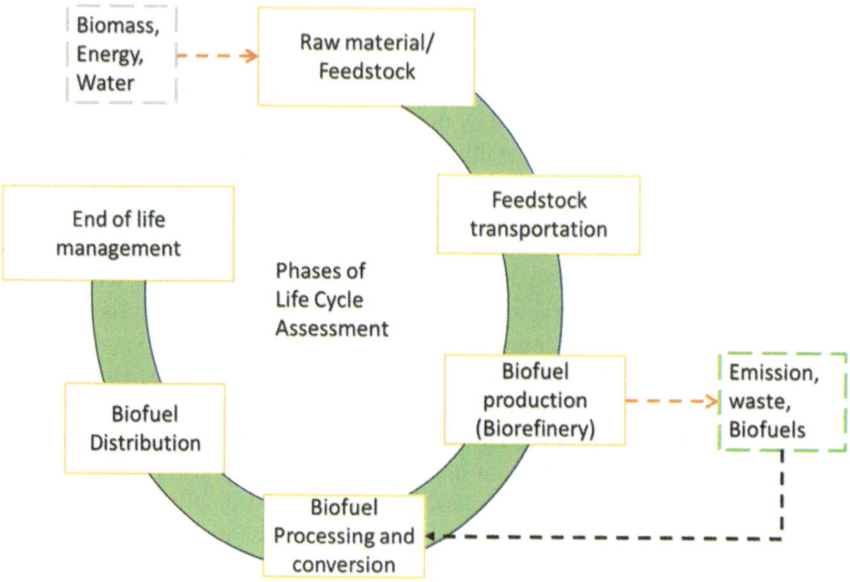

Fig. 8.1 Systematic represents the phases of life cycle assessment

types i.e. attributional and consequential. Attributional LCA assesses contemporary environmental burdens, recognizes the hotspots, and simplifies the comparisons with fossil fuels. It supports decision-making for policymakers, producers, and consumers. Consequential LCA on the other hand, evaluates the impacts of policy changes on the environment considering economy-wide effects through the incorporation of economic models (Gheewala, 2023).

The expansion of biofuel, by rising demand for food crops, has been connected with inflation of food prices which displaces food crops, resulting in both direct and indirect land-use changes. Direct land-use change (dLUC) occurs where biofuel feedstocks are cultivated, while indirect land-use change (iLUC) influences other regions globally, since the land is converted from forests, pastures, and peatlands to agricultural land to satisfy the increased demand for agri-products (Aoun & Gabrielle, 2017).

Defining the goal and scope is crucial in LCA to establish the objective of the study. At this stage, it is crucial to define the system boundaries and functional units. The functional units quantify the product performance which typically concentrates on bioenergy units or the biofuel production amounts. The system boundaries usually include biomass cultivation (Phase 1), conversion approaches with their environmental impacts (Phase 2), and ecological impacts of emissions along with co-products generated (Phase 3). Consequently, almost 25% of studies comprehensively estimate the whole biofuel lifecycle, including Phase 3 while 90% assessed only Phase 1 and Phase 2. Therefore, a holistic research approach should be followed to address this crucial gap and achieve the United Nations sustainable development goals (SDGs) (Osman et al., 2021).

8.2 Significance of LCA

Establishment of commercial utilization of bio-based resources is closed related to bioeconomy. Calculation will take into considerations the impact of harvesting the biomass as removing them from the environment may add to certain environmental disbalances for eg. generating land for plantation of crops for biofuel by removing original vegetation. This can challenge the existing trophic system as well as the food web. Consequently, monitoring, assessing, and forecasting ecological impacts are important to minimize negative influences on natural capital as the bioeconomy grows (Giuntoli et al., 2019).

LCA in biofuel research supports three major purposes viz (Giuntoli et al., 2019).

(i) Comparative analysis that deals with the comparison of biofuels with fossil fuels to determine the environmental performance, usually driven by policy requirements (such as GHG calculators etc.)
(ii) Impact assessment that identifies critical environmental influences of biofuels without direct comparison which are essential for research on novel feedstocks and procedures
(iii) Hotspot identification, pinpointing the hotspots which require transformations in biofuel supply chain to instruct R&D and therefore enhance sustainability.

The biofuel LCAs usually emphasize on climate impact of emission of greenhouse gases such as CO_2, CH_4, and N_2O. However, focus on temporary climate forcers like ozone-depleting substances, aerosols and their precursors should also be addressed. Significant variations occur in reports of global warming potentials for biofuels due to different methodological approaches, like the addressing iLUC, ecosystem carbon stock changes (comparing life cycle CO_2 emission equivalents with fossil fuels), and consideration of indirect N_2O emissions which is generated due to conversion of ammonia and nitrate leached in different ecological compartments (Hélias, 2019). For instance, biofuels derived from food crops can exhibit higher CO_2 equivalents than fossil fuels when considering these factors. Considering emerging biofuels such as microalgal biodiesel, LCA studies are acting as a promising potential for development of biofuels with significant GHG reductions. LCA studies provide insights on the importance on the selection of innovative feedstock, developed production processes, and carbon capture technologies in enhancing their environmental benefits (Tu et al., 2017).

Biofuel production, besides a promising candidate for renewable energy sources, also presents substantial water consumption problems. The studies indicate that biofuels such as biodiesel and bioethanol have significantly high green (rain water used by plant) and blue water (surface and ground water used by plants) footprints in comparison to fossil fuels, providing opportunities for innovation and sustainability to address water scarcity challenges (Pradinaud et al., 2019). The environmental impact broadens beyond water volume to overcome water scarcity which influences human health, ecosystems, and biodiversity. The ISO standards for water foot printing focus on the comprehensive evaluation, addressing the availability and scarcity of freshwaters, and ecological impacts for enhancing the reliability and importance of biofuel LCAs (Faist Emmenegger et al., 2011).

LCAs assess the environmental burdens of biofuel products which aids in informed alternatives. However, by transparently reporting assumptions, sensitivity analyses, and confidence levels, relative biofuel LCAs significantly enhances decision-making (Agostini et al., 2020). LCA assists identification of environmental trade-offs. Since forest-derived and food-based biofuels may have high EROI (Energetic Return on Energy Invested), they can display higher environmental influences compared to natural gas, especially in terms of eutrophication, water usage, and acidification (Havukainen et al., 2018). EROI basically is the ratio of the amount of energy that is produced by the system or technology compared to the amount of energy used to create it. It characterizes the economic efficiency of an energy source (Zhang et al., 2017) Thus, EROI is an excellent tool to evaluate in a comparative manner the efficacy of various fuels. This can be related to energy difference of a system and represented as net energy analysis (NEA). Ultimately it is connected to LCA as it will represent total energy inputs for a system (Gupta et al., 2018). In this context, the forest derived and food-based biofuels show better energy conversion proving their efficacy and sustainable use.

In a study by Wang et al. (2021), a comparative analysis of EROI was presented (Fig. 8.2) which shows that biofuels had higher average EROI as against fossil fuels. Even the pathways used had an impact on EROI with pyrolysis being most efficient. Considering factors such as driving conditions, engine type, and fuel blend composition, LCA guides to evaluate the trade-offs amongst different biofuels. Although lower blends offer a decline in CO and PM emissions, higher blends might increase acetaldehyde emissions. These insights help in the transformation and optimization of biofuel technologies for improved air quality (Jeswani et al., 2020). Therefore, the LCA approach contributes significantly to the development of biofuel such as co-designing which optimizes processes for minimal negative impact on the environment.

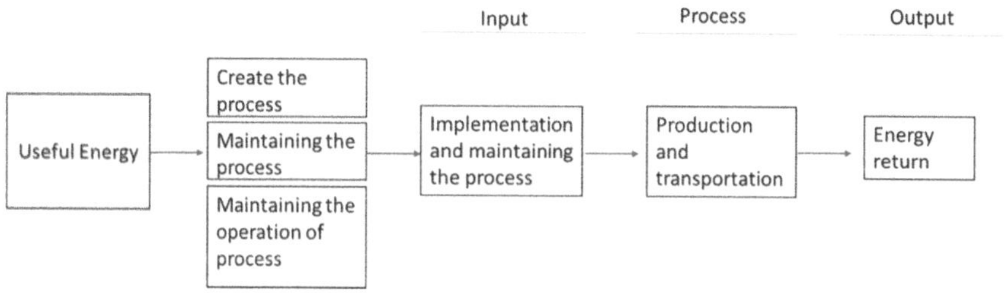

Fig. 8.2 Parameters involve in EORI calculation

8.3 Life Cycle Assessment of Bioethanol

Bioethanol is majorly produced from substrates namely, starch-based biomass, sugar-based biomass, and lignocellulosic biomass (Selvakumari et al. 2020). The production mainly depends on the feed stock type, design, conversion process, and local condition (Robak & Balcerek et al., 2018; Rezania et al., 2020; Devi et al., 2023). The process uses less energy and is cost efficient as it includes multistage conversion of biomass for breakdown of larger polymeric compounds such as lignin and cellulose (Stolarski et al., 2015; Morales et al., 2015). The algal biomass (microalgae, seaweed) is also used as a source for bioethanol production and is more efficient in terms of biofuel production, energy utilization, and CO_2 absorption during production (Jambo et al., 2016). Production of bioethanol majorly comprises of four stages i.e. (a) feedstock handling and pretreatment, (b) saccharification, (c) fermentation, and (d) distillation (Fig. 8.3) (Aditiya et al., 2016, Jambo et al., 2016). The type of feedstock used determines the conversion process, which is reflected in the LCA results. For example, lignin rich feedstock required more effective pretreatment resulting in higher energy consumption. So, pretreatment has been identified as the strategic stage for seamless recovery of bioethanol because it can influence other subsequent stages of bioconversion of biomass (Igwebuike et al., 2024). Thus, modified pretreatment techniques can significantly decrease environmental and economic impact eventually aiding in reduction of greenhouse gases (GHG) emission and global warming potential (GWP) (Igwebuike et al., 2024; Scacchi et al., 2010).

Various system boundaries have been studied by researchers for LCA of bioethanol production including cradle-to-gate, cradle-to-grave, and gate-to-gate depending upon biomass used in production (Table 8.1) (Rajaeifar et al., 2019). Zhang et al. (2017) concludes that the changes in boundaries of LCA results in changing the system boundary from the pretreatment stage to whole bioethanol production plant. In recent years, a wide range of LCA studies has been conducted worldwide to measure the bioethanol production. The difference in methodology hypothesis, technique results in variation in LCA outcomes (Igwebuike et al., 2024).

US is the world's dominant producer of bioethanol with 54% production worldwide followed by China, Brazil, European Union, and India (Kumar et al., 2024). Transportation sector of US is dominated by blended bioethanol (E30, E85) (Duffield et al., 2015). Clean Air Act authorizes that Environmental Protection Act (EPA) establish the annual volume requirement for renewable fuel standards (RFS) as E30 (30% bioethanol) and E85 (85% bioethanol) (Bracmort & Cowan et al., 2017). US utilizes high starch content feedstock such as sorghum and corn, among which sweet sorghum is evolving more in bioethanol production due to its high fermentable sugar extraction (Appiah-Nkansah et al., 2019). China is the second largest country in production of 2G bioethanol from its existing 12 plants with an annual capacity of 340 million tonnes utilizing renewable sources as feedstock. Chinese government in 2017, implemented legislation of employment of ethanol as a fuel, aiming with 10% blending (Hassan et al., 2021).

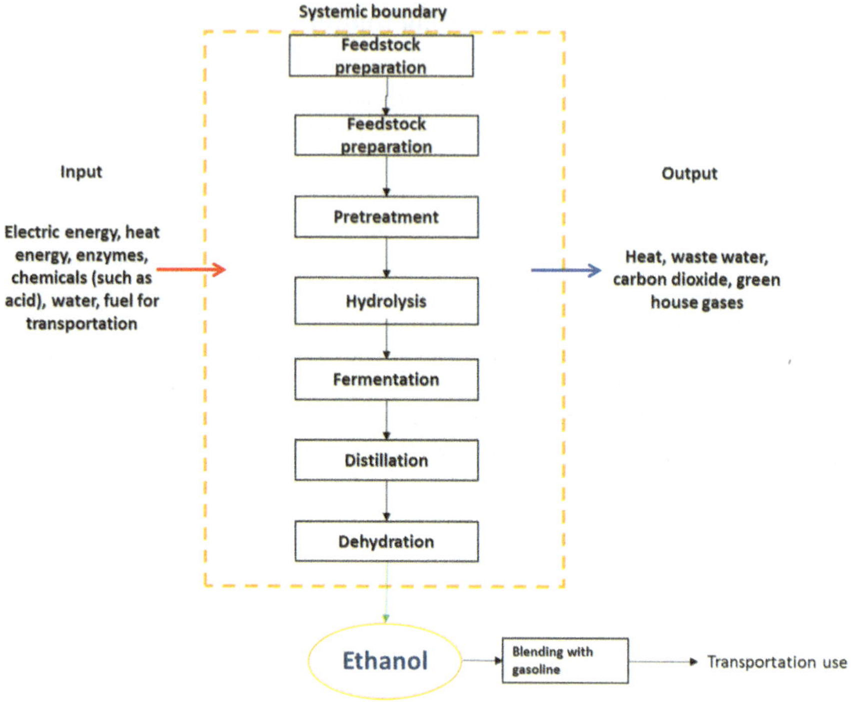

Fig. 8.3 The bioethanol production from the first, second and third generation feedstock

Table 8.1 System boundaries for LCA of bioethanol production

System boundaries	Description
Cradle-to-gate	Assessing a product from raw material to factory production facility
Cradle-to-grave	Assessing a life cycle of raw material to final disposal stage
Gate-to-gate	Assessing only the production process

Brazil is one of the steadiest countries with respect to bioethanol production from sugarcane and its use as vehicle fuel since 1970s (de Souza Abud & de Farias Silva et al., 2019). Two cases have been studied for LCA study by researchers. First, in which production is by sugarcane and electricity by bagasse (by product), and a second case involving bioethanol production from both sugarcane and bagasse along with electricity from process waste. Latter has been shown to have high efficiency and hence is economically appealing whereas the first case exhibits lower GHG emission (Santoyo-Castelazo et al., 2023).

For analyzing LCA of bioethanol production by molasses in India, researchers performed cradle-to-grave analysis. LCA comparison was also conducted, in which sugarcane base ethanol was taken into consideration which showed that northern regions (10.8%) of India had higher GHG emission as compared to western regions

(24.2%) on account of high energy consumption in sugarcane cultivation, production and molasses transportation in northern regions than western regions (Anand & Kaushal et al., 2021).

Indian government set in motion a national program on biofuel in 2001, wherein the government launched its ethanol-blend petrol program (EBPP) in 2003 and directed in 2009 the implementation of 10% bioethanol blending in gasoline across 20 states and 4 Union territories. Due to the energy security challenges, the government aimed for 20% blending with gasoline by 2017 across country. Data showed that minimal 3% blending was achieved which was planned for 20% by 2017. This objective was pushed back due to inadequate availability of infrastructure, distillation technology and advanced methodology. In the 21st conference of the Parties (COP21) of Convention of Biological Diversity held in 2015, Indian policymakers rejuvenated the blending mission of ethanol and aimed it for 20% ethanol blending by 2030. National policy on biofuels has been targeted for production of 1G, 2G and 3G bioethanol for reducing nation's dependence on import to meet its energy requirement. India has aimed for 5% blend of bioethanol by 2030 (Chandel et al., 2024). According to USDA report (Biofuel Annual, IN2024–0024), in India ethanol use for fuel consumption is 6245 million liters in 2023 (Fig. 8.4). Market penetrance of blend rate% in 2023 is 12.0 (million liters) which is projected for 11.5 (Million liters) in 2024.

Well-to-Tank analysis was also conducted in Nepal of molasses-based ethanol production (Anand & Kaushal et al., 2021). Thai renewable energy policy (Thailand) gives higher position to the usage of gasohol (Ethyl or Grain alcohol), a fuel for transportation with blending of 10% bioethanol and 90% gasoline. Molasses and cassava are the prominent source for bioethanol production in Thailand (Norasethasopon et al., 2010). According to renewable fuel association (RFA), currently United States holds first position in ethanol production with 15,580 Mgal following by Brazil with 8470 Mgal and India with 1510 Mgal by 2023.

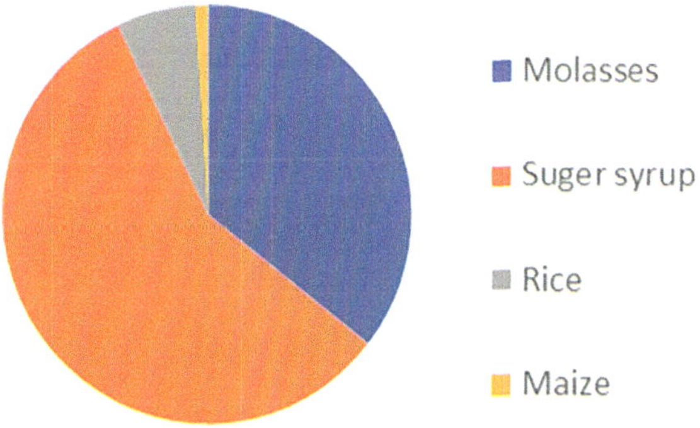

Fig. 8.4 Feedstock used for fuel ethanol production (million tonnes)

8.4 Life Cycle Assessment of Biohydrogen

Biohydrogen is a clean energy source which is being extensively used in various sectors as energy source. Among alternative fuels, hydrogen (H_2) is viewed as an important energy carrier of the future. It has the highest energy content per unit weight of any known fuel (122 MJ kg1 H_2), and it is the only fuel that is not chemically bound to carbon (Djomo & Blumberga, 2011). Various countries have launched various programs to promote the production and consumption of biohydrogen for vehicle fuel etc (Fig. 8.5).

Various routes have been studied for hydrogen production which majorly include, Green H_2 and Blue H_2 production. Blue H_2 may use fossil fuels as its feedstock for hydrogen production (Amaya-Santos et al., 2021). Biohydrogen can be produced through various sources such as water, coal, methane, algal biomass, and food and agriculture waste (Holladay et al., 2009; Amaya-Santos et al., 2021). Majorly two models have been used: steam methane reforming (SMR) and autothermal reforming (ATR) (Holladay et al., 2009). SMR uses natural gas (methane) feedstock supply and is more energy efficient than ATR model (Oni et al., 2022). These models are efficient when they are combined with strategy of Carbon Capture and Storage (CCS) (involves capturing CO_2 at emission sources with subsurface storage) (Muradov & Veziroğlu et al., 2005). In these techniques, natural gas reacts with steam and oxygen resulting in formation of hydrogen and higher carbon dioxide, carbon monoxide and other radicals, which negatively impact the environment (higher GHG emission and GWP) (Aasberg-Petersen et al., 2011). In some cases, carbon dioxide is captured by liquefication and compression method (SMR) and while in ATR is emitted to the

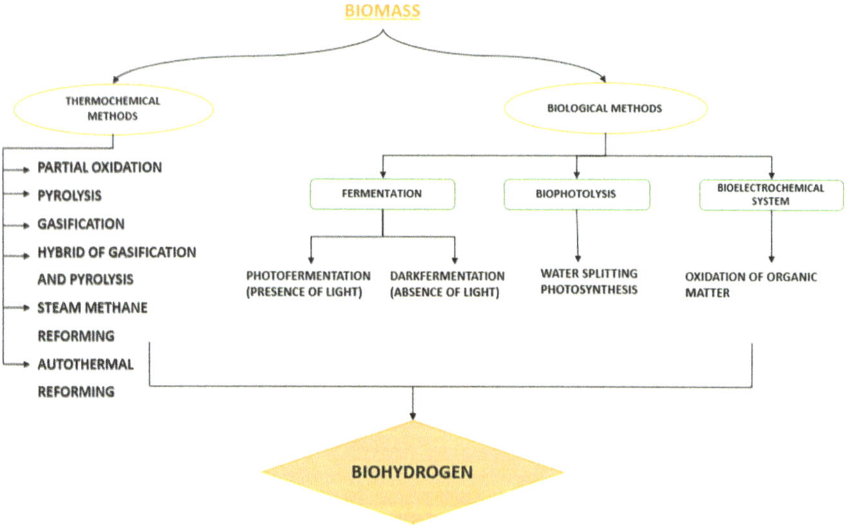

Fig. 8.5 The biomass is used by two pathways for the biohydrogen production

environment (Wang et al., 2024). Greenhouse gas emission for natural gas ranges between 22 g CO_2-eq.per MJ and 48 g CO_2-eq.per MJ when combined with CCS model and 90 g CO_2-eq.per MJ without CCS model (Amaya-Santos et al., 2021).

Green-H_2 uses electrolysis method for hydrogen production. Renewable source (water) is used, which accounts for 4% of total global hydrogen production. Green-H_2 is majorly applied in electric vehicle fuel cell, aviation and maritime industries and refineries (Lou et al., 2023; Lazaro & Hallenbeck et al., 2019). Electrolysis is divided further in two methods namely, Alkaline Water Electrolysis (AWE), which uses electrolyte such as sodium hydroxide, potassium hydroxide and sodium chloride (Dash et al., 2024) and the other one being Proton Exchange Membrane (PEM) electrolysis which employs proton exchange membrane for selective transfer of proton upon water breakdown to form hydrogen (Kumar & Himabindu, 2019). Green hydrogen is an emergent alternative for fossil-based fuel due to its energy efficiency and lesser environment impact. It has a drawback of high electricity input (~ 4974 MJ) but the oxygen produced is emitted to environment which has zero impact (Amaya-Santos et al., 2021).

Biohydrogen majorly uses biomass which is renewable such as agricultural waste, food waste and microalgae (Sharma et al., 2023). Biomass is targeted for pretreatment processing for breakdown of complex substances and production of fermentable compounds which is converted to biohydrogen by dark fermentation or photo fermentation (Dash et al., 2024; Jin et al., 2024). Dark fermentation is done by obligatory or facultative anaerobic bacteria for example *E. coli* which can convert organic substrates into hydrogen and organic acids in the absence of light. Various pretreatment techniques, such as heat, chemical, or pH treatments are introduced to prevent contamination which may lead to methane production (Sanghvi et al., 2024). In photo fermentation light works as energy source using photosynthetic bacteria with nitrogenase activity convert organic matter into hydrogen and carbon dioxide (Sanghvi et al., 2024). Various methods of biohydrogen production are represented in Fig. 8.6.

Indian government launched National Green Hydrogen Mission for energy transition and switching to renewable sources for energy production (Singh et al., 2024). This targets for making India a Global Hub for production and utilization of green hydrogen and aims for zero dependence for energy by 2047. India has announced the mission of Net Zero Emission by 2070, hence, launched National Green Hydrogen Mission for production target of 5 MMT (Million Metric Tonne) of Green Hydrogen with disbursement of Rs. 19,744 crores. This mission primarily aims for efficient green hydrogen production via electrolysis or other biochemical pathways along with its consumption for transportation and other industrial sectors by 2030 (Srinivasan et al., 2023).

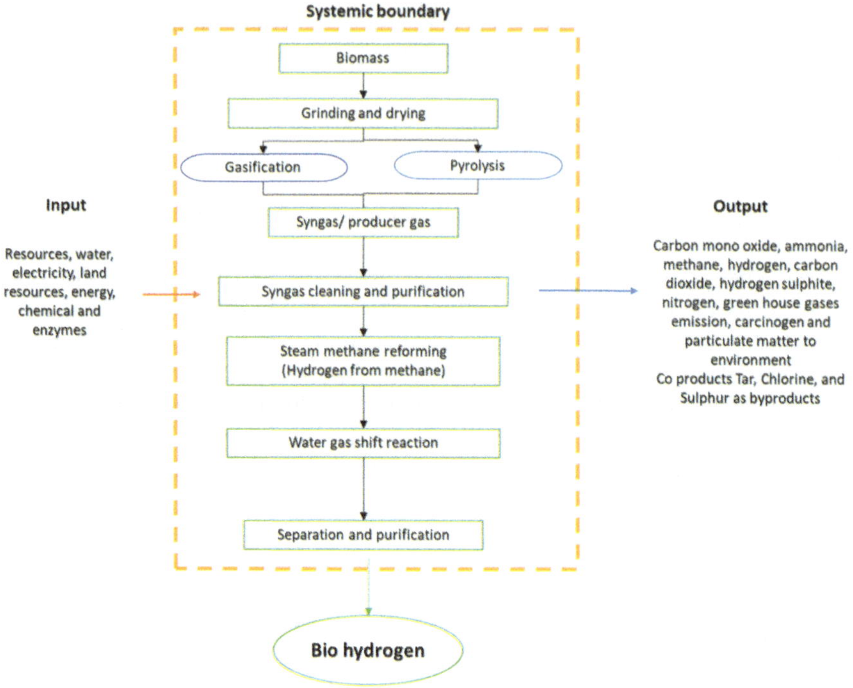

Fig. 8.6 The schematic represents the detail thermochemical method of biohydrogen production

8.5 Life Cycle Assessment of Biodiesel

Biodiesel production pathway is based on feedstock type which influences input and output of energy. Here, cradle-to-gate LCA model is discussed for utilization of waste cooking oil (WCO) (Fig. 8.5), lignocellulosic crop, and renewable biomass. Biodiesel is a methyl ester or ethyl ester of fatty acids which is majorly produced from non-edible vegetable oils, acid oil, used cooking oil (UCO), or animal fat (Neupane, 2022). Biodiesel is produced by three ways, namely, pyrolysis, micro-emulsion, and transesterification process (Daud et al., 2015; Neupane, 2022). Transesterification process (Fig. 8.7) is involved in production of biodiesel from oil [seed extracted oil, algal biomass, WCO]. System boundaries for transesterification process from European countries have been studied by researchers (Chung et al., 2019). This concluded that biodiesel from WCO is more efficient than jatropha oil-based biodiesel. The biodiesel production unit is included in system boundary (Tan et al., 2015). This applies to the additional steps for grinding and extraction of oil that demands more energy consumption, land requirement, labor, electricity, and chemical energy and will eventually result in emission of heat, carcinogen, respiratory organic substance, GHG and outflow of waste water. Hence, it has been concluded that utilization of

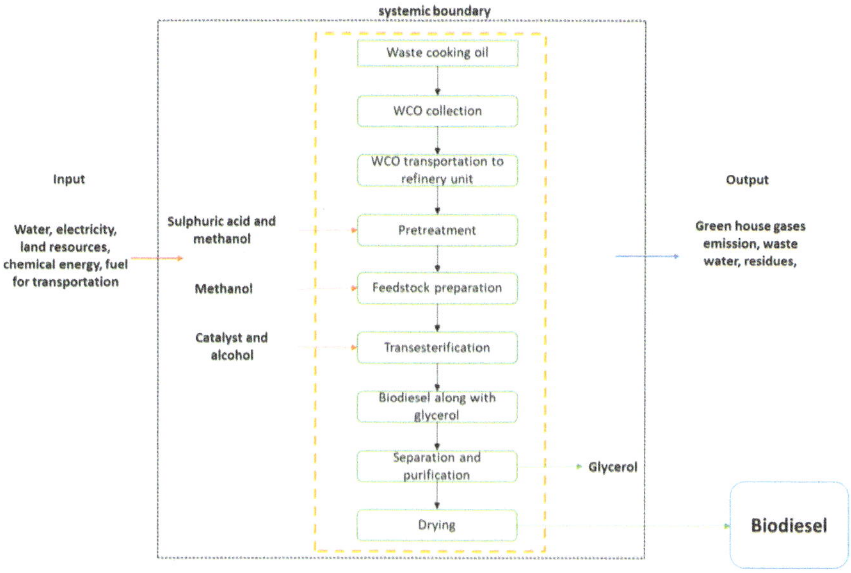

Fig. 8.7 The biodiesel production scheme represented the input, output and systemic boundary using waste cooking oil

WCO is more efficient than seed oil usage in context of human toxicity, environment health, and resource consumption (Hosseinzadeh-Bandbafha et al., 2022).

An economic analysis was conducted which stated that alkaline-catalyzed process with methanol and sodium hydroxide required an investment of 1,309,668 Euros (for seed extract oil), while the alkaline-catalyzed process with acid pretreatment required an input of 2,702,337 Euros (using WCO). The production of biodiesel from waste cooking oils is preferable from both economic and environmental perspectives since it presents a reasonable payback period (lower than Case I) despite the high fixed investment. Further, diesel consumption for transportation of raw material was lower for jatropha oil-based (0.00537 Pt) than WCO (0.027Pt) (Chung et al., 2019).

Biodiesel is categorized into three types based upon their raw material composition (Saladini et al., 2016; Sims et al., 2010; Chowdhury & Loganathan et al. 2019). In India, biodiesel is being produced from various sources such as vegetable oil, used cooking oil, animal fat, seed oil, etc. As reported by USDA report (Biofuel Annual, IN2024–0024), biodiesel consumption and production capacity in 2023 was 190 ML (Million Liters) and 200 ML which has increased in 2024 to 220 ML and 226 ML, respectively. It has been evaluated that in 2023, on-road biodiesel use was 40 ML which increased impressively to 105 ML by 2024. In China, LCA studies have been done for biodiesel in which multiple sources are used for biodiesel production such as, soyabean, jatropha seed, vegetable seed, castor seed, algae, waste cooking oil, and waste extraction oil (Liang et al., 2013). Study has shown that algal-derived biodiesel has higher energy demand than it contains while, WCO-derived biodiesel has higher human toxicity potential (Carneiro et al., 2017). Studies have shown in China, algal-based biodiesel had net energy yield as negative and economic benefit as zero. From studies it has been demonstrated that biodiesel based on biomass has higher input of energy, agricultural input and higher electricity demand and higher carcinogen and greenhouse gas and heat emission due to additional steps required for feedstock preparation, pretreatment process, extraction, and drying process (Chowdhury & Loganathan et al., 2019).

Countries have their own specific approaches and guidelines for biodiesel quality standards. Those standards hold the guidelines and limitations of biodiesel compositions namely methanol/ethanol content, blending proportion, etc. which are concluded by countries individually. The internation organizations established by countries for biodiesel standards are: India-BIS (Bureau of Indian Standards) [IS 15607:2005, IS 15067:2016, IS 15607:2022], European Union- CEN (European Committee for Standardization) [EN 14214 (specifies for FAME)], United States-ASTM (American Society for Testing and Material) [ASTM D6751], Brazil; ANP (Brazilian National Agency for Petroleum, Natural Gas, and Biofuel), Indonesia-BSN (National Standardization Agency of Indonesia) [SNI 7182:2015].

8.6 Life Cycle Assessment of Bio-Oil

Bio-oil is an emergent biofuel which is being used as liquid transportation fuel. But it cannot be used directly due to it high water and oxygen content. It must be converted to upgraded or advanced bio-oil by various treatment methodologies such hydrothermal liquification and pyrolysis (Hansen et al., 2020). These methodologies have an impact on environment as it requires high amount of energy input in form of electricity with heat and gas output having implications for global warming potential (GWP). In a case study presented by Steele et al. (2012) on LCA of biofuel from forest biomass, pyrolysis method led to the yield of liquid bio-oil, char (activated carbon), and syngas. Here, LCA includes transportation cost of forest biomass to refinery unit for pretreatment by using various chemical and thermochemical processes (van Schalkwyk et al., 2020). It is followed by drying process using high amount of heat flow. Pretreatment is fully dependent upon the biomass used and its particle size. After completion of pretreatment and drying, pyrolysis was done under anaerobic conditions which is completely endothermic (Salman et al., 2017). After pyrolysis, rapid quenching is followed which include high energy input in the form of electricity (Guo et al., 2020). In this process, bio-oil is extracted using series of condensation (Zheng et al., 2023). Other co-products which are produced along with bio-oil are destined for recovery process, as char is recovered and then used for heat generation (Gupta et al., 2021). Syngas produced from pyrolysis method (Fig. 8.8) is condensed and reused in industrial process (van Schalkwyk et al., 2020).

Various aspects are taken in account such as atmospheric emission (GWP, GHG emission and ozone layer depletion), land usage, terrestrial toxicity, aquatic toxicity, terrestrial and aquatic eutrophication, carcinogens, and acidification. Tortoioli et al.,

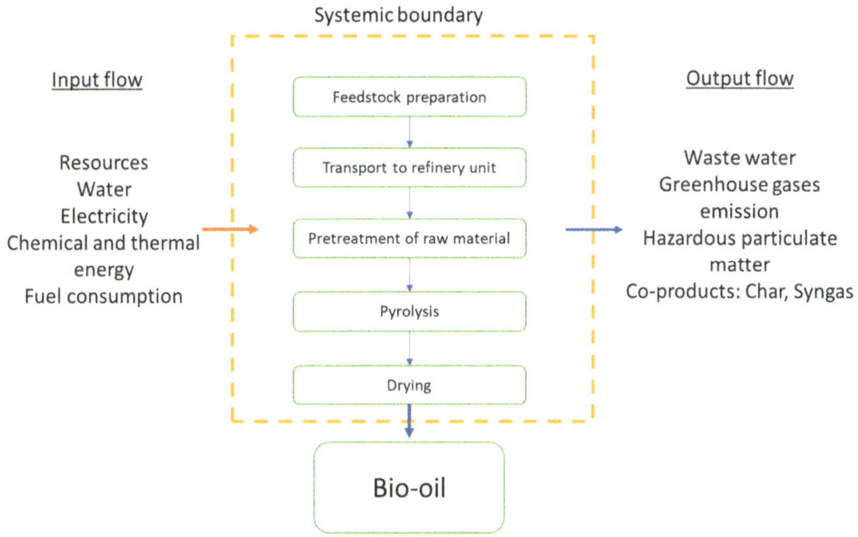

Fig. 8.8 The Life cycle assessment includes input flow, output flow systemic boundary for bio-oil production by pyrolysis

(2020), have studied environmental impact of bio-oil and concluded that collection of raw material and bio-oil distribution accounts for high amount of fuel consumption and GHG emission (CO, CO_2, SOx, NOx) that eventually deals with ozone layer depletion and GWP. Next stage that is feedstock preparation, has input of heat and electricity along with chemical and gases, and output flow of waste and heat. Drying and pyrolysis stage contributes to huge gases emission in atmosphere because of high energy demand (heat, electricity, and chemical energy). Carbon dioxide released during the process is mostly biogenic which result from combustion of biomass (Steele et al., 2012). Other compounds such as hazardous air pollutant, carbon monoxide, methane, nitrogen oxide, particulate matter, sulphur oxides, and organic compounds equally come up with GHG emission and GWP. Along with emission to air, emission to water has also been analyzed and total energy consumed during entire life cycle of cradle-to-grave (i.e. from production to use of biofuel) bio-oil production through pyrolysis is 0.5249 MJ/MJ from biomass (Steele et al., 2012).

8.7 Life Cycle Assessment of Biogas

Biogas is an energy source which is being consumed for various purposed. It contains methane, carbon dioxide and other gases such as nitrogen and ammonia. This can be used in its raw form or can be upgraded to enhance its methane content for further used in transportation and as Compressed Bio-Gas (CBG). Biogas production can be done by utilization of various biomass form such as agricultural biomass and organic waste. The biogas production processes have distinct environmental, resource, and human impact based upon their feedstock and method of production used. Biogas production is initiated by collection of suitable raw material biomass or organic waste. Majorly biogas is produced by anaerobic digestion of biomass and organic waste (from farm and animal waste) by means of anaerobic microorganism. System boundary for biogas production is depicted in Fig. 8.9. Major process of anaerobic digestion involves, hydrolysis phase in which polymers of biomass or organic waste is collected and broken down by hydrolysis and converted to monomers which are fermentable compounds (intermediate compounds such as amino acids and sugars) (Sárvári Horváth et al., 2016). Hydrolysis is followed by fermentation process in which these monomers are targeted for fermentation which further include acetogenesis and methanogenesis process (methane production) (Sárvári Horváth et al., 2016).

After anaerobic digestion there is separation of gas leading to the production of crude or raw biogas along with digestate. Hence, anaerobic digestion produces biogas along with digestate (Jameel et al., 2024). This raw form of biogas is now aimed for purification and upgradation (Awe et al., 2017). After production of biogas, it is followed by Combined Heat and Power (CHP) plant. After entering CHP plant, heat and electricity is generated which are further used in various sectors as energy source. Digestate produced as byproduct in biogas production is used as fertilizer in agricultural area (Koszel & Lorencowicz et al., 2015). In-situ and ex-situ methods are employed for upgrading biogas. In-situ methods are: CO_2 desorption, pressurized reactor, H_2 addition, and electro-methanogenesis, while ex-situ methods include

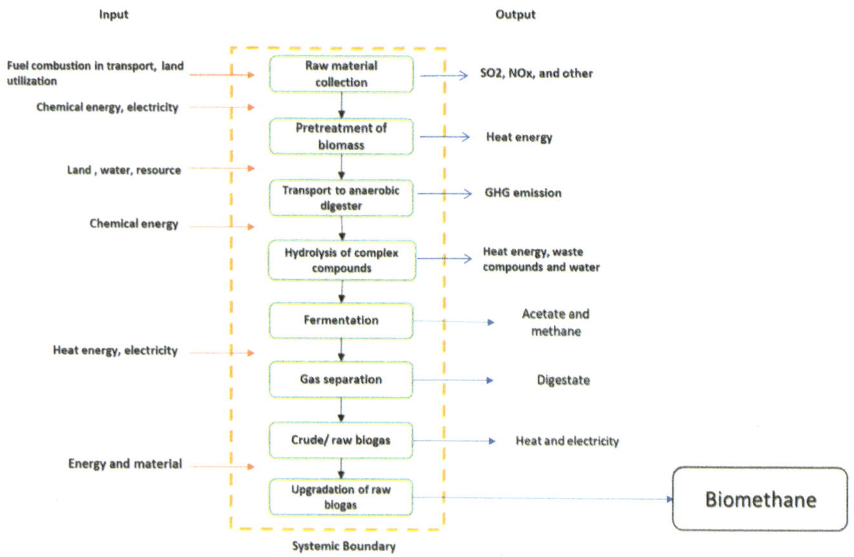

Fig. 8.9 The systemic boundary, input and output parameters for biogas production

catalytic conversion, membrane gas permeation, physical and chemical scrubbing, absorption, desulphurization, and adsorption (Upadhyay et al., 2022).

LCA of biogas production, includes GHG emission, GWP, eutrophication, acidification (AD), etc. These aspects are analyzed in every stage of life cycle of biogas production. Transportation of raw material has great impact on GHG emission which cannot be neglected. Distance between collection point and production unit plays crucial role in reducing GWP by adjusting combustion of fuel and emission of GHG which eventually effect climate change, ozone depletion potential, and AD. Another stage that is anaerobic digestion has role in effecting environment and human health. The process includes pretreatment processes, methodologies and technologies used in anaerobic digestion, type of digesters and substrate types used. Various studies showed that pretreatment and digestion phases exhibit highest contribution to eutrophication and acidification with GHG emissions like CO_2, CH_4, and SO_2. It also causes highest GWP contribution having the paramount impact on environment and resource (Fantin et al., 2015). After anaerobic digestion of biomass, combined heat and power (CHP) plant stage has high environmental impact. In this stage, there is GHG emission (majorly SO_2 and NO_x) during combustion process which contributes towards acidification potential and eutrophication potential. Biogas upgrading process involves high amount of energy input in form of heat and electricity and chemical utilization such as in scrubbing and catalytic conversion (Hijazi, et al., 2016). This results in emission of GHG and various toxic compounds in air and water.

In India, following National Biogas and Manure Management Programme (NBMMP), various biogas plants have been installed in Maharashtra, Andhra Pradesh, Karnataka, Gujarat and other places. Indian government has launched various policies, for promoting biogas production and utilization with respect to increasing demand of energy including Sustainable Alternative Towards Affordable Transportation (SATAT) programme (2018), Galvanising Organic Bio-Agro Resources (GOBARdhan) programme (2018), Waste to Energy Programme (2020), National Biogas Programme (NBP) (2020), and Blending Mandate Obligation (Nov, 2023). According to the International Energy Agency (IEA), China (81 TWh/y) and Germany (87 TWh/y) hold top position in biogas production with substantial number of biogas plants installations and are followed by other countries like Brazil, France, United Kingdom, India etc (Gustafsson et al., 2024).

8.8 Conclusion

Biomass mediated biofuel production contributes 9% (~ 51 EJ) of the global primary energy supply making it a very significant alternative fuel source. However, its successful displacement of traditional fuel depends not only on economic and technical feasibility but also involves environmental impacts. LCA allows for a complete evaluation of these biofuels in context of feedstock and production pathways wherein not only energy production is analyzed but environmental impact in terms of GHG and other pollutants is explored. There is great need of standardized features covering include functional unit, system boundaries, allocation methodology, and environmental indicators. This unified methodology will allow comparing biofuel LCA studies, which currently is not possible as different studies follow different rules. Whereas it is preferable that full LCA is carried out in academia, industry should have access to a simplified and cost-effective version of LCA. It is a powerful tool but needs to be continuously refined to enable better sustainable predictions.

Acknowledgments Authors are thankful to the J.C. Bose University of Science and Technology for providing necessary research support.

References

Aasberg-Petersen, K., Dybkjær, I., Ovesen, C. V., Schjødt, N. C., Sehested, J., & Thomsen, S. G. (2011). Natural gas to synthesis gas–catalysts and catalytic processes. *Journal of Natural Gas Science and Engineering, 3*(2), 423–459.

Aditiya, H. B., Mahlia, T. M. I., Chong, W. T., Nur, H., & Sebayang, A. H. (2016). Second generation bioethanol production: A critical review. *Renewable and Sustainable Energy Reviews, 66*, 631–653.

Agostini, A., Giuntoli, J., Marelli, L., & Amaducci, S. (2020). Flaws in the interpretation phase of bioenergy LCA fuel the debate and mislead policymakers. *The International Journal of Life Cycle Assessment, 25*, 17–35.

Amaya-Santos, G., Chari, S., Sebastiani, A., Grimaldi, F., Lettieri, P., & Materazzi, M. (2021). Biohydrogen: A life cycle assessment and comparison with alternative low-carbon production routes in UK. *Journal of Cleaner Production, 319*, 128886.

Anand, A., & Kaushal, P. (2021). Life cycle assessment of strategic locations to establish molasses-based bioethanol production facility in India. *Cleaner Environmental Systems, 3*, 100055.

Aoun, W. B., & Gabrielle, B. (2017). Life cycle assessment and land-use changes: effectiveness and limitations. In *Life-cycle assessment of biorefineries* (1st ed., p. 322).

Appiah-Nkansah, N. B., Li, J., Rooney, W., & Wang, D. (2019). A review of sweet sorghum as a viable renewable bioenergy crop and its techno-economic analysis. *Renewable Energy, 143*, 1121–1132.

Awe, O. W., Zhao, Y., Nzihou, A., Minh, D. P., & Lyczko, N. (2017). A review of biogas utilisation, purification and upgrading technologies. *Waste and Biomass Valorization, 8*, 267–283.

Bracmort, K., & Cowan, T. (2017). *The renewable fuel standard (RFS): Waiver authority and modification of volumes*. Congressional Research Service.

Carneiro, M. L. N., Pradelle, F., Braga, S. L., Gomes, M. S. P., Martins, A. R. F., Turkovics, F., & Pradelle, R. N. (2017). Potential of biofuels from algae: Comparison with fossil fuels, ethanol and biodiesel in Europe and Brazil through life cycle assessment (LCA). *Renewable and Sustainable Energy Reviews, 73*, 632–653.

Chandel, A. K., Arora, R., Gupta, R., Narisetty, V., & Kumar, V. (2024). Global biofuel Alliance and roadmap for ethanol blending program in India: Analysis and perspective. *Sugar Tech, 26*(4), 1005–1015.

Chowdhury, H., & Loganathan, B. (2019). Third-generation biofuels from microalgae: A review. *Current Opinion in Green and Sustainable Chemistry, 20*, 39–44.

Chung, Z. L., Tan, Y. H., San Chan, Y., Kansedo, J., Mubarak, N. M., Ghasemi, M., & Abdullah, M. O. (2019). Life cycle assessment of waste cooking oil for biodiesel production using waste chicken eggshell derived CaO as catalyst via transesterification. *Biocatalysis and Agricultural Biotechnology, 21*, 101317.

Dash, S., Singh, A., Jose, S., Elangovan, D., Surapraraju, S. K., & Natarajan, S. K. (2024). Advances in green hydrogen production through alkaline water electrolysis: A comprehensive review. *International Journal of Hydrogen Energy, 83*, 614–629.

Daud, N. M., Abdullah, S. R. S., Hasan, H. A., & Yaakob, Z. (2015). Production of biodiesel and its wastewater treatment technologies: A review. *Process Safety and Environmental Protection, 94*, 487–508.

de Souza Abud, A. K., & de Farias Silva, C. E. (2019). Bioethanol in Brazil: Status, challenges and perspectives to improve the production. *Bioethanol production from food crops*, 417–443.

Devi, A., Bajar, S., Sihag, P., Sheikh, Z. U. D., Singh, A., Kaur, J., Bishnoi, N. R., & Pant, D. (2023). A panoramic view of technological landscape for bioethanol production from various generations of feedstocks. *Bioengineered, 14*(1), 81–112.

Djomo, S. N., & Blumberga, D. (2011). Comparative life cycle assessment of three biohydrogen pathways. *Bioresource Technology, 102*(3), 2684–2694.

Duffield, J. A., Johansson, R., & Meyer, S. (2015). *US ethanol: An examination of policy, production, use, distribution, and market interactions*. US Department of Agriculture.

Faist Emmenegger, M., Pfister, S., Koehler, A., De Giovanetti, L., Arena, A. P., & Zah, R. (2011). Taking into account water use impacts in the LCA of biofuels: An Argentinean case study. *The International Journal of Life Cycle Assessment, 16*, 869–877.

Fantin, V., Giuliano, A., Manfredi, M., Ottaviano, G., Stefanova, M., & Masoni, P. (2015). Environmental assessment of electricity generation from an Italian anaerobic digestion plant. *Biomass and Bioenergy, 83*, 422–435.

Gheewala, S. H. (2023). Life cycle assessment for sustainability assessment of biofuels and bioproducts. *Biofuel Research Journal, 10*(1), 1810–1815.

Giuntoli, J., Bulgheroni, C., Marelli, L., Sala, S., & Pant, R. (2019). *Brief on the use of life cycle assessment (LCA) to evaluate environmental impacts of the bioeconomy*.

Guo, M. (2020). The global scenario of biofuel production and development. *Practices and perspectives in sustainable bioenergy: a systems thinking approach*, 29–56.

Gupta, M., Srivastava, M., Agrahari, S. K. & Detwal, P. (2018). Waste to energy technologies in India: A review. *Journal of Energy and Environmental Sustainability, 6*, 29–35.

Gupta, S., Mondal, P., Borugadda, V. B., & Dalai, A. K. (2021). Advances in upgradation of pyrolysis bio-oil and biochar towards improvement in bio-refinery economics: A comprehensive review. *Environmental Technology & Innovation, 21*, 101276.

Gustafsson, M., Meneghetti, R., Souza Marques, F., Trim, H., Dong, R., Al Saedi, T., Rasi, S., Thual, J., Kornatz, P., Wall, D., Berntsen, C., Saxegaard, S., Lyng, K.A., Nägele, H.J., Heaven, S., & Bywater, A.. (2024). A perspective on the state of the biogas industry from selected member countries. Gustafsson, M., Liebetrau, J. (Ed.) IEA Bioenergy Task 37:2.

Hansen, S., Mirkouei, A., & Diaz, L. A. (2020). A comprehensive state-of-technology review for upgrading bio-oil to renewable or blended hydrocarbon fuels. *Renewable and Sustainable Energy Reviews, 118*, 109548.

Hassan, M. K., Chowdhury, R., Ghosh, S., Manna, D., Pappinen, A., & Kuittinen, S. (2021). Energy and environmental impact assessment of Indian rice straw for the production of second-generation bioethanol. *Sustainable Energy Technologies and Assessments, 47*, 101546.

Havukainen, J., Nguyen, M. T., Väisänen, S., & Horttanainen, M. (2018). Life cycle assessment of small-scale combined heat and power plant: Environmental impacts of different forest biofuels and replacing district heat produced from natural gas. *Journal of Cleaner Production, 172*, 837–846.

Hélias, A. (2019). At the boundary between anthropogenic and environmental systems: The neglected emissions of indirect nitrous oxide. *The International Journal of Life Cycle Assessment, 24*(3), 412–418.

Hijazi, O., Munro, S., Zerhusen, B., & Effenberger, M. (2016). Review of life cycle assessment for biogas production in Europe. *Renewable and Sustainable Energy Reviews, 54*, 1291–1300.

Holladay, J. D., Hu, J., King, D. L., & Wang, Y. (2009). An overview of hydrogen production technologies. *Catalysis Today, 139*(4), 244–260.

Hosseinzadeh-Bandbafha, H., Nizami, A. S., Kalogirou, S. A., Gupta, V. K., Park, Y. K., Fallahi, A., Sulaiman, A., Ranjbari, M., Rahnama, H., Aghbashlo, M., & Peng, W. (2022). Environmental life cycle assessment of biodiesel production from waste cooking oil: A systematic review. *Renewable and Sustainable Energy Reviews, 161*, 112411. https://mopng.gov.in/en/refining/ethanol-blended-petrol

Igwebuike, C. M., Awad, S., & Andrès, Y. (2024). Renewable energy potential: Second-generation biomass as feedstock for bioethanol production. *Molecules, 29*(7), 1619.

Jambo, S. A., Abdulla, R., Azhar, S. H. M., Marbawi, H., Gansau, J. A., & Ravindra, P. (2016). A review on third generation bioethanol feedstock. *Renewable and Sustainable Energy Reviews, 65*, 756–769.

Jameel, M. K., Mustafa, M. A., Ahmed, H. S., Jassim Mohammed, A., Ghazy, H., Shakir, M. N., Lawas, A. M., Khudhur Mohammed, S., Idan, A. H., Mahmoud, Z. H., & Sayadi, H. (2024). Biogas: Production, properties, applications, economic and challenges: A review. *Results in Chemistry*, 101549.

Jeswani, H. K., Chilvers, A., & Azapagic, A. (2020). Environmental sustainability of biofuels: A review. *Proceedings of the Royal Society A, 476*(2243), 20200351.

Jin, M., Wei, X., Mu, X., Ren, W., Zhang, S., Tang, C., & Cao, W. (2024). Life-cycle analysis of biohydrogen production via dark-photo fermentation from wheat straw. *Bioresource Technology, 396*, 130429.

Koszel, M., & Lorencowicz, E. (2015). Agricultural use of biogas digestate as a replacement fertilizers. *Agriculture and Agricultural Science Procedia, 7*, 119–124.

Kumar, D., Dutt, S., Jaiswal, A. K., Kaundal, B., Kumar, D., & Singh, B. (2024). Bioethanol production from potatoes in India: A SWOT analysis. *Heliyon, 10*(23).

Kumar, S. S., & Himabindu, V. (2019). Hydrogen production by PEM water electrolysis–a review. *Materials Science for Energy Technologies, 2*(3), 442–454.

Lazaro, C. Z., & Hallenbeck, P. C. (2019). Fundamentals of biohydrogen production. In *Biohydrogen* (pp. 25–48). Elsevier.

Liang, S., Xu, M., & Zhang, T. (2013). Life cycle assessment of biodiesel production in China. *Bioresource Technology, 129*, 72–77.

Lou, Y., Fan, Z., Friedmann, J., Corbeau, A., Agrawal, M., & Khatri, A. (2023). *The potential role of biohydrogen in creating a net-zero world*. Columbia Center of Global Energy Policy. January.

Morales, M., Quintero, J., Conejeros, R., & Aroca, G. (2015). Life cycle assessment of lignocellulosic bioethanol: Environmental impacts and energy balance. *Renewable and Sustainable Energy Reviews, 42*, 1349–1361.

Muazu, R. I., Borrion, A. L., & Stegemann, J. A. (2021). Life cycle assessment model for biomass fuel briquetting. *Waste and Biomass Valorization*, 1–16.

Muradov, N. Z., & Veziroğlu, T. N. (2005). From hydrocarbon to hydrogen–carbon to hydrogen economy. *International Journal of Hydrogen Energy, 30*(3), 225–237.

Neupane, D. (2022). Biofuels from renewable sources, a potential option for biodiesel production. *Bioengineering, 10*(1), 29.

Norasethasopon, W. (2010). Effect of gasohol production on the sugarcane industry in Thailand. *Journal of Asia Pacific Studies, 1*(2).

Oni, A. O., Anaya, K., Giwa, T., Di Lullo, G., & Kumar, A. (2022). Comparative assessment of blue hydrogen from steam methane reforming, autothermal reforming, and natural gas decomposition technologies for natural gas-producing regions. *Energy Conversion and Management, 254*, 115245.

Osman, A. I., Mehta, N., Elgarahy, A. M., Al-Hinai, A., Al-Muhtaseb, A. A. H., & Rooney, D. W. (2021). Conversion of biomass to biofuels and life cycle assessment: A review. *Environmental Chemistry Letters, 19*, 4075–4118.

Pradinaud, C., Núñez, M., Roux, P., Junqua, G., & Rosenbaum, R. K. (2019). The issue of considering water quality in life cycle assessment of water use. *The International Journal of Life Cycle Assessment, 24*, 590–603.

Rajaeifar, M. A., Tabatabaei, M., Aghbashlo, M., Hemayati, S. S., & Heijungs, R. (2019). Biodiesel production and consumption: Life cycle assessment (LCA) approach. In *Biodiesel: From production to combustion* (pp. 161–192).

Rezania, S., Oryani, B., Cho, J., Talaiekhozani, A., Sabbagh, F., Hashemi, B., Rupani, P. F., & Mohammadi, A. A. (2020). Different pretreatment technologies of lignocellulosic biomass for bioethanol production: An overview. *Energy, 199*, 117457.

Robak, K., & Balcerek, M. (2018). Review of second-generation bioethanol production from residual biomass. *Food Technology and Biotechnology, 56*(2), 174.

Rocha, M. H., Capaz, R. S., Lora, E. E. S., Nogueira, L. A. H., Leme, M. M. V., Renó, M. L. G., & del Olmo, O. A. (2014). Life cycle assessment (LCA) for biofuels in Brazilian conditions: A meta-analysis. *Renewable and Sustainable Energy Reviews, 37*, 435–459.

Saladini, F., Patrizi, N., Pulselli, F. M., Marchettini, N., & Bastianoni, S. (2016). Guidelines for emergy evaluation of first, second and third generation biofuels. *Renewable and Sustainable Energy Reviews, 66*, 221–227.

Salman, C. A., Schwede, S., Thorin, E., & Yan, J. (2017). Enhancing biomethane production by integrating pyrolysis and anaerobic digestion processes. *Applied Energy, 204*, 1074–1083.

Sanghvi, A. H., Manjoo, A., Rajput, P., Mahajan, N., Rajamohan, N., & Abrar, I. (2024). Advancements in biohydrogen production–a comprehensive review of technologies, lifecycle analysis, and future scope. *RSC Advances, 14*(49), 36868–36885.

Santoyo-Castelazo, E., Santoyo, E., Zurita-García, L., Luengas, D. C., & Solano-Olivares, K. (2023). Life cycle assessment of bioethanol production from sugarcane bagasse using a gasification conversion process: Bibliometric analysis, systematic literature review and a case study. *Applied Thermal Engineering, 219*, 119414.

Sárvári Horváth, I., Tabatabaei, M., Karimi, K., & Kumar, R. (2016). Recent updates on biogas production-a review. *Biofuel Research Journal, 3*(2), 394–402.

Scacchi, C. C. O., González-García, S., Caserini, S., & Rigamonti, L. (2010). Greenhouse gases emissions and energy use of wheat grain-based bioethanol fuel blends. *Science of the Total Environment, 408*(21), 5010–5018.

Sharma, R. K., Nazari, M. A., Haydary, J., Singh, T. P., & Mandal, S. (2023). A review on advanced processes of biohydrogen generation from lignocellulosic biomass with special emphasis on thermochemical conversion. *Energies, 16*(17), 6349.

Sims, R. E., Mabee, W., Saddler, J. N., & Taylor, M. (2010). An overview of second generation biofuel technologies. *Bioresource Technology, 101*(6), 1570–1580.

Singh, A., Olsen, S. I., & Pant, D. (2013). Importance of life cycle assessment of renewable energy sources. In *Life cycle assessment of renewable energy sources* (pp. 1–11). Springer London.

Singh, A. P. (2024). Assessment of India's green hydrogen Mission and environmental impact. *Renewable and Sustainable Energy Reviews, 203*, 114758.

Srinivasan, M., Ghoge, K., Haldar, S., Bazaz, A., & Revi, A.. (2023). Climate finance in India.

Steele, P., Puettmann, M. E., Penmetsa, V. K., & Cooper, J. E. (2012). Life-cycle assessment of pyrolysis bio-oil production. *Forest Products Journal, 62*(4), 326–334.

Stolarski, M. J., Krzyżaniak, M., Łuczyński, M., Załuski, D., Szczukowski, S., Tworkowski, J., & Gołaszewski, J. (2015). Lignocellulosic biomass from short rotation woody crops as a feedstock for second-generation bioethanol production. *Industrial Crops and Products, 75*, 66–75.

Tahir, F., Mabrouk, A., Al-Ghamdi, S. G., Krupa, I., Sedlacek, T., Abdala, A., & Koc, M. (2021). Sustainability assessment and techno-economic analysis of thermally enhanced polymer tube for multi-effect distillation (med) technology. *Polymers, 13*(5), 681.

Tan, Y. H., Abdullah, M. O., & Nolasco-Hipolito, C. (2015). The potential of waste cooking oil-based biodiesel using heterogeneous catalyst derived from various calcined eggshells coupled with an emulsification technique: A review on the emission reduction and engine performance. *Renewable and Sustainable Energy Reviews, 47*, 589–603.

Tortoioli, S., Paolotti, L., Romagnoli, F., Boggia, A., & Rocchi, L. (2020). Environmental assessment of bio-oil transformation from thistle in the Italian context: An LCA study. *Rigas Tehniskas Universitates Zinatniskie Raksti, 24*(3), 430–446.
Tu, Q., Eckelman, M., & Zimmerman, J. (2017). Meta-analysis and harmonization of life cycle assessment studies for algae biofuels. *Environmental Science & Technology, 51*(17), 9419–9432.
Upadhyay, A., Kovalev, A. A., Zhuravleva, E. A., Kovalev, D. A., Litti, Y. V., Masakapalli, S. K., Pareek, N., & Vivekanand, V. (2022). Recent development in physical, chemical, biological and hybrid biogas upgradation techniques. *Sustainability, 15*(1), 476.
van Schalkwyk, D. L., Mandegari, M., Farzad, S., & Görgens, J. F. (2020). Techno-economic and environmental analysis of bio-oil production from forest residues via non-catalytic and catalytic pyrolysis processes. *Energy Conversion and Management, 213*, 112815.
Wang, F., Wang, R., & He, Z. (2021). The impact of environmental pollution and green finance on the high-quality development of energy based on spatial Dubin model. *Resources Policy, 74*, 102451.
Wang, L., He, L., & He, Y. (2024). Review on absorption refrigeration technology and its potential in energy-saving and carbon emission reduction in natural gas and hydrogen liquefaction. *Energies, 17*(14), 3427.
Zhang, J., Jia, C., Wu, Y., Xia, X., Xi, B., Wang, L., & Zhai, Y. (2017). Life cycle energy efficiency and environmental impact assessment of bioethanol production from sweet potato based on different production modes. *PLoS One, 12*(7), e0180685.
Zheng, J. L., Zhu, Y. H., Dong, Y. Y., Chen, Y., & Zhu, M. Q. (2023). Techno-economic analysis and life cycle assessment of industrial production of ammonia via bio-oil conversion. *Energy, 280*, 128223.

Chapter 9
Byproducts in the Biofuel Production Process and Their Environmental Application

Sakshi Jaiswal

Abstract The transition toward renewable energy sources has positioned biogas and biomethane as critical components in the global energy matrix. This comprehensive study investigates the environmental implications of biomethane production from landfill biogas through Life Cycle Assessment (LCA) methodology. The research identifies key process stages contributing to environmental burdens, with carbon dioxide removal accounting for 84% of global warming impacts and secondary biogas compression contributing 37% to fossil resource depletion. Our analysis demonstrates that biomethane deployment as a vehicular fuel offers substantial environmental benefits compared to conventional fossil fuels. The study proposes optimization strategies for critical process units and evaluates the potential of biomethane to mitigate greenhouse gas emissions in both light-duty and heavy-duty transportation sectors. These findings contribute to the growing body of evidence supporting biogas valorization as a sustainable pathway for waste management and renewable energy production.

Keywords Biogas production · Biomethane upgrading · Life cycle assessment · Environmental impact · Renewable energy · Landfill gas · Sustainable transportation · Waste valorization

9.1 Introduction

Energy availability represents a fundamental prerequisite for human well-being and socioeconomic development. The contemporary global energy landscape, characterized by heavy dependence on fossil fuel resources, presents unprecedented challenges, particularly for developing economies. The tripartite objectives of

S. Jaiswal (✉)
Department of Biotechnology, Mata Gujri Mahila Mahavidyalaya (Autonomous), Jabalpur, Madhya Pradesh, India
e-mail: jaiswal.sj08@gmail.com

V. Singh et al. (eds.), *Bioenergy and Value Added Byproducts from Biomass Waste*, Sustainable Environmental Waste Management Strategies,
https://doi.org/10.1007/978-3-032-22849-9_9

energy security, environmental sustainability, and economic growth constitute the primary drivers of energy policy across nations (Johnson & Williams, 2019; Smith et al., 2020). Projections indicate the potential depletion of conventional fossil fuel reserves—including coal, petroleum, and natural gas—within the next century, necessitating urgent diversification of energy sources (Anderson et al., 2021; Chen et al., 2020; Roberts & Thompson, 2019).

Biogas emerges as a particularly promising renewable energy vector due to several advantageous characteristics: abundant and cost-effective feedstock availability, versatility in application domains (including thermal energy generation, electricity production, transportation fuel, and chemical feedstock), and potential for producing high-value products such as hydrogen, purified carbon dioxide, and advanced biofuels (Martinez et al., 2019). Global biogas-based electricity generation capacity has demonstrated remarkable expansion, increasing by approximately 90% from 65 GW in 2010 to 120 GW in 2019, underscoring the technology's growing significance in energy transition pathways (International Energy Agency, 2020) (Fig. 9.1).

Biogas composition typically comprises methane (CH_4) and carbon dioxide (CO_2) as primary constituents, along with trace components including water vapor, hydrogen sulfide (H_2S), siloxanes, ammonia (NH_3), carbon monoxide (CO), and various volatile organic compounds (Kumar et al., 2018). The energetic value of biogas derives principally from its combustible components—methane, hydrogen, and carbon monoxide—which facilitate diverse thermal and electrical applications (Zhang et al., 2014). While methane itself is non-toxic, it functions as a potent asphyxiant in confined spaces through oxygen displacement and represents a significant greenhouse gas with a global warming potential approximately 20 times greater than carbon dioxide over a 20-year timeframe (IPCC, 2021; Lee & Park, 2020). This

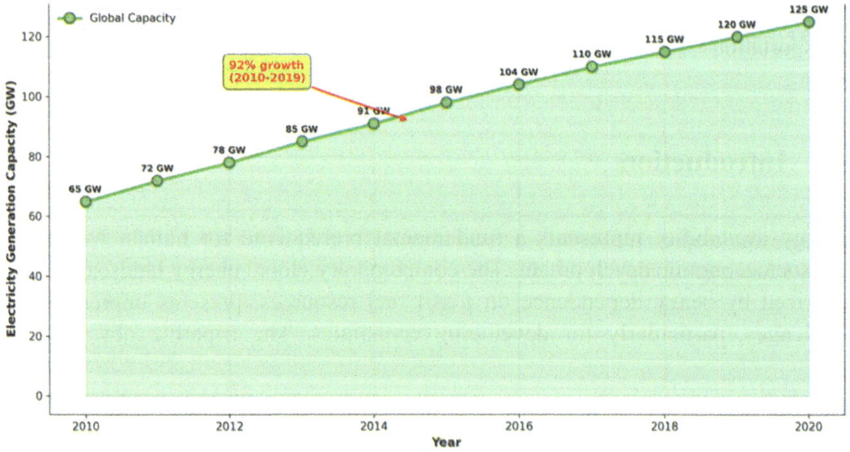

Fig. 9.1 Global biogas-based electricity generation capacity growth (2010–2020)

environmental consideration underscores the critical importance of controlled biogas production and utilization systems (Godoy et al., 2024).

9.1.1 Biogas Composition and Properties

The biochemical characteristics of biogas exhibit substantial variability depending on feedstock composition and operational parameters of the anaerobic digestion process. Beyond the principal components of methane and carbon dioxide, biogas contains various trace constituents that significantly influence its utility as an energy carrier (Table 9.1). Specific impurities, particularly hydrogen sulfide and carbon dioxide, require removal to ensure optimal performance in downstream applications (Wilson et al., 2018) (Fig. 9.2).

Table 9.1 Typical composition of raw biogas

Component	Concentration range	Unit
Methane (CH_4)	50–75	% v/v
Carbon dioxide (CO_2)	25–50	% v/v
Hydrogen sulfide (H_2S)	0–10,000	ppm
Ammonia (NH_3)	0–100	ppm
Water vapor (H_2O)	1–5	% v/v
Siloxanes	0–50	mg/m^3

Source Adapted from Angelidaki et al. (2018) and Sun et al. (2015)

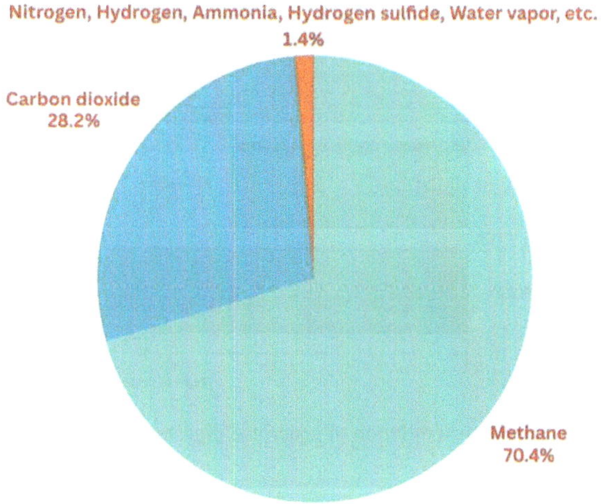

Fig. 9.2 Typical composition of raw biogas from landfill sources

9.1.2 Anaerobic Digestion Process

Biogas production occurs through anaerobic digestion (AD), a sophisticated biochemical process involving sequential metabolic stages orchestrated by diverse microbial consortia (Kougias et al., 2016; Tian et al., 2015). The process encompasses four primary phases:

- Hydrolysis: Complex organic polymers (proteins, carbohydrates, lipids) undergo enzymatic breakdown into simpler monomers and oligomers
- Acidogenesis: Fermentative bacteria convert hydrolysis products into volatile fatty acids, alcohols, hydrogen, and carbon dioxide
- Acetogenesis: Acetogenic bacteria transform intermediate products into acetate, hydrogen, and carbon dioxide
- Methanogenesis: Methanogenic archaea convert acetate and hydrogen/carbon dioxide into methane gas

Process efficiency and biogas composition are significantly influenced by operational parameters including temperature regime (mesophilic vs. thermophilic), hydraulic retention time, organic loading rate, pH maintenance, and feedstock characteristics (An et al., 2017; Zhang et al., 2017). Optimization of these parameters is essential for maximizing methane yield while minimizing process inhibition from factors such as ammonia accumulation, volatile fatty acid buildup, or long-chain fatty acid toxicity (Ma et al., 2015; Moestedt et al., 2016) (Fig. 9.3).

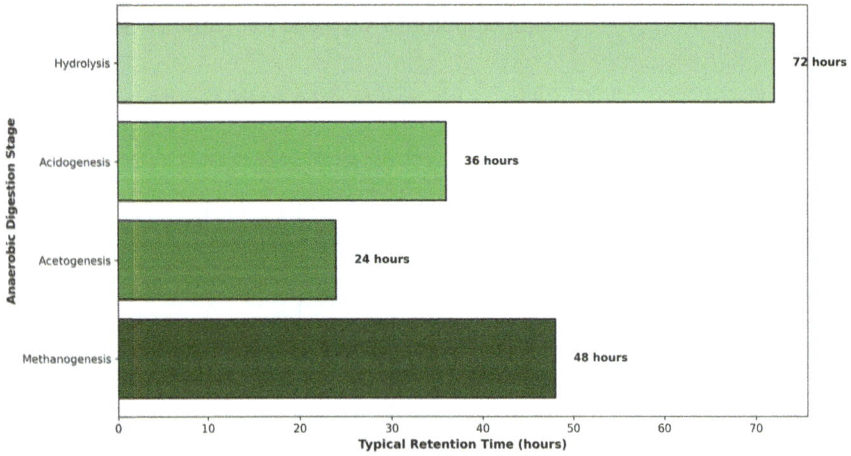

Fig. 9.3 Typical duration of anaerobic digestion process stages

9.2 Biogas Upgrading Technologies

The transformation of raw biogas into biomethane—a refined product suitable for injection into natural gas grids or utilization as vehicular fuel—necessitates the removal of carbon dioxide and trace contaminants through upgrading processes (Angelidaki et al., 2018). Various physicochemical technologies have been developed and commercialized for this purpose, each characterized by distinct advantages, limitations, and operational requirements (Muñoz et al., 2015; Ryckebosch et al., 2011).

9.2.1 Classification of Upgrading Technologies

Contemporary biogas upgrading methodologies can be categorized into several primary technological families (Table 9.2):

Each technology presents unique operational characteristics, capital requirements, and suitability for specific production scales and purity specifications. The selection of an appropriate upgrading system depends on multiple factors including feedstock composition, desired product quality, economic constraints, and environmental considerations (Aieamsam-Aung et al., 2023; Kapoor et al., 2019) (Fig. 9.4).

9.3 Environmental Assessment Through Life Cycle Analysis

The environmental performance of biomethane production systems requires comprehensive evaluation through Life Cycle Assessment (LCA) methodology, which systematically quantifies environmental impacts across the entire value chain from feedstock collection through end-use application (Lansche & Müller, 2017). This holistic approach enables identification of environmental hotspots and optimization opportunities within the production pathway (Godoy et al., 2024).

Table 9.2 Comparison of biogas upgrading technologies

Technology	CH_4 purity (%)	Energy input	Primary application
Water scrubbing	95–98	Medium	Large-scale plants
Chemical scrubbing	96–99	High	Industrial scale
Pressure swing adsorption	95–98	Medium–High	Medium-scale facilities
Membrane separation	90–96	Low-Medium	Small-scale operations
Cryogenic separation	> 99	Very high	High-purity requirements

Source Compiled from Tuinier and van Sint Annaland (2012) and Jensen et al. (2021)

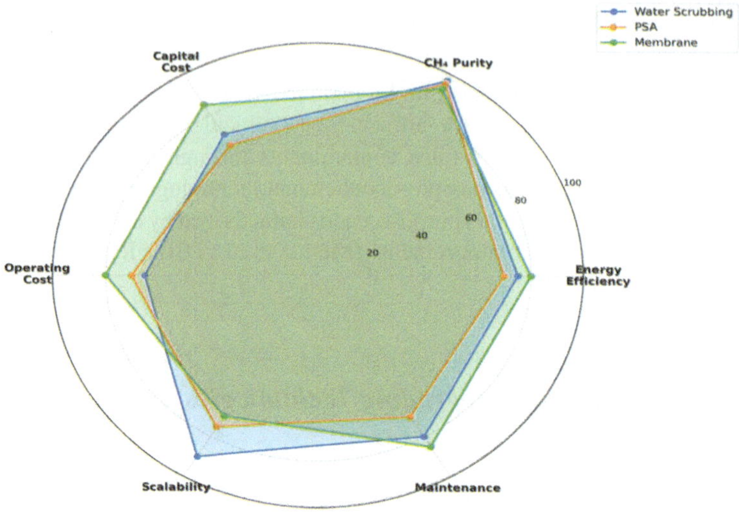

Fig. 9.4 Multi-criteria performance comparison of upgrading technologies (scale: 0–100, higher is better)

9.3.1 System Boundaries and Functional Unit

The LCA framework employed in this investigation encompasses the following system boundaries:

- Landfill biogas collection and preliminary treatment
- Primary compression and conditioning
- Carbon dioxide removal through upgrading technology
- Secondary compression to distribution pressure
- Storage and dispensing infrastructure.

The functional unit for environmental impact assessment was defined as 1 MJ of delivered biomethane energy, enabling direct comparison with conventional fossil fuel alternatives (Rajendran et al., 2012).

9.3.2 Impact Category Results

Quantitative assessment revealed significant disparities in environmental contribution across different process stages (Table 9.3). The carbon dioxide removal operation emerged as the predominant contributor to climate change impacts, accounting for 84% of total global warming potential. This substantial contribution derives primarily from the electrical energy requirements of separation equipment and associated methane slip during the upgrading process (Godoy et al., 2024).

Table 9.3 Environmental impact distribution by process stage

Process stage	Global warming (%)	Resource depletion (%)
Biogas collection	4.2	8.5
Primary compression	6.8	12.3
CO_2 removal	**84.0**	28.6
Secondary compression	3.5	**37.0**
Storage and distribution	1.5	13.6

Note Highlighted cells indicate process stages contributing > 30% to respective impact categories

Secondary compression operations demonstrated the highest contribution (37%) to fossil resource scarcity, attributable to the substantial electrical energy demands for achieving distribution-grade pressure specifications. These findings underscore the critical importance of energy efficiency optimization in upgrading infrastructure design and operation.

9.4 Biomethane as Transportation Fuel

The application of biomethane as a vehicular fuel presents substantial opportunities for greenhouse gas emission mitigation in the transportation sector. Comparative life cycle analysis demonstrates significant environmental advantages relative to conventional fossil-derived fuels across multiple vehicle categories (Table 9.4).

The substantial emission reductions observed for biomethane (70–75% for light-duty vehicles; 66–71% for heavy-duty vehicles) derive from multiple factors including avoided landfill methane emissions, displacement of fossil fuel extraction and processing, and the renewable nature of the feedstock. These findings support the strategic deployment of biomethane infrastructure for transportation decarbonization initiatives (Godoy et al., 2024) (Fig. 9.5).

Table 9.4 Comparative well-to-wheel GHG emissions by fuel type

Fuel type	Light-duty vehicles	Heavy-duty vehicles
Gasoline/petrol	168 g CO_2-eq/km	N/A
Diesel	152 g CO_2-eq/km	845 g CO_2-eq/km
Fossil natural gas	138 g CO_2-eq/km	728 g CO_2-eq/km
Biomethane (landfill)	**42 g CO_2-eq/km**	**245 g CO_2-eq/km**
Emission reduction	**70–75%**	**66–71%**

Source Life cycle assessment data from Perin et al. (2020) and Dobers (2019)

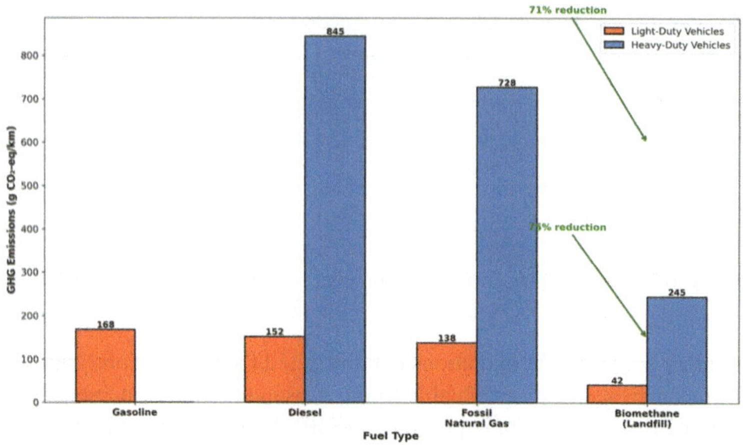

Fig. 9.5 Comparatively to wheel CHG emissions by fuel type and vehicle category

9.5 Optimization Strategies and Recommendations

Based on the environmental hotspot analysis, several optimization strategies are recommended to enhance the sustainability profile of biomethane production systems:

9.5.1 CO_2 Removal Process Improvements

- Integration of renewable electricity sources for upgrading equipment operation
- Implementation of advanced separation technologies with reduced methane slip
- Heat recovery systems to minimize parasitic energy consumption
- Process intensification through hybrid upgrading configurations
- Utilization of separated CO_2 for value-added applications (e.g., algae cultivation, greenhouse enrichment).

9.5.2 Compression System Optimization

- Deployment of high-efficiency compression technologies
- Implementation of variable-speed drive systems for demand-responsive operation
- Waste heat recovery from compression processes
- Optimization of compression staging to minimize specific energy consumption.

9.5.3 System-Level Integration

Holistic system design should consider:

- Co-location of upgrading facilities with renewable energy infrastructure
- Integration with existing natural gas distribution networks where feasible
- Development of regional biogas-to-biomethane clusters to achieve economies of scale
- Implementation of advanced process control and monitoring systems.

9.6 Future Perspectives and Research Directions

The expanding role of biogas and biomethane in global energy systems necessitates continued research and development across multiple domains.

9.6.1 Technological Advancement

- Development of novel upgrading technologies with enhanced energy efficiency
- Advanced biological upgrading processes utilizing specialized microbial consortia
- Integration of artificial intelligence for process optimization and predictive maintenance
- Novel materials for improved separation efficiency and durability.

9.6.2 Feedstock Diversification

Expansion beyond traditional waste streams to include:

- Agricultural residues and energy crops with minimal land-use competition
- Industrial organic waste streams from food processing and manufacturing
- Emerging feedstocks such as microalgae and macroalgae
- Co-digestion strategies to optimize biogas yield and composition.

9.6.3 Policy and Market Development

- Enhanced policy frameworks supporting biomethane integration into energy systems
- Development of certification schemes ensuring sustainability compliance
- Market mechanisms for carbon credit generation from waste-derived biomethane

- Infrastructure investment programs facilitating biomethane distribution.

9.7 Conclusions

This investigation demonstrates that biomethane production from landfill biogas represents a viable pathway for sustainable energy generation with substantial environmental benefits. Life cycle assessment reveals that CO_2 removal operations and compression systems constitute primary environmental hotspots, contributing 84% and 37% to global warming potential and fossil resource depletion, respectively. Implementation of recommended optimization strategies—including renewable energy integration, process efficiency enhancement, and heat recovery—can significantly improve the environmental performance of biomethane production systems.

Comparative analysis demonstrates that biomethane deployment as a transportation fuel offers emission reductions of 70–75% for light-duty vehicles and 66–71% for heavy-duty vehicles relative to conventional fossil fuels. These substantial benefits derive from avoided landfill methane emissions, displacement of fossil fuel consumption, and the inherently renewable nature of the feedstock.

The continued expansion of biogas and biomethane production systems requires concerted efforts across technological innovation, policy development, and infrastructure investment. As global energy systems transition toward decarbonization, waste-derived biomethane will play an increasingly critical role in achieving climate mitigation targets while simultaneously addressing waste management challenges. Future research should focus on process optimization, feedstock diversification, and system-level integration to maximize the sustainability benefits of this renewable energy pathway.

Acknowledgements Author of this manuscript is thankful to the Institute for providing research support.

References

Aieamsam-Aung, P., et al. (2023). Advanced biogas upgrading technologies: Performance evaluation and economic analysis. *Scientific Reports, 13*(1), 9342. https://doi.org/10.1038/s41598-023-36425-5

An, D., Wang, T., Zhou, Q., Wang, C., Yang, Q., Xu, B., & Zhang, Q. (2017). Effects of total solids content on performance of sludge mesophilic anaerobic digestion and dewaterability of digested sludge. *Waste Management, 62*, 188–193.

Anderson, J. K., Thompson, R. W., & Miller, S. D. (2021). Global fossil fuel reserve projections and energy transition pathways. *Energy Policy, 145*, Article 111728.

Angelidaki, I., Treu, L., Tsapekos, P., Luo, G., Campanaro, S., Wenzel, H., & Kougias, P. G. (2018). Biogas upgrading and utilization: Current status and perspectives. *Biotechnology Advances, 36*(2), 452–466.

Chen, L., Wang, Y., & Liu, H. (2020). Renewable energy transition: Challenges and opportunities for developing economies. *Renewable and Sustainable Energy Reviews, 128*, Article 109912.

Dobers, G. M. (2019). Acceptance of biogas plants taking into account space and place. *Energy Policy, 135*, Article 110987.

Godoy, V., Martín-Lara, M. Á., Garcia-Garcia, G., Arjandas, S., & Calero, M. (2024). Environmental impact assessment of the production of biomethane from landfill biogas and its use as vehicle fuel. *Renewable Energy, 237*, Article 121685. https://doi.org/10.1016/j.renene.2024.121685

International Energy Agency. (2020). *Outlook for biogas and biomethane: Prospects for organic growth*. World Energy Outlook Special Report, IEA Publications, Paris.

IPCC. (2021). *Climate change 2021: The physical science basis. Contribution of Working Group I to the Sixth Assessment Report of the Intergovernmental Panel on Climate Change*. Cambridge University Press.

Jensen, M. B., et al. (2021). Pressure swing adsorption for biogas upgrading: Process optimization and performance analysis. *Biochemical Engineering Journal, 166*, Article 107869.

Johnson, P. R., & Williams, K. M. (2019). Energy security in the 21st century: Geopolitical dimensions and policy responses. *Energy Research & Social Science, 54*, 78–89.

Kapoor, R., et al. (2019). Biogas upgrading technologies: Status, challenges and economic perspectives. *Environmental Science and Pollution Research, 26*, 11631–11661.

Kougias, P. G., Treu, L., Campanaro, S., Zhu, X., & Angelidaki, I. (2016). Dynamic functional characterization and phylogenetic changes due to long chain fatty acids pulses in biogas reactors. *Scientific Reports, 6*, 28810.

Kumar, S., Singh, R., & Mishra, P. (2018). Biogas composition variability: Influence of feedstock and operational parameters. *Renewable Energy, 125*, 641–652.

Lansche, J., & Müller, J. (2017). Life cycle assessment of biogas versus dung combustion household cooking systems in developing countries—A case study in Ethiopia. *Journal of Cleaner Production, 165*, 828–835.

Lee, S. H., & Park, J. K. (2020). Methane as a greenhouse gas: Environmental implications and mitigation strategies. *Environmental Monitoring and Assessment, 192*, 458.

Ma, J., Zhao, Q. B., Laurens, L. L. M., Jarvis, E. E., Nagle, N. J., Chen, S., & Frear, C. S. (2015). Mechanism, kinetics and microbiology of inhibition caused by long-chain fatty acids in anaerobic digestion of algal biomass. *Biotechnology for Biofuels, 8*, 141.

Martinez, E. J., Raghavan, V., Gonzalez-Andres, F., & Gomez, X. (2019). New trends in biogas production from macroalgae and microalgae. *Resources, 8*(2), 91.

Moestedt, J., Müller, B., Westerholm, M., & Schnürer, A. (2016). Ammonia threshold for inhibition of anaerobic digestion of thin stillage and the importance of organic loading rate. *Microbial Biotechnology, 9*(2), 180–194.

Muñoz, R., et al. (2015). A review on the state-of-the-art of physical/chemical and biological technologies for biogas upgrading. *Reviews in Environmental Science and Bio/technology, 14*, 727–759.

Perin, J. K. H., Borth, P. L. B., Torrecilhas, A. R., da Cunha, L. S., Kuroda, E. K., & Fernandes, F. (2020). Optimization of methane production parameters during anaerobic co-digestion of food waste and garden waste. *Journal of Cleaner Production, 272*, Article 123130.

Rajendran, K., Aslanzadeh, S., & Taherzadeh, M. J. (2012). Household biogas digesters—A review. *Energies, 5*(8), 2911–2942.

Roberts, D. A., & Thompson, M. J. (2019). Energy transition pathways: A global perspective on fossil fuel phaseout. *Global Environmental Change, 58*, Article 101966.

Ryckebosch, E., Drouillon, M., & Vervaeren, H. (2011). Techniques for transformation of biogas to biomethane. *Biomass and Bioenergy, 35*(5), 1633–1645.

Smith, J. A., Williams, R. T., & Brown, K. L. (2020). Global energy security: Challenges and pathways to sustainability. *Energy for Sustainable Development, 54*, 45–58.

Sun, Q., Li, H., Yan, J., Liu, L., Yu, Z., & Yu, X. (2015). Selection of appropriate biogas upgrading technology—A review of biogas cleaning, upgrading and utilisation. *Renewable and Sustainable Energy Reviews, 51*, 521–532.

Tian, Z., Zhang, Y., Li, Y., Chi, Y., & Yang, M. (2015). Rapid establishment of thermophilic anaerobic microbial community during the one-step startup of thermophilic anaerobic digestion from a mesophilic digester. *Water Research, 69*, 9–19.

Tuinier, M.J., & van Sint Annaland, M. (2012). Biogas purification using cryogenic packed-bed technology. *Industrial & Engineering Chemistry Research, 51*(15), 5552–5558.

Wilson, C. A., et al. (2018). Trace contaminant removal from biogas: Impact on downstream applications and environmental performance. *Renewable Energy, 127*, 945–957.

Zhang, C., Su, H., Baeyens, J., & Tan, T. (2014). Reviewing the anaerobic digestion of food waste for biogas production. *Renewable and Sustainable Energy Reviews, 38*, 383–392.

Zhang, W., Heaven, S., & Banks, C. J. (2017). Continuous operation of thermophilic food waste digestion with side-stream ammonia stripping. *Bioresource Technology, 244*(1), 611–620.

Chapter 10
Global Bioenergy Production and Its Role in Sustainable Development Goals (SDGs)

Priyanka Yadav, Diksha Singh, Shivani Mishra, and Kusum Yadav

Abstract The world has relied heavily on fossil fuels over the past few decades, but environmental concerns have compelled the world to consider biofuel as a viable alternative for steady economic growth. Additionally, biofuel lessens environmental issues and helps maintain national energy security. In order to identify the barriers preventing biofuel from becoming a more sustainable and alternative fuel, the current study focuses on a thorough examination of bioenergy policy initiatives implemented by numerous government agencies across several nations. A comprehensive examination of policy assistance programs, blending targets, and feedstock utilisation has been conducted. To better understand the present state of the bioenergy market, the potential of commercial companies to produce bioenergy is also highlighted. The chapter ends with the writers' perspectives on the ongoing problems with the production of bioenergy worldwide.

Keywords Fossil fuels · Biofuels · Bioenergy · Environmental issues

10.1 Introduction

The urgent problem of supplying energy to meet human demands both now and in the future while adhering to regulations to avoid unintended repercussions in the future is a complicated one that calls for a significant structural and economic shift, or "energy transition." There is a fair amount of agreement that the "energy transition" needs to address issues with greenhouse gas emissions and forest degradation while maintaining global food security. One significant energy source that is known to produce a relatively favorable—even negative—carbon net balance is bioenergy. The subject of how bioenergy will work with other future human demands, primarily food and industrial raw materials (such textiles, etc.), is brought up by projections

P. Yadav · D. Singh · S. Mishra · K. Yadav (✉)
Department of Biochemistry, University of Lucknow, Lucknow, India
e-mail: anukusum@gmail.com

V. Singh et al. (eds.), *Bioenergy and Value Added Byproducts from Biomass Waste*, Sustainable Environmental Waste Management Strategies,
https://doi.org/10.1007/978-3-032-22849-9_10

of an increase in energy demand. The evaluation of the amount of usable land that is available, followed by the conditions and potential demand for such resources, is the fundamental problem. About 55.6 EJ, or about 10% of the world's primary energy supply, is responded to by bioenergy today. Table 3 displays the current bioenergy supply's approximate makeup. It reveals that biomass residues comprise about 23% and firewood alone accounts for 67%. The available arable land area, technological and economic performance, and environmental and spatial factors are the primary determinants of each bioenergy source's (biosource) supply level (Blair et al., 2021).

Although it is generally acknowledged that the supply of bioenergy may be expanded, it is still unclear how and to what extent. The first consideration while growing energy crops, such as woody crops or other crops like corn, soybeans and sugarcane, is the amount of land that is accessible. Land for food production must be included in estimates of future land availability for bioenergy products in all scenarios.

SG 7: Human as well as economic development is hampered by a lack of access to the energy sources and transformation systems. Numerous renewable and non-renewable energy sources, including solar, wind, hydropower, geothermal, biofuels, natural gas, coal, petroleum, and uranium, are found in the environment (Kabeyi & Olanrewaju, 2022b).

Global climate change will result from increased usage of fossil fuels without steps to reduce greenhouse gas emissions. Disaster risk reduction as well as climate change mitigation are aided by energy efficiency and the growing usage of renewable energy sources. Ecosystem preservation and upkeep enable the use as well as advancement of hydropower and bioenergy sources (Wuebbles & Jain, 2001).

Since 2015, numerous energy access projects have received backing from governments, businesses, and civil society organisations under the auspices of the Sustainable Development Goal 7 (SDG 7) of the UN. Notably, funds and expenditures allocated to renewable energy are seen as supporting the achievement of other SDGs pertaining to poverty alleviation, food security, health, as well as the climate change, in addition to providing "energy for all." It is appropriate to reflect and analyse how energy access is propelling development projects, affecting local populations, and influencing governance processes, as there are less than ten years remaining to achieve the SDG 7 targets (Brauch et al., 2022). Some important Facts are;

- For cooking and heating, 3 billion people use coal, wood, charcoal, or the animal dung.
- Energy is the primary cause of climate change, contributing around 60% of all greenhouse gas emissions worldwide.
- Global carbon dioxide emissions have risen by over 46% since 1990.
- With 16% of global electricity coming from hydropower at competitive costs, it is currently the greatest renewable electricity source. In a number of developed, growing, and developing nations, it controls the majority of the electrical mix.
- With 10% of the world's primary energy supply coming from bioenergy, it is currently the greatest renewable energy source.

10.1.1 Targets Linked to the Environment:

- **Target 7.1**: By 2030, guarantee that everyone has access to reasonably priced, dependable, and contemporary energy services. Transitioning to clean energy is critical for lowering greenhouse gas emissions, improving air quality, and alleviating the effects of the three global crises: climate change, biodiversity loss, and pollution and waste. Ongoing conflicts, especially in Ukraine, and global economic instability continue to produce major volatility in energy prices, forcing some countries to invest more in renewables while others increase their reliance on coal.
- **Target 7.2**: Significantly raise the proportion of renewable energy in the world's energy mix by 2030. While renewable energy has grown at an unparalleled rate over the previous decade, its share of overall final energy consumption has remained stable, at roughly 17%, while global energy consumption has grown at a comparable rate. According to the 2021 Tracking SGD7 Report, renewables are most dynamic in the power sector, reaching approximately 25% in 2018, whereas progress in the heat and transportation sectors has been significantly slower.
- **Target 7.3**: Double the pace of energy efficiency advancement worldwide by the year 2030. Energy intensity, or the amount of energy used to produce one unit of economic output, is used as a proxy for energy efficiency and to monitor progress towards SDG 7.3. Global progress towards SDG 7.3 is measured by the annual change in total energy supply and the GDP of the country or region under consideration. A decrease in energy intensity shows an increase in energy efficiency, as less energy is used to create the same amount of economic production, or in other words, more economic output is produced for each unit of energy expenditure.
- **Target 7.a**: By 2030, increase global collaboration to make clean energy research and technology more accessible, including energy efficiency, renewable energy, and cleaner, more sophisticated fossil fuel technologies. Encourage investment in clean energy technology and energy infrastructure.
- **Target 7.b**: In line with their individual support programs, develop infrastructure and modernise technology by 2030 to provide modern, sustainable energy services to all developing nations, especially landlocked, small island developing states, as well as least developed nations (McCollum et al., 2017).

10.1.2 Indicators Linked to the Environment:

7.1.1 Proportion of the population with an access to electricity.

7.1.2 Proportion of the population with primary reliance on the clean fuels as well as technology.

7.1.3 The intensity of energy were measured in the terms of the primary energy as well as GDP.

7.A.1 The International financial flows to the developing countries in the support of the cleans energy research as well as development and also the production of energy, such as hybrid systems.

7.B.1 The installed renewable energy-generating capacity in the developing as well as developed countries (in watts per capita) (Chepeliev et al., 2024, p. 269).

10.2 Global Bioenergy Production Trends

The term global bioenergy market is considered to be an essential as well as dynamic component of the larger renewable energy landscape, which are driven by the requirement for the sustainable, lower carbon energy sources. Therefore Bioenergy can harness the power of some organic materials like biogas, biomass as well as biofuels, for the production of transportation fuels and electricity. So, this growth of the market is underpinned by different key factors (Barot, 2022).

Firstly, bioenergy can play a major part in the transition towards the renewable energy sources as well as the mitigations of the climatic changes. Worldwide the government and the industries were incrementally changing to bioenergy as an eco-friendly as well as viable alternative to that of fossil fuels, and significantly it reduces the greenhouse gas emission along with curbing reliance on finite resources (Souza et al., 2017).

Secondly, the term bioenergy market may benefit from a well-established as well as versatile infrastructure. Several forms of the bioenergy, such as biogas, the biomass combustion and biofuels were already integrated into the existing energy supply chains, by making them accessible and adaptable for industries as well as for consumers (Yue et al., 2014).

Thirdly, the presence of diverse feedstock sources can contribute to market resilience. The agricultural residues, organic waste, the forest biomass, as well as energy crops offer a range of the feedstock options, by decreasing the risk of the supply shortages as well as fostering sustainability (Ghatak, 2011).

Additionally, technological advancements as well as ongoing research as well as development efforts may continue to increment or improve the economic validity alongwith the efficiency of the production of bioenergy.

The innovations in the fermentation processes, the biomass conversion technologies as well as the utilization of co-product are driving progress in these fields.

Lastly, the supportive policies as well as incentives like renewable targets, the carbon pricing mechanisms, as well as subsidies, encourage the investment in the bioenergy projects as well as foster market growth (Baumber, 2017).

10.3 Categories of Bioenergy Production

There are several categories for the bioenergy production. Some important categories are mentioned in the Fig. 10.1

10.3.1 Solid Biomass

Consists of organic waste, wood, and crop residues. mostly used for direct combustion or conversion into pellets and briquettes for industrial energy requirements, heating, and electricity production (Japhet et al., 2019).

10.3.2 Advanced Biofuels

Next-generation biofuels made from waste, algae, or non-food biomass. These provide more sustainability and efficiency and include cellulosic ethanol, green diesel, and synthetic biofuels (Kumar et al., 2019).

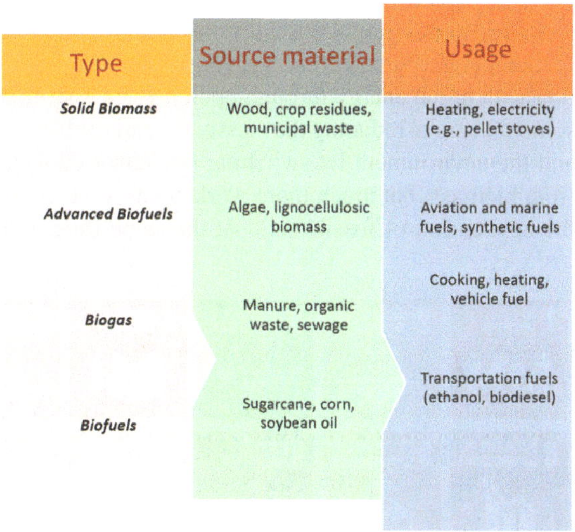

Fig. 10.1 Categories for the bioenergy production

10.3.3 Biogas

Made by anaerobic digestion of organic waste, including food scraps, sewage, and manure. It can be converted to biomethane for injection into gas grids or utilised as fuel for automobiles, or it can be used straight for heating and cooking (Kiyasudeen et al., 2016).

10.3.4 Biofuels

Transportation-related liquid fuels such as bioethanol (made from corn and sugarcane) and biodiesel (made from vegetable and animal fats). They are essential to decarbonising aircraft and road travel (Longo et al., 2024).

10.4 Bioenergy and Its Role in Sustainable Development

Bioenergy is the source of clean and green energy. It is eco-friendly and can be replace the fossil fuel in future. It is also play important role in the sustainable environmental development (Fig. 10.2).

In numerous impoverished countries, traditional biomass continues to be the predominant source of energy, providing nearly one-third of the world's population with home energy through open fires or traditional cookstoves. There has been some success in reducing the massive negative effects on human health, the economy, and the environment by switching to cleaner cookstoves and cleaner fuels derived from biomass, but much more work needs to be done, perhaps including expanding the application of fossil fuels. At the same time, biomass is growing quickly as a

Fig. 10.2 Bioenergy and its role in sustainable development

commercial energy source, particularly for transportation fuels. Although bioenergy can help achieve climate goals and support rural livelihoods, if it is not used correctly, it may worsen ecosystem, land, and water degradation, decrease food security, and raise greenhouse gas (GHG) emissions. It will take agriculturally sustainable practices and marketplaces that offer better livelihood prospects and fair terms of trade for large-scale commercial biofuels to support sustainable development. Converting the opportunity into reality is the difficult part (Sagar & Kartha, 2007).

10.4.1 Energy Access and Poverty Alleviation (SDG 1—No Poverty, SDG 7—Affordable and Clean Energy)

Economic growth and poverty alleviation are greatly aided by having access to contemporary energy services. For everyone, especially in rural and underserved areas, bioenergy—which is made from biomass, biogas, and biofuels—can be a game-changer in providing contemporary, affordable, dependable, and sustainable energy (Kaygusuz, 2011).

10.4.2 Improved Cookstoves and Clean Cooking Solutions

More than 2.3 billion people still cook and heat their homes using traditional biomass, including as wood, manure, and crop leftovers, in many poor nations (Thakur et al., 2025). This dependence results in:

- Every year, indoor air pollution results in over 3.2 million preventable deaths.
- significant time commitments for gathering fuelwood, particularly for women and girls.
- soil deterioration and deforestation.

Solutions based on bioenergy, like: Better biomass cookstoves (ICS), digesters that use biogas (using kitchen leftovers or animal waste), Stoves that use briquettes or pellets (Mapelli & Mungwe, 2013).

Case Study: India's Ujjwala Yojana reduced energy poverty and improved the health and empowerment of women by providing LPG (and concurrently, biogas) connections to nearly 90 million households.

10.4.3 Decentralized Bioenergy Systems for Rural Electrification

Bioenergy mini-grids and standalone systems offer a scalable and reasonably priced way to acquire electricity in off-grid and energy-poor areas (Niraula, 2015).

Schools, clinics, small enterprises, and houses are powered by biomass gasifiers and biogas generators.

- Energy crops, animal manure, and agricultural wastes can all power systems.
- Encourage the efficient use of energy in cottage businesses, food processing, and irrigation.
- For instance, Husk Power Systems (India and Nigeria) provide underprivileged areas with round-the-clock electricity by using rice husks to power mini-grids.

10.4.4 Socioeconomic Benefits of Bioenergy Access

1. **Income Generation**: The collection of feedstock, installation, operation, and maintenance of systems are all jobs generated locally by bioenergy (Bioenergy I.E.A., 2009).
2. **Women's Empowerment**: By easing the load of gathering fuelwood, more time may be spent on learning, earning a living, and engaging in communal activities (Shailaja, 2000).
3. **Health Benefits**: Lower levels of respiratory and cardiovascular diseases are associated with cleaner indoor air (Kumar et al., 2023).
4. **Educational Outcomes**: Having access to electricity allows schools to function, promotes digital learning, and prolongs study hours. (Sovacool & Ryan, 2016).

10.4.5 Key Facts

- **SDG 7 Target 1**: Make sure that everyone has access to modern, reasonably priced, and dependable energy services by 2030.
- **Employment in Bioenergy**: More than 3.6 million employment worldwide in 2022 (Trypolska et al., 2021).
- **Cost-effectiveness**: In distant locations, community biogas systems can provide electricity at a price of US$0.08–0.12 per kWh, which is competitive with fossil fuel alternatives (Asghar et al., 2022).

10.5 Climate Change Mitigation (SDG 13)

SDG 13, which demands immediate action to prevent climate change and its effects, places a strong emphasis on mitigating its effects. Bioenergy is essential because it substitutes carbon–neutral or even carbon-negative energy sources for fossil fuels. Carbon capture and utilisation (CCU) technology and biofuels for transportation are two important areas where bioenergy makes a contribution (Louman et al., 2019).

10.6 Biofuels and Transportation

Approximately 24% of worldwide CO_2 emissions are caused by the transport sector (IEA, 2023). Making the switch to biofuels is a crucial step in decarbonising this industry, particularly in heavy-duty transportation, shipping, and aircraft, where electrification is still difficult (Küfeoğlu, 2024).

10.6.1 First- and Second-Generation Biofuels

- Bioethanol derived from corn and sugarcane is mixed with petrol (such as E10 and E20).
- Blended with diesel (such as B10 or B20), biodiesel is made from vegetable or animal fats.
- Second-generation biofuels, which are made from lignocellulosic biomass as well as agricultural leftovers, have lifecycle emissions that are 60–90% lower than those of fossil fuels.

Sustainable Aviation and Marine Fuels

- Aviation emissions can be reduced by up to 80% with Sustainable Aviation Fuels (SAF), which are made from waste oils, algae, or forest leftovers.
- Adoption is being fuelled by incentives and mandates (such as the EU ReFuel Aviation and the US SAF Grand Challenge).

Global Biofuel Leaders

- **Brazil**: more than half of all passenger cars are powered by sugarcane ethanol.
- The USA is the world's greatest producer of ethanol from maize.
- India's National Bio-Energy Policy calls for a 20% ethanol blend by 2025.
- **Impact**: Between 2010 and 2022, biofuels avoided more than 1.4 billion tonnes of CO_2 emissions worldwide (IRENA, 2023).

10.6.2 Carbon Capture and Utilization (CCU) with Bioenergy

One of the rare "negative emissions" technologies is bioenergy with carbon capture as well as storage (BECCS), which extracts more CO_2 from the atmosphere than it releases (Kreuter & Lederer, 2021).

How It Works

- As biomass grows, it absorbs CO_2.
- The CO_2 generated during fermentation or combustion is caught and stored underground or used again in bioenergy power plants or biorefineries.
- A net carbon-negative energy cycle is therefore produced.

Carbon Utilization

Captured CO_2 can be reused in: **Building materials** (carbon-infused cement), **Synthetic fuels, as well as the Algae cultivation** (Oluremi, 2025).

10.6.3 Global BECCS Projects

In order to become carbon-negative, the Drax Power Station in the UK is switching to biomass using BECCS (Errendal, 2021).

CO_2 is captured from the manufacturing of ethanol by the Illinois Industrial CCS Project (USA) (Gollakota & McDonald, 2012).

Potential: By 2050, BECCS might assist achieve the objectives of the Paris Agreement by removing up to 5 Gt CO_2 annually worldwide (Chiquier et al., 2022).

10.7 Benefits of Bioenergy for Climate Action

Figure 10.3 depicts the Benefits of bioenergy for the climatic action.

10.8 Sustainable Agriculture and Rural Development (SDG 2—Zero Hunger & SDG 8—Decent Work and Economic Growth)

In addition to helping to mitigate climate change and provide clean energy, bioenergy is essential for supporting rural livelihoods, sustainable agriculture, and regional economic growth. Key elements of SDGs 2 (Zero Hunger) and 8 (Decent Work

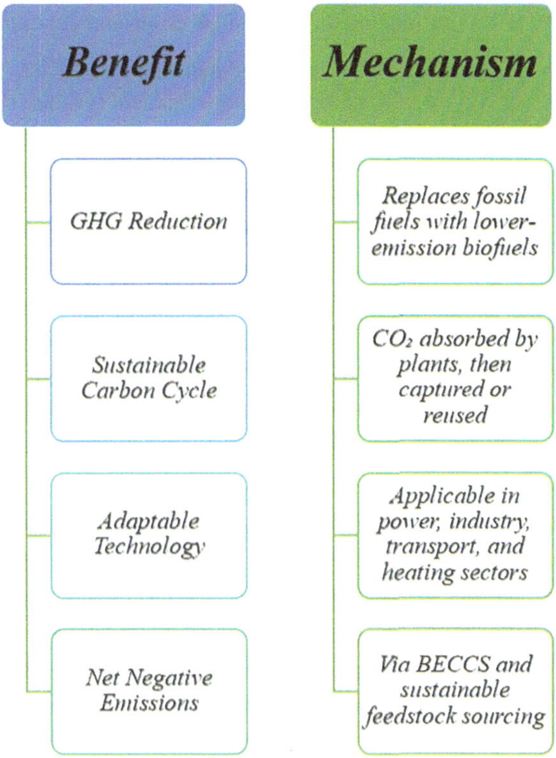

Fig. 10.3 Benefits and mechanism of Bioenergy

and Economic Growth) are addressed by bioenergy through its integration with agricultural systems and rural infrastructure (Creutzig et al., 2015).

10.9 Bioenergy Crops and Sustainable Farming

Through the production of energy crops and the integrated use of agricultural leftovers, bioenergy promotes sustainable agriculture (Muller, 2009).

Energy Crops

In order to minimise conflicts between food and fuel, dedicated crops such as Jatropha, switchgrass, Miscanthus, sugarcane, and sweet sorghum are grown on marginal or degraded soils. These crops need little in the way of chemical inputs and are frequently drought-tolerant (Rao et al., 2019).

Benefits:

- Using perennial root systems to enhance soil and sequester carbon
- increases biodiversity and land productivity

- allows farmers to diversify their crops and get extra revenue.

As an example, in India and sub-Saharan Africa, intercropping food crops with bioenergy crops maximises land usage and revenue (Wicke et al., 2011).

10.10 Waste-to-Energy and Circular Agriculture

Rice husk, sugarcane bagasse, animal manure, and crop leftovers are examples of agricultural and food industry waste that can be used to produce (Gupta et al., 2023):

Anaerobic digestion produces biogas.
Biopower (by gasification or burning of biomass)
Biofertilizers (made from slurry or digestate).

Advantages

- Enhances garbage management in rural regions and lowers air pollution and open burning
- Increases agricultural yields by returning nutrients to the soil.
- Reduces expenses and emissions by replacing chemical fertilisers.

Case Study: SATAT Scheme (India): With buyback assurances from oil firms, the Sustainable Alternative Towards the Affordable Transportation initiative encourages the establishment of compressed biogas (CBG) plants using agro-waste. The objective of this scheme is to utilize biomass waste to produce CBG; To reduce dependency on fossil fuels and cut down crude oil imports; To promote entrepreneurship and rural employment in bioenergy; To help manage agricultural and organic waste effectively; and To contribute to India's climate goals and Sustainable Development Goals (SDGs) (Singh & Kumar, 2023).

10.11 Employment and Income Generation in Rural Areas

There are several job prospects in rural areas thanks to the bioenergy value chain:

- Harvesting and raising feedstock
- Gathering and moving biomass
- Building and maintaining bioenergy facilities
- Distribution of biogas and biofertilizer.

According to IRENA (2023):

Globally, the bioenergy industry supports more than 3.6 million employment. Several case studies show that rural bioenergy initiatives have raised household income by 10% to 30%. Decentralised bioenergy initiatives in Brazil and Kenya have

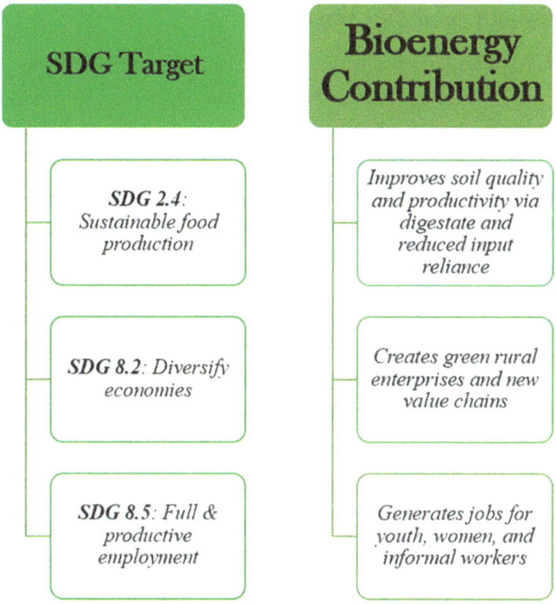

Fig. 10.4 SDG Targets and their bioenergy contributions

spawned regional economic models centred on pellet fuel production, mini-grids, and biodigester upkeep (Reid et al., 2020).

Synergies with SDG Targets

Figure 10.4 represents the SGD targets with their bioenergy contributions. The above SDGs include SDG 2.4, SDG 8.2, and SDG 8.5 along with their contributions.

10.12 Energy Security and Economic Development (SDG 7 and SDG 9)

Enhancing energy security, fostering economic growth, and spurring industrial innovation and infrastructure expansion are all made possible by bioenergy. Bioenergy promotes inclusive industrialisation and creates robust, low-carbon energy systems by utilising regional biomass resources and renewable technology, achieving SDGs 7 and 9 (Kothari et al., 2023).

Job Creation in the Bioenergy Sector

From the collection of feedstock to the manufacturing and distribution of fuel, the bioenergy value chain provides substantial job possibilities at every level, directly supporting SDG 8 (decent work) and indirectly supporting SDG 9 (Singh, 2015).

Employment Pathways

- **Agriculture**: Producing and harvesting bioenergy crops, such as energy grass, bamboo, and sweet sorghum
- **Logistics**: Gathering, moving, and storing biomass waste
- **Industry**: Building and running biomass-based power plants, biogas facilities, and biofuel refineries
- **Services**: Technical assistance, sales, and upkeep of bioenergy systems

Global Impact

In 2023, bioenergy supported more than 3.6 million employment worldwide (IRENA) (Gehring & Cand, 2024).

Compressed biogas (CBG) units have made thousands of rural jobs possible in India thanks to the SATAT program (Prasad et al., 2022).

For instance, under SATAT, a 5-ton/day CBG factory can provide 25–30 direct and indirect jobs.

Infrastructure Development and Industrial Innovation—By encouraging the development and growth of sustainable infrastructure, bioenergy boosts regional economies and lessens reliance on imported energy.

Infrastructure Contributions

- Bio-refineries are establishments that transform biomass into a variety of goods, such as energy, bioethanol, and bioplastics.
- Biogas facilities: Both commercial (municipal/agro-industrial) and domestic (household-scale)
- Mini-grids: Rural electrification systems fuelled by biomass that assist local businesses and farmers.

Technological Innovation

- Sophisticated conversion techniques (such as gasification, enzymatic hydrolysis, and pyrolysis)
- Systems of hybrid energy: Using biomass and solar energy together to create reliable off-grid power
- Intelligent bioenergy solutions: Optimising plant operations and feedstock utilisation with IoT and AI.

Impact on Energy Security

- Bioenergy lessens dependency on imported fossil fuels. The bioenergy may able to offer an alternative to those of the conventional energy sources, decrementing the reliance on coal, oil as well as gas imports.
- Encourages the use of locally accessible biomass to diversify energy sources, protecting economies from fluctuations in world prices. It utilizes the locally available biomasses that helps to diversity the energy mix, and self-sufficiency as well as shields the national economy from the global energy price volatility.

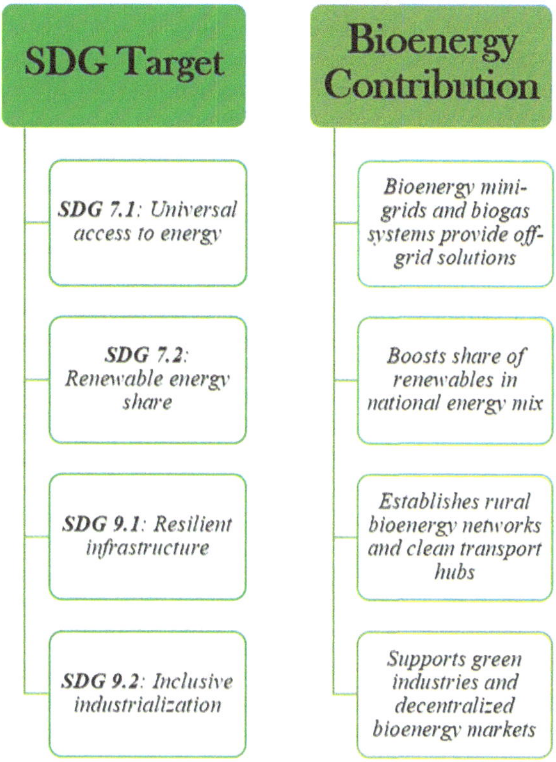

Fig. 10.5 SDG targets and their bioenergy contributions

10.13 Contribution To SDG Target

Figure 10.5 represents the SGD targets with their bioenergy contributions. The above SDGs include SDG 7.1, SDG 7.2, SDG 9.1, SDG 9.2 along with their contributions.

10.14 Challenges and Trade-Offs

Although bioenergy has a lot of potential advantages, there are drawbacks and compromises. The goal of bioenergy is to lower greenhouse gas (GHG) emissions and help achieve global climate change mitigation goals. However, bioenergy raises a number of sustainability issues, particularly in relation to the effects of exploiting land for the cultivation of energy crops. Depending on the circumstances, growing energy crops may result in trade-offs or synergies between lowering GHG emissions and other sustainability benefits. The primary trade-offs and synergies related to land usage for dedicated energy crop production were determined using the framework of the Sustainable Development Goals (SDGs) of the United Nations. Additionally, the context-specific factors that influence such synergies and trade-offs were discovered, including biomass feedstock, past land use, climate, soil type, and agricultural management. GHG emission reduction (SDG 13) and other SDGs were compared pairwise, and the most recent literature was examined. Four hundred and twenty-seven observations were categorised as synergy (170), trade-off (176), or no effect (81). Perennial crops grown on arable land, pasture, or marginal land in the "cool temperate moist" temperate zone and "high activity clay" soils showed the greatest synergies with SDGs linked to the environment, such as biodiversity conservation and water quality. The majority of trade-offs concerned the availability of water and food security. Climate zone and soil type have less of an impact on synergies and trade-offs than previous land usage and feedstock type (Halsnæs et al., 2024).

10.15 Policy and Technological Innovations for Sustainable Bioenergy

Energy generated from biological resources, including energy crops, forestry and agricultural waste, post-consumer waste, and other organic materials, is known as bioenergy (Järvinen et al., 2025). Biogas, liquid biofuels, solid biomass, and biomethane are all included. Through the use of biomass boilers or district heating networks, bioenergy can be used to generate power, heat buildings and industrial processes, and supply fuel for transportation by land, sea and air (Kabeyi & Olanrewaju, 2022a, 2022b). Additionally, chemical businesses can use bio-based resources as feedstock (such as biomethanol) to create industrial goods, such as bio-based polymers. Combining bioenergy with carbon capture and storage technology can also result in negative emissions (Parkhey, 2022).

About half of all renewable energy consumption in 2019 came from bioenergy, which now accounts for 9% of the world's energy demand (Kober et al., 2020). In developing nations, where biomass is used in crude and ineffective stoves, almost half of its present applications are conventional, such as cooking and heating. This leads to significant indoor air pollution, which has detrimental health effects (Mehetre et al., 2017). In many places, gathering fuel wood and making charcoal for cooking also contributes to deforestation and biodiversity loss. Recent years have seen the development of modern, effective uses of bioenergy, which now supply 2% of power generation, over 3% of transportation energy needs, and 8% of all energy demand in buildings and industries. Only a small portion of the feedstocks used to produce chemicals and plastics—less than 1% of overall production—are currently derived from bio-based sources (Bioenergy, I.E.A., 2009).

Modern biomass has a great deal of promise to increase energy supplies in countries like China, India, and Brazil that have large populations and growing demand. It can be transformed into alternatives to gas or oil, or it can be burned directly for heating or power generation. Biomass is digested anaerobically to produce biogas.

The transportation industry is the primary use for liquid biofuels, a practical renewable alternative to petrol. With the largest fleet of flexible-fuel vehicles that can run on bioethanol, an alcohol primarily produced by fermenting the carbohydrates in sugar or starch crops like corn, sugarcane, and sweet sorghum, Brazil leads the world in liquid biofuels (Nogueira et al., 2020). The energy transformation that is compatible with achieving the 1.5 °C climate targets heavily relies on bioenergy. By 2030, modern bioenergy's share of the energy and feedstock demand must increase by a ratio of three. By 2050, it will account for 17% of total final energy consumption, supporting all end-use sectors and particularly facilitating emission reductions in industries with few alternatives, such as aviation fuel and feedstock in the chemicals industry (Mandley et al., 2022).

To limit negative effects and encourage sustainability, regulations must be carefully crafted. Competition for land, emissions from changing land uses, deforestation, biodiversity loss, and air pollution are some of the possible effects. Though they are very context-specific based on location, socioeconomic conditions, and local political and regulatory capabilities, appropriate solutions and actions to assure sustainability are available (Smith et al., 2016).

Additionally, a thorough and cohesive policy framework that tackles the biggest deployment obstacles is needed. Clear and long-term goals and plans, financial and fiscal incentives, mandates and responsibilities, support for R&D, infrastructure and commercialisation, training and skill development, steps to provide trustworthy information and increase public awareness, and sustainability governance backed by certifications and regulations are all examples of appropriate policies.

The role of bioenergy in the energy transition, deployment obstacles, and policy solutions to these obstacles have all been examined by IRENA. Above all, IRENA has been studying the sustainability elements of bioenergy and finding best practice-backed policy frameworks to guarantee its sustainability (Asmelash et al., 2020).

10.16 Conclusion

The production of bioenergy on a global scale has the potential to revolutionise the pursuit of a low-carbon, inclusive, and sustainable future. It supports a number of SDGs when handled sustainably, including those pertaining to responsible production, renewable energy, economic development, and climate action. To reach its full potential, bioenergy expansion must be balanced with social justice, environmental integrity, and food security.

References

Asghar, N., Amjad, M. A., Ur Rehman, H., Munir, M., & Alhajj, R. (2022). Achieving sustainable development resilience: Poverty reduction through affordable access to electricity in developing economies. *Journal of Cleaner Production, 376*, Article 134040.

Asmelash, E., Prakash, G., Gorini, R., & Gielen, D. (2020). Role of IRENA for global transition to 100% renewable energy. *Accelerating the transition to a 100% renewable energy era* (pp. 51–71). Springer International Publishing.

Barot, S. (2022). Biomass and bioenergy: Resources, conversion and application. In *Renewable energy for sustainable growth assessment* (pp. 243–262).

Baumber, A. (2017). Enhancing ecosystem services through targeted bioenergy support policies. *Ecosystem Services, 26*, 98–110.

Bioenergy, I.E.A. (2009). *Bioenergy—A sustainable and reliable energy source*. International Energy Agency Bioenergy.

Blair, M. J., Gagnon, B., Klain, A., & Kulišić, B. (2021). Contribution of biomass supply chains for bioenergy to sustainable development goals. *Land, 10*(2), 181.

Brauch, M. D., Toledano, P., Mehranvar, L., Iliopoulos, T., Sasmal, S., Aydos, M. (2022). Scaling investment in renewable energy generation to achieve sustainable development goals 7 (affordable and clean energy) and 13 (climate action) and the Paris Agreement: Roadblocks and drivers.

Chepeliev, M., Diachuk, O., Podolets, R., Semeniuk, A., & Trypolska, G. (2024). Net-zero transition in Ukraine: Implications for sustainable development goal 7. In: *Aligning the Energy Transition with the Sustainable Development Goals* (p. 269).

Chiquier, S., Fajardy, M., & Mac Dowell, N. (2022). CO_2 removal and 1.5°C: What, when, where, and how? *Energy Advances, 1*(8), 524–561

Creutzig, F., Ravindranath, N. H., Berndes, G., Bolwig, S., Bright, R., Cherubini, F., Chum, H., Corbera, E., Delucchi, M., Faaij, A., & Fargione, J. (2015). Bioenergy and climate change mitigation: An assessment. *GCB Bioenergy, 7*(5), 916–944.

Errendal, S. (2021). *BECCS—A climate friend or foe? A discourse depiction of the negative emissions technology BECCS in the UK* (Master Thesis Series in Environmental Studies and Sustainability Science).

Gehring, J. C., & Cand, B. M. (2024). The sustainable energy imperative: A future generations perspective on technologies leading the clean energy transition.

Ghatak, H. R. (2011). Biorefineries from the perspective of sustainability: Feedstocks, products, and processes. *Renewable and Sustainable Energy Reviews, 15*(8), 4042–4052.

Gollakota, S., & McDonald, S. (2012). CO_2 capture from ethanol production and storage into the Mt Simon Sandstone. *Greenhouse Gases: Science and Technology, 2*(5), 346–351.

Gupta, A. P., Upadhyay, P., Sen, T., & Dutta, J. (2023). Agricultural waste as a resource: the lesser travelled road to sustainability. In *Agriculture waste management and bioresource: The circular economy perspective* (pp. 1–20).

Halsnæs, K., Some, S., & Pathak, M. (2024). Beyond synergies: Understanding SDG trade-offs, equity and implementation challenges of sectoral climate change mitigation options. *Sustainability Science, 19*(1), 35–49.

Japhet, J. A., Tokan, A., & Kyauta, E. E. (2019). A review of pellet production from biomass residues as domestic fuel. *International Journal of Environment, Agriculture and Biotechnology, 4*(3), 835–842.

Järvinen, M., Paulomäki, H., van Kasteren, H., Oksanen, E., Herzon, I., Forsius, M., Akujärvi, A., Holmberg, M., Junttila, V., Pekkonen, M., & Schulz, T. (2025). Bioenergy and waste. In *Designing renewable energy systems within planetary boundaries* (pp. 461–544). Springer.

Kabeyi, M. J. B., & Olanrewaju, O. A. (2022a). Biogas production and applications in the sustainable energy transition. *Journal of Energy, 2022*(1), 8750221.

Kabeyi, M. J. B., & Olanrewaju, O. A. (2022b). Sustainable energy transition for renewable and low carbon grid electricity generation and supply. *Frontiers in Energy Research, 9*, Article 743114.

Kaygusuz, K. (2011). Energy services and energy poverty for sustainable rural development. *Renewable and Sustainable Energy Reviews, 15*(2), 936–947.

Kiyasudeen S. K., Ibrahim, M. H., Quaik, S., & Ahmed Ismail, S. (2016). An introduction to anaerobic digestion of organic wastes. In *Prospects of organic waste management and the significance of earthworms* (pp. 23–44).

Kober, T., Schiffer, H. W., Densing, M., & Panos, E. (2020). Global energy perspectives to 2060–WEC's World Energy Scenarios 2019. *Energy Strategy Reviews, 31*, Article 100523.

Kothari, R., Goria, K., Bharti, A., Singh, H. M., Pathak, V. V., Pathak, A., & Tyagi, V. V. (2023). Sustainable development goals (SDGs-7) for bioeconomy with bioenergy sector. In *Sustainable butanol biofuels* (pp. 29–56). CRC Press.

Kreuter, J., & Lederer, M. (2021). The geopolitics of negative emissions technologies: Learning lessons from REDD+ and renewable energy for afforestation, BECCS, and direct air capture. *Global Sustainability, 4*, Article e26.

Küfeoğlu, S. (2024). Transportation sector emissions. In *Net zero: Decarbonizing the global economies* (pp. 493–554). Springer Nature Switzerland.

Kumar, N., Sonthalia, A., Pali, H. S., & Sidharth. (2019). Next-generation biofuels—Opportunities and challenges. In *Innovations in sustainable energy and cleaner environment* (pp.171–191).

Kumar, P., Singh, A. B., Arora, T., Singh, S., & Singh, R. (2023). Critical review on emerging health effects associated with the indoor air quality and its sustainable management. *Science of the Total Environment, 872*, Article 162163.

Longo, A., Calabrese, C., Nobre, C., Poravou, C. A., Yangin-Gomec, C., Grottola, C. M., di Perta, E. S., Battista, G., Ariemma, J. M. G., Michel, J. B., & Valentino, L. (2024). Key enabling applications to drive the development of biorefineries.

Louman, B., Keenan, R. J., Kleinschmit, D., Atmadja, S., Sitoe, A. A., Nhantumbo, I., de Camino Velozo, R., & Morales, J. P. (2019). SDG 13: Climate action—Impacts on forests and people. In *Sustainable development goals: Their impacts on forests and people* (pp.419–444).

Mandley, S., Wicke, B., Junginger, H., van Vuuren, D., & Daioglou, V. (2022). Integrated assessment of the role of bioenergy within the EU energy transition targets to 2050. *GCB Bioenergy, 14*(2), 157–172.

Mapelli, F., & Mungwe, J. N. (2013). Modern energies services for cooking: From improved cook-stoves to domestic and community biogas based systems. *Renewable energy for unleashing sustainable development: Blending technology, finance and policy in low and middle income economies* (pp. 43–74). Springer International Publishing.

McCollum, D., Gomez Echeverri, L., Riahi, K., & Parkinson, S. (2017). Sdg7: Ensure access to affordable, reliable, sustainable and modern energy for all.

Mehetre, S. A., Panwar, N. L., Sharma, D., & Kumar, H. (2017). Improved biomass cookstoves for sustainable development: A review. *Renewable and Sustainable Energy Reviews, 73*, 672–687.

Muller, A. (2009). Sustainable agriculture and the production of biomass for energy use. *Climatic Change, 94*(3), 319–331.

Niraula, A. (2015). *Scaling up of off-grid solar micro grids*. University of Edinburgh.

Nogueira, L. A. H., Souza, G. M., Cortez, L. A. B., & de Brito Cruz, C. H. (2020). Biofuels for transport. In *Future energy* (pp. 173–197). Elsevier.

Olurcmi, D. (2025). Waste management meets carbon capture: A sustainable approach.

Parkhey, P. (2022). Biomethanol: Possibilities towards a bio-based economy. *Biomass Conversion and Biorefinery, 12*(5), 1877–1887.

Prasad, S., Singh, A., Rathore, D., & Rakshit, A. (2022). Biogas technology for improving livelihoods and agricultural sustainability. In *Innovation in small-farm agriculture* (pp. 107–119). CRC Press.

Rao, P. S., Vinutha, K. S., Kumar, G. A., Chiranjeevi, T., Uma, A., Lal, P., Prakasham, R. S., Singh, H. P., Rao, R. S., Chopra, S., & Jose, S. (2019). Sorghum: A multipurpose bioenergy crop. *Sorghum: A State of the Art and Future Perspetives, 58*,399–424.

Reid, W. V., Ali, M. K., & Field, C. B. (2020). The future of bioenergy. *Global Change Biology, 26*(1), 274–286.

Sagar, A. D., & Kartha, S. (2007). Bioenergy and sustainable development? *Annual Review of Environment and Resources, 32*(1), 131–167.

Shailaja, R. (2000). Women, energy and sustainable development. *Energy for Sustainable Development, 4*(1), 45–64.

Singh, V. P., & Kumar, N. (2023). Financing India's 2030 NDC targets and beyond. In *The Role of Coal in a Sustainable Energy Mix for India* (pp. 283–298). Routledge.

Singh, J. (2015). Overview of electric power potential of surplus agricultural biomass from economic, social, environmental and technical perspective—A case study of Punjab. *Renewable and Sustainable Energy Reviews, 42*, 286–297.

Smith, P., House, J. I., Bustamante, M., Sobocká, J., Harper, R., Pan, G., West, P. C., Clark, J. M., Adhya, T., Rumpel, C., & Paustian, K. (2016). Global change pressures on soils from land use and management. *Global Change Biology, 22*(3), 1008–1028.

Souza, G. M., Ballester, M. V. R., de Brito Cruz, C. H., Chum, H., Dale, B., Dale, V. H., Fernandes, E. C., Foust, T., Karp, A., Lynd, L., & Maciel Filho, R. (2017). The role of bioenergy in a climate-changing world. *Environmental Development, 23*, 57–64.

Sovacool, B. K., & Ryan, S. E. (2016). The geography of energy and education: Leaders, laggards, and lessons for achieving primary and secondary school electrification. *Renewable and Sustainable Energy Reviews, 58*, 107–123.

Thakur, I., Sharma, R. P., & Kumar, S. (2025). Agricultural wastes: Their implications and management. In Organic Production of Vegetable Crops (pp. 267–288). Apple Academic Press.

Trypolska, G., Kryvda, O., Kurbatova, T., Andrushchenko, O., Suleymanov, C., & Brydun, Y. (2021). Impact of new renewable electricity generating capacities on employment in Ukraine in 2021–2030. *International Journal of Energy Economics and Policy, 11*(6), 98–105.

Wicke, B., Smeets, E., Watson, H., & Faaij, A. (2011). The current bioenergy production potential of semi-arid and arid regions in sub-Saharan Africa. *Biomass and Bioenergy, 35*(7), 2773–2786.

Wuebbles, D. J., & Jain, A. K. (2001). Concerns about climate change and the role of fossil fuel use. *Fuel Processing Technology, 71*(1–3), 99–119.

Yue, D., You, F., & Snyder, S. W. (2014). Biomass-to-bioenergy and biofuel supply chain optimization: Overview, key issues and challenges. *Computers & Chemical Engineering, 66*, 36–56.

Chapter 11
Advances in Catalysis for Biomass Conversion to Biofuels

Abhishek Anand, Shyam Sunder Rao, Anamika Maurya, Parmanand Maurya, and Rahul Ranjan

Abstract Sustainable energy solutions are imperative for the world's growing energy demands and environmental challenges. Biofuels represent a viable and promising alternative to fossil fuels. Converting biomass into biofuels offers a feasible approach to renewable energy production, utilising diverse and abundant feedstocks such as algae, lignocellulosic biomass, and agricultural residues. Catalysis enhances these biomass-to-biofuel conversion processes' efficiency, selectivity, and economic viability. This chapter explores recent advancements in homogeneous and heterogeneous catalysts tailored for critical biomass conversion pathways, including hydrolysis, pyrolysis, gasification, and hydrodeoxygenation. Particular focus is placed on developing bifunctional and multifunctional catalysts capable of facilitating simultaneous reactions and catalysts incorporating metals, metal oxides, and zeolites with specific porosities and surface chemistry. Additionally, the chapter examines the impact of innovative catalytic materials, such as metal–organic frameworks (MOFs), hybrid systems, and nanostructured catalysts, on enhancing process efficiency and sustainability. Integrating catalytic processes with green technologies, such as electrocatalysis, supercritical fluids, and microwave-assisted conversion, offers a promising pathway to achieving carbon neutrality and energy efficiency. This chapter addresses emerging solutions alongside challenges such as feedstock variability, catalyst deactivation, and scale-up limitations. By combining advanced catalytic approaches with innovative reactor designs and process intensification strategies, biomass conversion holds tremendous potential to revolutionise the biofuel industry and contribute to a sustainable energy future.

Keywords Biomass · Biofuels · Biodiesel · Catalyst · Biodiesel

---

A. Anand · S. S. Rao · A. Maurya · P. Maurya
Department of Chemical Engineering and Technology, IIT (BHU) Varanasi, Varanasi, Uttar Pradesh, India

R. Ranjan (✉)
School of Biochemical Engineering, IIT (BHU) Varanasi, Varanasi, Uttar Pradesh, India
e-mail: rahulranjan.rs.bce19@itbhu.ac.in

V. Singh et al. (eds.), *Bioenergy and Value Added Byproducts from Biomass Waste*, Sustainable Environmental Waste Management Strategies,
https://doi.org/10.1007/978-3-032-22849-9_11

11.1 Introduction to Catalysis in Biomass Conversion

The imperative to identify renewable sources of energy and chemicals has become increasingly urgent due to pressing environmental challenges, such as global warming and the depletion of fossil fuels, which have significantly influenced global research agendas (Sadorsky, 1999). Biomass is anticipated to emerge as a pivotal source of fuels and chemicals in the foreseeable future. Recent advancements in biomass conversion processes have facilitated the development and implementation of innovative technologies capable of transforming diverse biomass feedstocks into a spectrum of valuable products (Shafiee & Topal, 2009). These technological advancements have profoundly impacted the sustainable fuel production landscape and established essential building blocks for tomorrow's chemicals.

Bioethanol and biodiesel are presently the two most popular first-generation biofuels globally. Bioethanol is generally produced from corn and sugarcane in several countries, besides using other sources of sugar (Hamza et al., 2021). The hydrolysis of biomass is the critical step in bioethanol production, which involves the breakdown of carbohydrates into simple sugars. Biomass needs to be pre-treated for subsequent enzymatic degradation. The cellulose then undergoes enzymatic hydrolysis to produce glucose, which fermentation can convert into biofuels and chemicals. Biodiesel can be produced from various plant seeds such as palm, jatropha, colza, sunflower, rapeseed, and soybean-based oils. Transesterification is one of the standard methods applied for the production of biodiesel. In contrast, catalytic cracking is a direct route applied to produce kerosene, gasoline, and diesel.

This chapter investigates the diverse processes involved in converting biomass into biofuels, particularly on utilising lignocellulosic biomass. While we do not underestimate the significance of other potential feedstocks and processes, our primary focus will be upgrading strategies for producing fuels from aqueous solutions containing carbohydrates derived from lignocellulose. Through the examples presented herein, we aim to elucidate various alternatives for biomass processing, offering options that facilitate the most efficient processing of each feedstock.

Consistent combustion of conventional fuels in transportation and various industrial applications primarily contributes to environmental pollution. The rise in concentrations of greenhouse gases is a significant factor driving dramatic climatic changes, resulting in the widely recognised phenomenon of "global warming." This rapid shift in climate presents substantial risks to biodiversity and human life. Consequently, there is an urgent imperative to investigate alternatives to traditional fossil fuels through the development and implementation of renewable energy sources.

A significant paradigm shift has occurred in recent decades, moving from fossil fuels towards renewable energy sources. Renewable energy is increasingly recognised as a viable solution for achieving sustainable energy growth while minimising adverse environmental impacts. It represents the most effective alternative to fossil fuels in numerous sectors, including power generation, transportation, and rural energy supply. Projections indicate that renewable energy resources will fulfil considerable global energy demand. According to the 2017 Renewable Energy Policy

Network report, global investment in renewable energy increased by a remarkable 4.4 times between 2004 and 2016. Notably, among the leading countries, China has made substantial investments in producing renewable energy derived from diverse renewable fuels (Table 11.1).

Biofuels exhibit significant diversity owing to the extensive variety of potential feedstocks. Biomass can be transformed into valuable biofuels, such as biohydrogen, bioethanol, and biodiesel. As renewable and sustainable energy sources, biofuels are cost-effective and vital in reducing greenhouse gas emissions. Biofuels are categorised according to the feedstock utilised in their production. Primary biofuels originate from wood-based feedstocks, including firewood and wood chips, utilized directly as fuel. First-generation biofuels are derived from food-based feedstocks, such as corn, sugarcane, and soybeans. Second-generation biofuels employ non-edible feedstocks, including agricultural residues, forestry waste, and energy crops, thereby avoiding competition with food supply. Third-generation biofuels originate from aquatic feedstocks, particularly algae, known for their rapid growth rates and high biomass yield. Fourth-generation biofuels are derived from feedstocks obtained from non-arable lands, incorporating genetically engineered crops and other novel sources, emphasising sustainability and reducing land-use competition (Figs. 11.1 and 11.2).

Biofuels are vastly diverse due to their infinite sources of feedstock. Biomass is converted into valuable biofuels such as biohydrogen, bioethanol, and biodiesel. Biofuels have emerged as renewable, sustainable, cost-efficient, and vital in producing fewer greenhouse gases (Ullah et al., 2015). Biofuels are categorised based on their feedstock type as primary (wood as feedstock), first generation (food-based feedstock), second generation (non-edible feedstock), third generation (aquatic source feedstock), and fourth generation (non-arable land feedstock). Various categories of biofuels are presented in Fig. 11.2. According to the World Energy Forum, conventional fuels are estimated to be consumed in the next ten decades (Kumar et al., 2020). Globally, energy consumption increases daily, and the population will increase by up to 25% from 2010 to 2040 (Singh et al., 2016). This upsurge in energy demand causes the depletion of renewable fuel reserves at a speedy rate and significantly impacts climate change. To meet this energy demand, sustainable, clean and green alternative options come into play (Encinar et al., 2002). Energy obtained from various renewable sources such as sun, wind, tidal, and biomass collectively offers a path to future energy necessities.

Bioenergy derived from biomass is increasingly recognised as a viable energy source with a lower environmental impact. The natural decomposition of biomass generates substantial quantities of methane. This greenhouse gas is approximately twenty times more potent than carbon dioxide in terms of its harmful effects on climate change (Pang et al., 2019). Biomass is a fuel source and critical in generating heat, which is applicable in various industrial and domestic processes. Furthermore, carbon dioxide is a vital component for biomass growth; the CO_2 utilised during biomass synthesis is sourced from the ambient environment. The combustion of biofuels results in the release of carbon dioxide, which is subsequently reabsorbed by biomass for its metabolic activities. This cyclical exchange renders the

Table 11.1 Comparison of various renewable energy sources

Renewable energy sources	Constraints	Life cycle assessment (LCA method)	Outcome	References
Hydropower energy	May cause flooding in nearby areas Only restricted to areas where water is available The best locations for dam construction present various challenges	Impact 2002 + IPCC and ReCiPe 2016	The research focused on alpine and non-alpine region-based hydropower plants. Alpine regions are better than non-alpine regions in terms of CO_2 emissions and global warming potential. Comparison with other sources such as fossil fuel, wind, biomass, and photovoltaic-based power plants shows that hydropower plants are more suited for the environment	Mahmud et al. (2019)
Geothermal Energy	Installation and startup of geothermal plants is expensive Corrosion issues that increase its maintainable cost	ILCD 2011 and ReCiPe 2016	Geothermal power plants showed better results than the existing national energy mix. However, when compared with solar PV and wind power plants, geothermal power plants cause greater environmental impacts	Basosi et al. (2020)
Solar energy	Economically less competent Sustainability based on the presence of sunlight	Eco indicator 99	Perovskite solar cells show less stable CO_2 emissions. Sensitive analysis on these solar cells indicates that, if the performance and lifespan increase over time, these cells will become the most environmentally friendly among other solar cells	Gong et al. (2015)

(continued)

Table 11.1 (continued)

Renewable energy sources	Constraints	Life cycle assessment (LCA method)	Outcome	References
Wind energy	Nede's continuous and large amount of wind required an energy storage system	CML 2001	Impacts of wind, nuclear, and hydro-power-plants were compared. The study finds that wind power plants have the most impact on nature. Moreover, the manufacturing phase of wind and hydro-power plants was significant whereas the decommissioning phase of nuclear power plants was of greatest impact	Wang et al., 2019)
Biomass energy	Plentiful as well as renewable sources of energy Help to reduce waste and make the environment green Most economical sources to harvest energy	TRACI, ReCiPe 2008, and IMPACT 2000+	Feedstock production creates a significant impact. Apart from the smog and ozone layer depletion, other categories were less impactful. Transportation distances also play a role in greenhouse gas emissions	Weldu, (2017)

overall process carbon neutral. This natural carbon utilisation approach promotes a carbon–neutral footprint. Biofuels serve as a sustainable and environmentally friendly alternative to conventional energy sources, exhibiting net-zero emissions of greenhouse gases. Bioenergy represents a viable, low-cost fuel source critical to the bioeconomy (Smets et al., 2014). With the increasing prices of fossil fuels, bioenergy production is anticipated to receive heightened attention over the next three decades (Pang et al., 2019).The production of fuels and chemicals from lignocellulosic biomass has historically been achieved through two primary platform technologies. The first is the biochemical process, wherein biomass is converted into reducing sugars through steps, including pretreatment and enzymatic hydrolysis, followed by microbial fermentation to yield fuel products. The second platform is the thermochemical process, which involves the treatment of biomass via gasification or pyrolysis to generate intermediates such as syngas or bio-oil, which can subsequently be upgraded into drop-in fuels. An emerging alternative to these conventional biomass conversion methods is the implementation of hybrid processes that integrate both

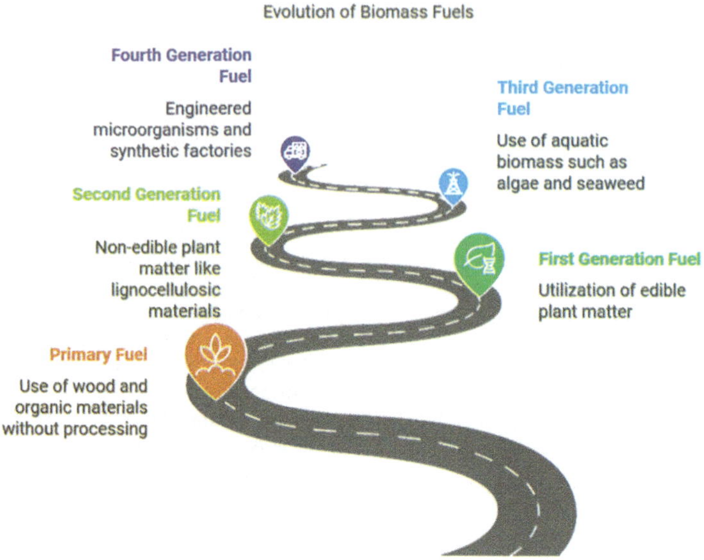

Fig. 11.1 Roadmap for the evolution of biofuels from biomass

Fig. 11.2 Various categories of biofuels

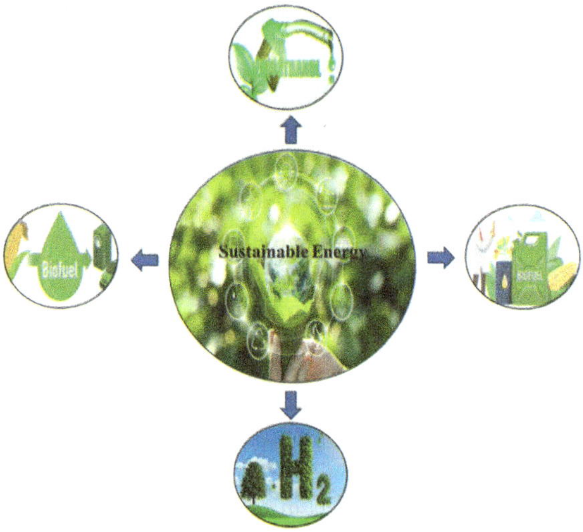

thermochemical and biochemical approaches. These hybrid processes are particularly advantageous because they can enhance carbon conversion efficiency, thereby optimising the production of targeted fuel products (Brown, 2007).

11.2 Importance of Catalysis in Biofuel Production

Biofuel production has emerged as critical in the global transition towards a more sustainable energy landscape, offering a viable alternative to traditional fossil fuels and addressing the pressing environmental concerns that have plagued the world for decades. The increasing energy demands driven by industrialisation and modernisation, coupled with the depletion of natural fuel reserves, have prompted the search for clean energy technologies to meet future energy needs while mitigating the detrimental environmental impact of greenhouse gas emissions (Kumar et al., 2020). Microalgae have been identified as a promising biofuel production feedstock, as they can efficiently convert sunlight, carbon dioxide, and other nutrients into a diverse array of valuable renewables, including biofuels (Encinar et al., 2002; Singh et al., 2016). Compared to first and second-generation biofuels, microalgae-based biofuels offer several advantages, such as higher growth rates, the ability to thrive in a wide range of environments, and the potential to utilise waste streams as feedstock, thereby reducing the competition with food production (Encinar et al., 2002; Sun et al., 2011). Furthermore, producing biofuels from microalgal biomass is environmentally sustainable and promises economic viability. Microalgal-derived biofuels, such as bioethanol and biodiesel, are carbon–neutral, as the amount of CO_2 released during their combustion equals the amount fixed by the biomass during photosynthesis.

11.3 Overview of the Biomass Conversion Pathways

Biomass refers to organic material from living or recently living organisms, classified as either non-lignocellulosic or lignocellulosic. Non-lignocellulosic biomass includes materials without significant lignin, such as food waste and algae, which can be easily converted into biofuels. Lignocellulosic biomass, found in wood, straw, and other plant materials, contains lignin and is more complex to convert, but is abundant and has significant potential for renewable energy (Aberilla et al., 2019; Kaloudas et al., 2021) Forms of biomass include woody, herbaceous, aquatic debris, farming manure, and industrial byproducts. It's use can help reducing greenhouse gas emissions and promote a sustainable, circular economy. Biomass can be transformed into valuable fuels or chemicals through a variety of technologies. These include gasification, which converts organic material into syngas; combustion, where biomass is burned to release energy; pyrolysis, which thermally decomposes biomass in the absence of oxygen; enzymatic hydrolysis, where enzymes break down complex carbohydrates into sugars; and fermentation, a biological process that converts sugars into ethanol or other byproducts. Each method offers unique advantages for harnessing the energy stored in biomass. This study explores the integration of hydrothermal and biochemical methods in biomass utilisation within the framework of a circular bioeconomy (Sadorsky, 1999). Thermochemical processes typically require substantial energy input and often necessitate the addition of solvents or catalysts. In contrast, biochemical methods can be lengthy and demonstrate reduced efficiency in breaking down recalcitrant biomass materials. Combining these two approaches presents a promising opportunity to harness the strengths of both methods in biofuel production. The hydrothermal methods serve as a pretreatment stage to enhance the suitability of biomass feedstock for subsequent biological processes, thereby improving overall process efficiency and final product yields (Aberilla et al., 2019). The integration of thermochemical and biochemical processes presents significant challenges. For instance, the use of catalysts or solvents in thermal methods may result in the contamination or destruction of microorganisms or produce various inhibitors that can adversely affect biological pathways. Furthermore, this integration may incur additional costs. The identification of biofuel sources, such as biodiesel and biochar, presents a valuable opportunity to mitigate the environmental impacts associated with fossil fuels (Balajii & Niju, 2019; Gunarathne et al., 2019). Moreover, adopting biofuels can contribute to reducing the growing dependence on fossil resources, thereby alleviating the pressures on non-renewable energy sources. It is imperative to utilize rigorous and scientifically sound methodologies to evaluate the advantages of biofuels relative to conventional energy sources (Chamkalani et al., 2020). Life Cycle Assessment (LCA) is recognized as a comprehensive evaluation framework (Astrup et al., 2015) that assesses the environmental impacts associated with the entire production chain of biofuels (Collotta et al., 2019).

Biomass combustion is not the best option for producing energy because it can harm the environment and does not recover all the energy stored in the material (Ullah et al., 2015) A better way to use biomass is to convert it into solid, liquid and gas forms. This method is a more efficient and eco-friendly energy source for heating, power, and transportation fuels (Kargbo et al., 2021; Wang et al., 2020a). There are two main ways to convert biomass: thermochemical and biochemical methods. Thermochemical methods use heat and controlled oxygen levels to change the entire biomass into different energy forms. In contrast, biochemical methods utilise enzymes, bacteria, and other engineered organisms to convert raw materials into liquid fuels, including drop-in biofuels (Singh et al., 2016). In recent decades, the synthesis process for biomass-derived fuels has progressed from first-generation biofuels to fourth-generation biofuels, encompassing second and third generations along the way. First-generation biofuels are derived from food crops, second-generation biofuels utilise inedible biomass, third-generation biofuels are produced from macro and microalgal biomass, and fourth-generation biofuels are based on genetically engineered algal and microbial systems (Ben-Iwo et al., 2016). Various biomass-to-liquid processes have produced numerous biomass-based fuels, chemicals, and organic compounds. These include methane, ethane, propane, butane, ethylene, methanol, ethanol, butanol, dimethyl ether, ammonia, acetic acid, formaldehyde, gasoline, diesel, wax, paraffin, bio-jet fuels, and more, all of which are readily available in markets worldwide (Demirbas & Demirbas, 2010).

Various thermochemical technologies, including direct combustion, torrefaction, hydrothermal liquefaction, pyrolysis, and gasification, have been developed to convert biomass into liquid fuels. These processes involve the controlled decomposition of biomass, which results in the generation of solid, liquid, and gaseous products, including synthesis gas (syngas). This syngas typically requires an additional catalytic process to be converted into liquid fuels known as drop-in biofuels. A significant advantage of these technologies is their capability to utilize a diverse range of biomass types as feedstock, distinguishing them from biochemical conversion methods (Osman, 2020). Thermal-based processes for biomass conversion can be divided into two categories: low-temperature processes, which generally operate below 300 °C, and high-temperature processes, which function above 300 °C. Low-temperature methods include torrefaction and hydrothermal carbonisation, while high-temperature methods encompass gasification, combustion, and pyrolysis (Song et al., 2021). Direct combustion of biomass, whether through flaming or smouldering, typically occurs at temperatures ranging from 1000 to 2000 °C in the presence of air. However, these processes can lead to emissions of nitrogen oxides (NOx) and other pollutants. As a result, gasification which operates within the range of 800–1300 °C, is viewed as a promising alternative for producing energy and chemicals with reduced emissions (Smets et al., 2014). The synthesis gas produced from gasification can be utilised for electricity generation or converted into liquid fuel through the Fischer–Tropsch process. Moreover, pyrolysis is considered an attractive option because it requires a lower temperature, typically reaching up to 500 °C, compared to gasification.

11.4 Homogeneous Catalysis in Biomass Conversion

Homogeneous catalysts have been developed for dehydrogenating alcohols, exhibiting chemo selectivity and activity under relatively mild conditions. However, recent advances suggest that an increasing number of heterogeneous catalysts are competitive with their homogeneous counterparts in terms of activity and selectivity. This chapter aims to provide a comprehensive and systematic comparison of both types of catalysts employed in the acceptor less dehydrogenation of ethanol and glycerol. The distinct advantages of each category of catalysts and identify the shortcomings that need to be addressed to develop more cost-effective, robust, and efficient catalytic systems. Furthermore, we highlight the challenges associated with designing catalysts that comply with the principles of green chemistry and the necessity of employing suitable metrics to evaluate the impacts of circular processes (Subramaniam et al., 2021).

Over the last decade, there has been a notable increase in the literature concerning ethanol dehydrogenation utilising both homogeneous and heterogeneous catalysts. Among the homogeneous catalysts, compounds from group VIII have garnered the most extensive research attention for this application. Milstein et al. demonstrated that the PNP–Ru(II) catalyst can facilitate ethanol dehydrogenation at low temperatures (−30 °C). However, the resultant aldehydes were sequestered by the catalyst, generating an aldehyde adduct with the ruthenium centre and the PNP pincer ligand (Montag et al., 2012).

11.5 Bioethanol Production Challenges

Bioethanol production from lignocellulosic biomass encounters significant obstacles due to the intricate multi-stage conversion process. The inherent recalcitrance of lignin, coupled with the variability in biomass composition, presents considerable challenges for pretreatment and enzymatic hydrolysis. These factors contribute to high costs and, consequently, limit the commercial viability of bioethanol production. The economic viability of bioethanol production is significantly hindered by high research and production costs. Additionally, the necessity for expensive enzyme blends and the use of energy-intensive pretreatment processes present substantial challenges that impede the widespread adoption of this technology.

11.5.1 Biorefinery Challenges

Integrating various processes within a biorefinery, particularly the synthesis of bioethanol and biogas production, presents considerable complexity. Achieving an optimal balance between energy input and output to enhance overall efficiency remains a significant challenge in this field. Membrane separation processes used in biorefineries face issues such as membrane fouling and concentration polarisation, which affect the purification of biofuels like bioethanol and biogas. Enhancements in membrane engineering are needed to overcome these limitations.

11.5.2 Biofuels and Biogas Challenges

The conversion of biomass-derived syngas into biofuels is impeded by inadequate mass transfer properties and suboptimal ethanol yields, significantly limiting commercialisation efforts. To overcome these challenges, conducting further research into biocatalysts and refining fermentation conditions is essential. The production of biofuels may result in detrimental environmental consequences due to the utilisation of synthetic chemicals and energy carriers. Moreover, waste and wastewater generation present substantial sustainability challenges that require careful consideration and management. The conversion of biomass-derived syngas into biofuels is limited by poor mass transfer properties and low ethanol yield, which hinder commercialisation. Addressing these issues requires further research into biocatalysts and fermentation conditions. Integration and Efficiency: Integrating different processes within a biorefinery, such as combining bioethanol and biogas production, is complex. Achieving a balance between energy input and output to improve efficiency remains a significant challenge.

11.5.3 Heterogeneous Catalysis for Biofuel Production

A functional substance that continuously forms active sites with its reactants under reaction circumstances is known as a heterogeneous catalyst. These catalysts alter the localized reactants rates of chemical reactions without altering the material's thermodynamic equilibrium (Shafiee & Topal, 2009). Heterogeneous catalysis, involving a gas or liquid-phase reactant and a solid catalyst, occurs through adsorbing one or more reactant molecules onto the catalyst's surface. This process increases the local concentration of reactants at the surface and facilitates their activation. The catalytic mechanism begins with the adsorption of the reactant(s), followed by the chemical transformation of adsorbed species, and concludes with the desorption of the product (Hamza et al., 2021) (Fig. 11.3).

11.5.3.1 Types of Heterogeneous Catalysts (Zeolites, Metal Oxides, Supported Metals, etc.)

Solid heterogeneous catalysts fall into one of three categories (Fig. 11.3):

Desulfurization reactions occur on a sulphide catalyst, oxidation reactions are catalysed by oxides. Insulator oxides catalyse dehydration, whereas acid/base solids interact with carbocation/carbanionic intermediates. These characteristics have given rise to the notion that the reactant and the catalyst molecules are somewhat compatible (Hamza et al., 2021).

Fig. 11.3 Various solid heterogeneous catalysts

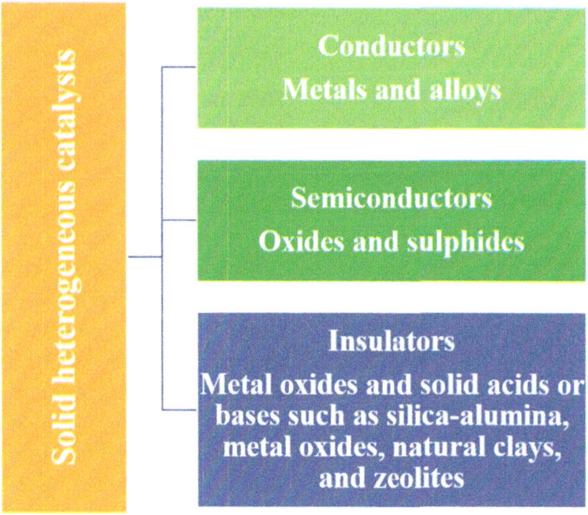

11.6 Role of Heterogeneous Catalysts in Pyrolysis, Gasification, and Hydro Processing

11.6.1 Role of Heterogeneous Catalysts in Pyrolysis

Catalytic pyrolysis of biomass can be performed in two ways: in-situ, using a single reactor with lower costs but poor heat transfer and faster catalyst deactivation, or ex-situ, using dual reactors for selective product control but with higher complexity and costs. Various catalysts are studied for specific product yields. Table 11.2 and Fig. 11.4 depict the common catalysts employed in the pyrolysis of various biomass feedstock. As of right now, zeolite-based catalysts, particularly ZSM-5 varieties of zeolites.

Wang et al. examined the physiochemical alterations in ZSM-5 and reported catalytic pyrolysis of lignocellulosic biomass (lignin, cellulose, and Xylan) over ZSM-5 (Mahmud et al., 2019). The findings demonstrated that the ZSM-5 catalyst generated olefins and aromatic hydrocarbons with selectivity. Following the reaction, ZSM-5's crystal structure and morphology remained unchanged, but the amount of

Table 11.2 Role of heterogeneous catalysts in pyrolysis of different biomass

Sources of feed	Catalysts	Conditions	Key findings	Reference
Lignin, cellulose and xylan	ZSM-5	Pyrolysis, catalytic upgrading, and ex-situ mode at 500 °C for 30 s	ZSM-5's inclusion aided in the synthesis of olefins and aromatic hydrocarbons	Mahmud et al. (2019)
Torrefied cedarwood	Ni, Cr modified Mo_2N/ HZSM-5, HZSM-5	20 °C/min, 600–800 °C, 20 s resident time, He is a gas carrier	When 5% Ni–Mo_2N/HZSM-5 is catalysed at 700 °C and the catalyst-to-biomass ratio is 5:1 (w/w), the yields of xylene, toluene, and benzene can be raised to 2.10, 3.94, and 5.54%, respectively	Basosi et al. (2020)
Black liquors	Ga, Mo or Zn modified H-ZSM-5	20 °C/s from 40 to 600 °C in-situ. Ex-situ: catalysed at 575 °C, pyrolysis at 550 °C, and 50 kPa	In both in-situ and ex-situ pyrolysis, Ga-modified HZSM-5 produced up to 7 weight percent of BTX (based on the organic fraction of the black liquors)	Gong et al. (2015)
Lignin waste	Al-SBA-15, HZSM-5, Al-SBA-15/ HZSM-5	45 min of resident time, 450, 550, and 650 °C, and 10 °C/min	At 550 °C, the concentration of mono-aromatic hydrocarbons was 42.57% utilizing Al-SBA-15/HZSM-5 (mass ratio: 1:3)	Wang et al. (2019)

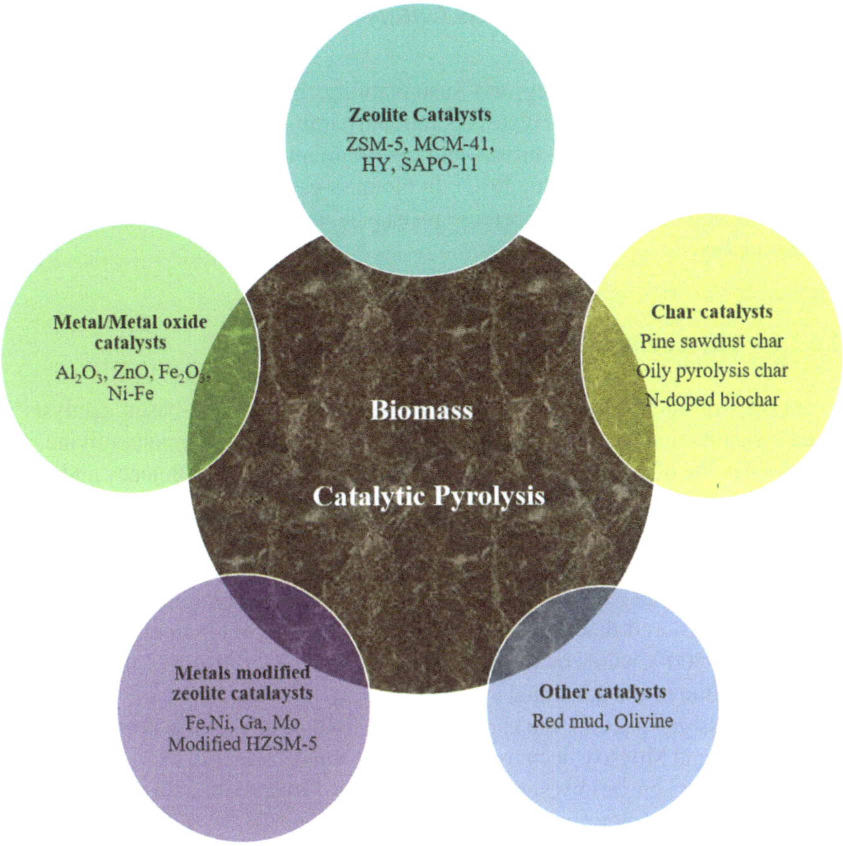

Fig. 11.4 A few common uses for catalysts in the pyrolysis of biomass

total acid, Brønsted acid, and Lewis's acid all dropped. At the same time, the coke that was deposited on ZSM-5 produced new micropores and raised their specific surface area. A series of metal-modified ZSM-5 catalysts, such as Fe, Ni, Ga, Mo, Cr, Zn, etc., have been shown to enhance the quality of pyrolysis products in addition to pure ZSM-5. For catalytic rapid pyrolysis of torrefied cedarwood, Liu et al. assessed Ni or Cr modified Mo2N/HZSM-5 catalysts. They suggested that torrefaction in conjunction with catalytic pyrolysis might aid in the biomass's selective conversion into aromatic hydrocarbons (Basosi et al., 2020).

11.6.2 Role of Heterogeneous Catalysts in Gasification

In order to improve efficiency and sustainability, biomass gasification research focuses on technology, gasifier types, operational parameters, feedstocks, target products, catalysts, models, mechanisms and environmental performance (Pang et al., 2019; Ullah et al., 2015; Weldu, 2017). In biomass gasification, catalysts are essential to lower activation energy, decrease resistance, increase efficiency, and improve product quality.

11.6.2.1 The Catalyst Used in the Gasification Process

Recent uses of carbon-based, Fe-based, and alkali metal oxide-based catalysts in biomass gasification are compiled in Fig. 11.5 and Table 11.3. Cement, olivine, and ZSM-5 are some of the other catalysts that are employed. High-efficiency and stable catalyst design is still difficult in extreme gasification settings.

Gasifier pine sawdust at 600, 700, and 800 °C in a fixed-bed reactor with N_2 and CO_2 gamifying agent, using the alkali metal oxide CaO as a catalyst and CO_2 absorber (Smets et al., 2014). The CaO and sawdust were combined in a 1:1 ratio. This study discovered that the catalytic process of CaO could significantly reduce 25% of CO_2 at 700 °C while increasing the production of H_2 (up to 10%) and lowering the heating value of the generated syngas.

For biomass gasification, the earth-available, low-cost, and non-noble transition metals Fe, Ni, and Mo have been frequently used. Kumar et al. (2020) created several Ni-based catalysts on various supports (AC, CNTs, ZrO_2, Al_2O_3, SiO_2, and Al_2O_3–SiO_2) and evaluated how well they catalysed soybean straw SCWG. It was also determined that the addition of promoters (Na, K, and Ce) to the Ni/ZrO_2 and Ni/Al_2O_3 catalysts improved H_2 yield and selectivity in the following order: ZrO_2 > Al_2O_3 > AC > CNT > SiO_2 > Al_2O_3–SiO_2. Increasing Ni dispersion and the metallic surface area, the promoter (Ce in particular) improved H_2 yields and reduced coke deposition.

Furthermore, Ni-modified MgO nano-catalysts (Mg1-xNixO) were used in a number of investigations by Mastuli et al. On the other hand, the $Mg_{0.80}Zn_{0.20}O$ catalyst generated the highest total gas yield (213.5 ml/g) and a 438.1% increase in H_2 yield, suggesting that metal oxide doped materials Mg1-xZnxO could be a new catalyst structure system for biomass SCWG. Another more recent study using Mg1-xNixO catalysts found that the mesoporous $Mg_{0.80}Ni_{0.20}O$ achieved the highest overall gas yield of 193.5 mL/g with a 361.7% increase in H_2 yield when compared to non-catalytic SCWG.

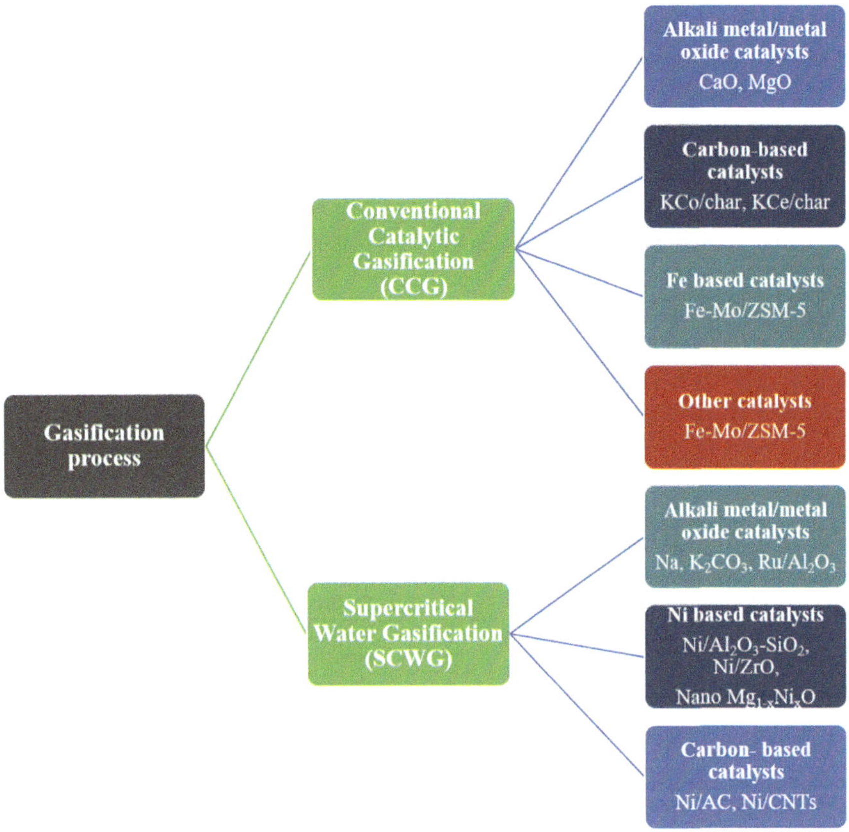

Fig. 11.5 Common catalysts used in biomass catalytic gasification

11.6.2.2 Role of Heterogeneous Catalysts in Hydro Processing

Bio-oils produced by hydro processing of biomass often have a more excellent heating value (HHV) and less oxygenated chemicals, making them of superior quality than those produced by pyrolysis (Encinar et al., 2002; Singh et al., 2016). The creation of improved catalysts to raise the yields and Caliber of liquid products has been the primary emphasis of hydro processing research up to this point (Sun et al., 2011).

11.6.2.3 Catalyst Used in Hydro Processing

As seen in Fig. 11.6 and described in Table 11.4, many types of homogeneous and heterogenous catalysts have been added to hydro processes to enhance liquefaction efficiency and bio-oil quality (Aberilla et al., 2019). Only homogeneous or heterogeneous catalysts are assessed in a single process in certain investigations. However, to

Table 11.3 Catalysts used for gasification of different biomass types

Sources of feed	Catalysts	Conditions	Key findings	Reference
Pine sawdust	CaO	CCG: 600, 700, 800 °C	The catalyst raised the syngas's H_2 concentration by up to 10% while lowering CO_2 emissions by 58.7%	Smets et al. (2014)
Long-strip compressed industrial sawdust char	Char supports KCo, KNi, KFe, or KCe	CCG: N_2, CO_2 atmosphere, 700–800 °C	Carbon conversion rose from 14.38 to 99% thanks to KCo catalyst. KCo > KNi > KFe > KCe was the catalysts order of catalytic performance	Pang et al. (2019)
Shells from coconuts and oil palm fronds	Dolomite, limestone, and Portland cement	CCG: 700–900 °C, 0–30wt.% catalyst loading, and air environment	Limestone produced the most syngas, followed by dolomite and Portland cement	Brown (2007)
Soybean straw	With promoters Na, K, and Ce, Ni is supported on AC, CNTs, ZrO_2, and Al_2O_3-SiO_2	SCWG: 45 min, 500 °C, 1:10 feedstock/water ratio, 23–25 MPa	The maximum H_2 yield (8.1 mmol/g) and selectivity were demonstrated by Ni/ZrO_2. All of the promoters decreased the production of coke while increasing H_2 output	Kumar et al. (2020)

compare homogeneous and heterogeneous catalysts, both homogeneous and heterogeneous catalysts are simultaneously studied. Furthermore, a dual system containing both homogeneous and heterogeneous catalysts is investigated in some recent studies to examine their synergistic effects.

In biomass Hydro process, homogenous catalysts have been employed extensively. Using a homogeneous Na_2CO_3 catalyst and a glycerol-water mixed solvent, Kashimalla et al. reported the Hydro process of rice straw (Kaloudas et al., 2021). Due to their high activity and ease of recovery from liquid products, heterogeneous catalysts are currently the subject of extensive study in biomass Hydro process, which lowers the high cost of the entire bio-crude production chain. Zeolite catalysts appear to have the best catalytic performance among them. HZSM-5, MCM-41, and γ-Al_2O_3 catalysts were used by Yang et al. for the Hydro process of antibiotic residue in ethanol–water mixed solvents (Balajii & Niju, 2019). The dual catalytic systems exhibited clear synergistic effects that efficiently promoted the low-temperature degradation of corn stalks to liquid products rich in amino phenyl compounds. With $ZnFe_2O_4$-NaOH dual catalysts, the maximum concentration (84%) of monophenyl compounds was achieved at 220 °C.

The main industrial applications of heterogeneous catalysis are:

(i) Bulk chemicals
(ii) Energy and transportation, oil refining, textiles, fine chemicals, pharmaceuticals, medical chemicals, food feed, polymers, and detergents

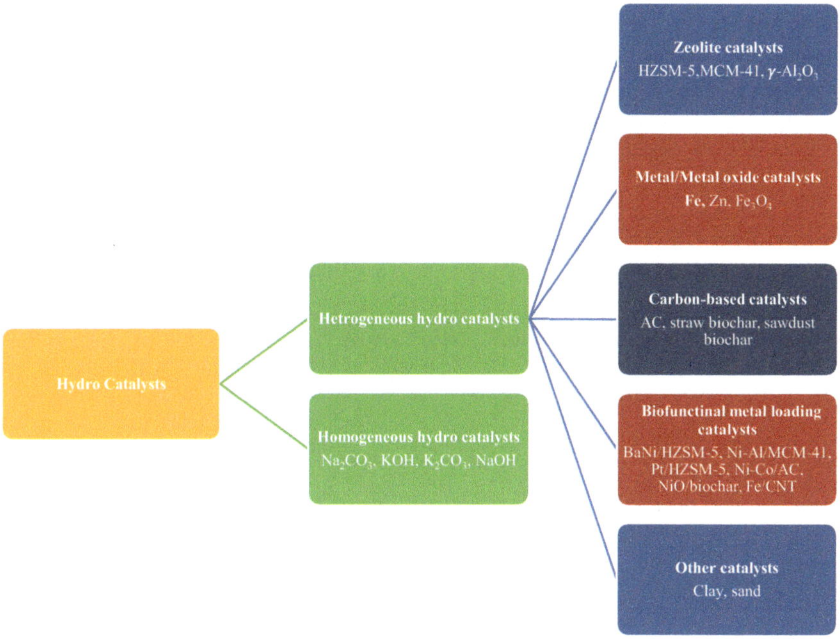

Fig. 11.6 Typical homogeneous and heterogenous catalysts used in biomass Hydro process

(iii) Catalyst design, subsequent catalyst and catalytic process development, plant design/engineering and realisation
(iv) Commercial manufacturing of adequate amounts of catalysts
(v) Plant operations and chemical reaction monitoring and control
(vi) Environmental problems (Hamza et al., 2021)

11.7 Advances in Catalyst Stability and Reusability

Catalyst stability and reusability improvements are essential for catalytic process optimisation, especially in industrial settings where durability and effectiveness are critical. Some recent advancements concentrate on enhancing the stability of different catalyst systems using breakthroughs in reaction conditions and materials science.

Table 11.4 Lists recent catalyst applications in hydro process for different biomass

Biomass	Catalysts	Solvents	Conditions	Key findings	References
Rice straw	Na_2CO_3	Water + glycerol	N_2 for 15–60 min, 21.2–60 bar, and 220–280 °C	The ideal circumstances for operation: 12.5 weight percent of Na_2CO_3 at 260 °C for 60 min at a glycerol-to-water ratio of 1:1 (v/v)	Kaloudas et al. (2021)
Antibiotic residue	HZSM-5, MCM-41, and γ-Al_2O_3	Ethanol + water	150 min at 280 °C, Ar at 0.69 MPa	Higher levels of carbon (up to 71.44%), hydrogen (up to 9.37%), and HHV are found in the catalytic bio-oil	Balajii and Niju (2019)
Empty fruit bunch	BaNi, BaLa and BaCe modified CHZSM5	Water	30 min at 250 °C with 10 bar N_2	Bio-oil yield and biomass conversion were enhanced by bimetallic modified CHZSM5 catalysts, with Ba1La2/CHZSM5 being the most successful catalyst	Gunarathne et al. (2019)
Municipal water sludge	$CoMo/Al_2O_3$, $NiMo/Al_2O_3$ and activated carbon	Water + formic acid	30 min at 325 °C, 0.2 MPa H_2 or N_2	The quantity and quality of bio-oils were enhanced by the combined use of a catalyst and formic acid	Chamkalani et al. (2020)

11.8 Conventional Catalyst Recovery Techniques

Filtration uses porous materials to separate liquid catalysts from solid catalysts. Variants such as pressure, vacuum, and gravity filtering are used. While vacuum and pressure filtration enable higher rates and efficiency, gravity filtration is slower and best suited for low-viscosity liquids. Among the difficulties are partial separation and liquid adhesion to particles, necessitating further procedures such drying and washing (Astrup et al., 2015).

This technique uses centrifugal force to separate particles based on density and speed up sedimentation. Although it is not cost-effective for industrial applications, it is quicker than natural sedimentation. It is difficult to recover smaller catalysts, such as nanoparticles fully (Collotta et al., 2019; Kargbo et al., 2021).

11.9 Limitations

Both approaches cause weight loss and raise financial issues at industrial scales because they are time-consuming and ineffective for nanoscale catalysts.

Magnetic separation provides an effective substitute to solve the problems of time, energy, and catalyst loss (Ben-Iwo et al., 2016; Wang et al., 2020a).

11.9.1 Magnetic Separation

By employing an external magnet to effectively separate solid catalysts, magnetic separation promotes green chemistry, the reuse of costly catalysts, and lowers waste and energy consumption.

11.9.2 Magnetic Nanoparticles (MNPs)

Magnetic nanoparticles (MNPs) are used in catalysis due to their superparamagnetic, high surface area, and ease of recovery via external magnetic fields. They offer excellent catalytic activity and reusability. Iron oxide-based MNPs, such as Fe_3O_4, are particularly valued for their low cost, nontoxicity, and ability to be functionalised with other materials.

11.9.3 Magnetically Recoverable Transition Metal Ferrite Nanoparticles

$CuFe_2O_4$ and other magnetic spinel ferrite nanoparticles (MFe_2O_4) are widely valued for their mechanical hardness, high surface area, and chemical stability, which make them perfect for heterogeneous catalysis Ali et.,al. proved that $CuFe_2O_4$ catalysts, which are produced by hydrothermal and co-precipitation techniques, effectively catalyse the conversion of used frying oil into biodiesel. Following five cycles, these nanoparticles maintained their strong catalytic efficacy with no discernible loss of activity, demonstrating exceptional reusability.

11.9.4 *Magnetically Recoverable Orthoferrites of Rare-Earth Nanoparticles*

Lanthanides, scandium, and yttrium are examples of rare earth elements (REEs), which are essential in a variety of fields, including metallurgy and electronics. Catalytic applications, such cerium-based catalysts for CO_2 hydrogenation and Fischer–Tropsch synthesis, are made possible by their changeable valence and electronic structures. Orthoferrites with high catalytic activity and magnetic recovery, such as holmium orthoferrites, are also employed as magnetic acid–base catalysts for redox processes.

11.9.5 *Surface Modification of Magnetic Nanoparticles*

Agglomeration of magnetic nanoparticles (MNPs) frequently lowers catalytic activity. To stop this, MNPs are coated with silica or carbon to create core–shell structures. This enhances stability, stops oxidation, and makes it easier for catalytic species, such as ionic liquids, to connect. Additionally, metal oxides like titania or cerium improve thermal resistance.

11.9.6 *Ionic Liquid Coated Magnetic Nanoparticles*

Excellent catalytic activity, simple recovery through magnetic separation, and enhanced recyclability are all displayed by ionic liquids (ILs) immobilised on magnetic nanoparticles (MNPs). By improving catalytic activity through polymer grafting or nanocomposites like metal–organic frameworks (MOFs), supported IL catalysts get beyond the disadvantages of IL, such as high viscosity and cost, and allow for effective, environmentally friendly processes with numerous reuses.

11.10 Nanotechnology in Catalyst Design

The goal of catalysis research is perfect selectivity while reducing waste and energy consumption. Protein-formed active sites in flexible cavities, which allow for precise reactant binding and transformation, are one example of this in enzymes. Using enzymes as inspiration, designing heterogeneous catalysts entails controlling active sites, binding sites, and pathways at the atomic to nanoscale to maximise efficiency and selectivity.

11.10.1 Nanostructured Catalysts for Enhanced Activity and Selectivity

Nanomaterials are extremely useful in the field of catalysis because of their special qualities, which include large surface area, customized catalytic active sites, and enhanced mass transfer (Demirbas & Demirbas, 2010; Osman, 2020). Because of their inherent catalytic active sites, graphene-based materials have been thoroughly investigated as metal-free catalysts, demonstrating their capacity to promote a variety of chemical reactions, such as oxidation, reduction, and coupling reactions (Song et al., 2021). Catalysts based on nanomaterials have also shown promise in the synthesis of biofuels, which is an essential component of sustainable energy production. In order to improve the catalytic activity and selectivity of the conversion of feedstocks produced from biomass into biofuels, nanocomposites which are composed of metal or metal oxide nanoparticles embedded in a support matrix have been used. Nanomaterials for water treatment and environmental cleanup Nanomaterials are appealing options for a variety of environmental cleanup applications due to their high surface area and reactivity.

Heavy metals, organic pollutants, and pathogens are just a few of the many contaminants that can be removed from water and wastewater using nanoparticles, such as those made of metal oxides or carbon-based compounds (Demirbas & Demirbas, 2010; Osman, 2020; Subramaniam et al., 2021). Effective and reasonably priced water treatment methods have been made possible by the special qualities of these nanomaterials, such as their potential for adsorption, catalytic activity, and disinfection. Nanomaterials sustainability has been further improved and the environmental impact of their manufacture has been decreased with the development of green and bio-inspired synthesis techniques. While retaining their intended qualities and functionality, these biogenic nanomaterials provide economical and sustainable substitutes for traditional nanomaterial fabrication. In conclusion Nanomaterials adaptability and special qualities have made them essential for solving a variety of global issues, especially those pertaining to environmental remediation, biofuels generation, and catalysis. A more sustainable future is very likely if nanomaterials, such as nanoparticles, nanocomposites, and graphene-based catalysts, are developed and used further (Demirbas & Demirbas, 2010; Montag et al., 2012; Osman, 2020; Subramaniam et al., 2021).

11.10.2 Role of Nanomaterials (Nanoparticles, Nanocomposites, and Graphene-Based Catalysts)

Catalysts based on graphene, nanoparticles, and nanocomposites. The development of green and bio-inspired synthesis routes for nanomaterials has emerged as a promising approach to enhance their sustainability and reduce the environmental impact associated with their production. This is because of their unique physicochemical properties, such as high surface area, reactivity, and catalytic activity, which have made them invaluable in addressing pressing global challenges related to sustainable energy and environmental reinteresting.

Synthesized from biological organisms or natural resources, biogenic nanomaterials provide economical and eco-friendly substitutes for traditional nanomaterial synthesis while preserving their intended characteristics and functionality (Osman, 2020; Schlögl, 2015).

In conclusion, nanomaterials such as nanoparticles, nanocomposites, and graphene-based catalysts have a variety of uses in environmental remediation, biofuel production, and catalysis, and their research and development are still ongoing.

11.10.3 Case Studies Showcasing Nano Catalyst Performance in Biofuel Synthesis

Nano catalysts are made from inexpensive materials such as leftover eggshells, animal bones, and molluscs. Promising for sustainable biodiesel generation, advanced catalysts including $CaO/CuFe_2O_4$, KF/CaO, and $SO_4/Mg–Al-Fe_3O_4$ have large active areas, high catalytic activity, and resistance to saponification. They also produce outstanding biodiesel yields. Because of their high activity, low cost, and reusability, heterogeneous nano catalysts like CaO are frequently utilized in the manufacture of biodiesel. By heating and grinding waste materials like egg and snail shells, CaO can be produced. Significant biodiesel yields with nano-CaO have been reported in studies: In (Védrine, 2017) obtained a 97% yield from soybean oil, (Védrine, 2017) obtained a 96% yield from leftover cooking oil, and (Yaşar, 2019) obtained a 96.81% yield from rapeseed oil. Transesterification is improved by the porous structure of nano-CaO. High biodiesel yields are also demonstrated by MgO nano catalysts, and in industrial applications, mixing MgO with TiO_2 or Al_2O_3 enhances catalytic performance while resolving separation issues.

Also, the biodiesel conversion yield (BCY) can be calculated using the following formula (Wang, 2021) (Table 11.5).

$$BCY(\%) = \frac{weight\ of\ biodiesel(g)}{weight\ of\ oil(g)}. \tag{11.1}$$

11.11 Biocatalysts in Biomass Conversion

An effective, sustainable, and eco-friendly method of converting renewable raw materials into value-added products like biofuels, chemicals, and materials is provided by biocatalysts, which are essential to biomass conversion.

Table 11.5 The yield of producing biodiesel from vegetable oil employing different kinds of nano-catalysts is shown in following table

Type of nano-catalyst	Feed	Ideal conditions	Percentage of BCY	References
KF/CaO-Fe_3O_4	Stillingia oil	M/O ratio = 12:1, catalyst = 4 wt%, 65 °C, 3 h	95	Liu et al. (2020)
KOH/ ZSM-5-Fe_3O_4	Canola oil	M/O ratio = 12.3:1, catalyst = 9.03 wt%, 65 °C, 3.26 h	93.6	Heeres et al. (2018)
Li/ZnO-Fe_3O_4	Rapeseed oil	M/O ratio = 12:1, catalyst = 0.8 wt%, 35 °C, 0.58 h	99.8	Wang et al. (2021)
CaO/Al/Fe_3O_4	Rapeseed oil	M/O ratio = 15:1, catalyst = 6 wt%, 3 h	98.7	Wang (2021)
CaO–CeO_2	Refined palm	M/O ratio = 20:1, catalyst = 5 wt%, 85 °C, 3 h	95	Liu et al. (2020)
MgO	Moringa oleifera seed	M/O ratio = 12:1, catalyst = 1 wt%, 45 °C, 4 h	93.6	Okolie et al. (2019)

11.11.1 Enzymes and Microbes as Catalysts in Biochemical Processes

Enzymes and microbes as catalyst are used in many different sectors and applications because of their well-established role as metabolic catalysts. Industrial enzymes have a very broad end-use market with many industrial and commercial uses (Narnaware & Panwar, 2021). The increasing need for sustainable solutions is driving a steady increase in the demand for industrial enzymes. One of the most abundant and practical sources of numerous enzymes has been and still is microbes (Yang et al., 2020). Waste production, unfavourable circumstances, and inefficiency are common problems in industrial chemical synthesis. With their high catalytic efficiency, stereoselectivity, and regioselectivity while functioning in mild circumstances, enzymes provide a sustainable substitute. They produce precise, well-defined products with few impurities, greatly speed up processes, and treat unconventional substrates without the need for substrate protection. In order to improve stability, substrate specificity, and activity, enzymes can be chemically or genetically altered. Despite limitations such cofactor needs, there are workarounds like whole-cell catalysts and cofactor recycling. Enzymes or microbial cells are used in about 150 industrial processes. In the very competitive global market, Novozymes is the market leader, followed by DSM and DuPont. Consumption is dominated by North America and Europe, with Asia–Pacific demand particularly in China, Japan, and India growing quickly.

Long employed in the food and beverage sectors, biocatalysts are now being applied in industrial chemistry. Microbial cells, extracts, and pure or immobilized enzymes are all used in biocatalysis. Its quick expansion is due to developments in metabolic engineering, directed evolution, protein expression, genome sequencing, high-throughput screening, and structural biology, which make a variety of effective catalytic processes possible (Gao et al., 2021).

11.11.2 Applications in Bioethanol and Biobutanol Production

For the manufacture of bioethanol and biobutanol, lignocellulosic-based biofuels, which are low in lignin and high in cellulose and hemicellulose, are an economical and sustainable option. ABE fermentation is aided by the high monosaccharide content of agro-waste substrates such as sugarcane bagasse, corn stover, rice straw, and cassava waste. Digestibility is influenced by pretreatment and cellulose crystallinity, which facilitate waste consumption and effective conversion to bioethanol and biobutanol (Jiao, 2020) (Fig. 11.7).

In the field of producing biofuel, biocatalysts-enzymes or microorganisms that can catalyse particular chemical reactions have drawn much interest, especially in the synthesis of bioethanol and biobutanol. The intrinsic sustainability, energy efficiency, and potential to reduce waste creation of biocatalysts make them superior

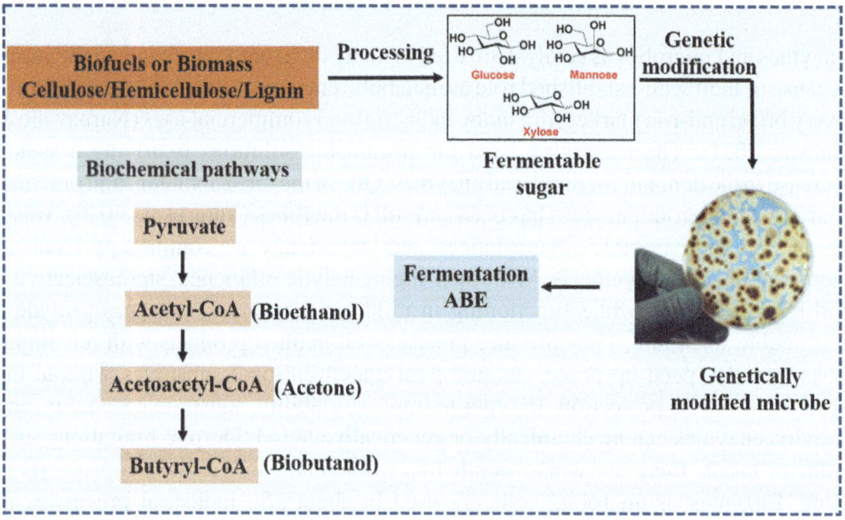

Fig. 11.7 Anaerobic ABE (Acetone-butanol-ethanol) fermentation of sustainable biofuels or biomass to produce biobutanol and bioethanol

to conventional chemical processes (Inayat et al., 2020). Biocatalysts are essential for converting feedstock-such as lignocellulosic biomass or agricultural waste into fermentable sugars during bioethanol production. Enzymatic saccharification, in which enzymes convert complex cellulosic and hemicellulose compounds into simpler sugars, is usually used in this procedure. Additionally, the efficiency of these saccharification phases can be increased using membrane-integrated processes, such as membrane bioreactors, which will boost bioethanol production overall (Okolie et al., 2021). Due to the inefficiency of traditional pretreatment and saccharification procedures and the excessive electricity consumption of common fermentation and purification processes, bioethanol production is still regarded as one of the most energy-intensive and difficult processes. As a result, it became essential to enhance the bioethanol production process by lowering energy needs. Similarly, biocatalysts can be used to help turn biomass or agricultural waste into the necessary product in the manufacturing of biobutanol. When compared to conventional chemical techniques, biocatalysts can provide benefits such as increased substrate specificity, lower energy requirements, and greater process sustainability (Gao et al., 2021; Inayat et al., 2020; Okolie et al., 2021).

11.11.3 Recent Advances in Enzyme Engineering for Biofuel Synthesis

Enzymatic biodiesel production is gaining attention due to novel approaches improving cost-effectiveness and efficiency. Advances include immobilization carriers, reactions in ionic liquids, solvent-free media, or supercritical CO_2. Sustainable feedstocks like waste oils and microalgae are explored (Table 11.6).

11.12 Catalyst Development for Specific Biofuels

11.12.1 Transesterification Reaction

The transesterification reaction of vegetable oils is depicted in Scheme 1. This reaction typically requires the presence of catalysts or the application of supercritical conditions to proceed efficiently. A range of catalysts has been developed for the production of biodiesel, encompassing both homogeneous and heterogeneous types. These catalysts include bases, acids, enzymes, and carbon-based catalysts derived from sources such as sugar, starch, and cellulose.

A comprehensive review conducted (Encinar et al., 2002), addresses several conventional and non-conventional base-catalysed transesterification methods. Within this framework, various catalysts, including boiler ashes and potassium

Table 11.6 The benefits and difficulties of certain cutting-edge methods and current developments in enzyme engineering for biofuel synthesis (Haarlemmer et al., 2016)

Current techniques	Benefits	Difficulties
Combining lipases from several sources	Results in increased yields, decreased reaction times, and broad substrate specificity	Making an enzyme cocktail or using genetic engineering is a laborious procedure
Using ionic liquids as a solvent	Enhanced enzyme activity, stability, and selectivity	Costly method
Transesterification using enzymes for inexpensive and abundant free fatty acid feedstocks	Lowers feedstock costs and waste management through the manufacturing of biodiesel	Careful difficulties with logistics and collection
Transesterification by enzymes in a supercritical CO_2 environment	Enhances diffusion and reaction rate, allows for the engineering of salvation ability, applicable to lipid extraction, and facilitates simple product separation	Expensive method that calls for complex equipment

hydroxide (KOH), have demonstrated considerable effectiveness in the methanolysis of palm and coconut oils, achieving yields of up to 90%. Furthermore, research indicates that methyl and ethyl esters can be produced with a yield of 90% from palm and coconut oil derived from press cake and waste generated by oil mills and refineries. This process utilizes ashes resulting from the fibres, shells, and husks of these oil seeds, in conjunction with catalysts such as lime, clay, and zeolites.

It has been observed that the methanolysis of palm oil under reflux conditions for two hours, employing coconut-shell ash and additional ashes obtained from the combustion of agricultural plant waste containing potassium and sodium carbonate, can yield esters in the range of 96–98%. Moreover, a catalyst system incorporating calcium oxide and magnesium oxide has been identified as the most effective catalyst, functioning optimally at temperatures between 60 and 63 °C. This system surpasses the performance of other catalysts, including potassium carbonate, sodium carbonate, iron (III) oxide, sodium methoxide, sodium aluminate, zinc, copper, tin, lead, and zinc oxide, in the methanol transesterification of low-erucic rapeseed oil.

Generally, the mechanism of the base-catalysed transesterification of vegetable oils involves four steps. The first step is the base's reaction with the alcohol, producing an alkoxide and the protonated catalyst. The second step is the alkoxide's nucleophilic attack at the triglyceride's carbonyl group, generating a tetrahedral intermediate (Schuchardt, 1994). The third step involves the formation of the alkyl ester and the corresponding anion of diglyceride. The final step consists in deprotonating the catalyst, thus regenerating the active species, which can now react with a second alcohol molecule, starting another catalytic cycle. The exact mechanism converts diglycerides and monoglycerides to a mixture of alkyl esters and glycerol.

The analysis of transesterification reactions was conducted using several feedstocks to enhance biodiesel quality through improvements in its physicochemical properties. The mechanism underlying the transesterification reaction is delineated in Eqs. (11.2a), (11.2b), and (11.2c). This process encompasses a series of consecutive and reversible reactions. In the presence of methanol and a catalyst, triglycerides are systematically converted into diglycerides (DG), monoglycerides (MG), and glycerol. At each stage of this process, one mole of fatty acid methyl ester is liberated.

$$\begin{array}{llcll}
CH_2-O-CO-R_1 & & & & CH_2-OH \\
| & & \text{Catalyst} & & | \\
CH-O-CO-R_2 + & CH_3OH & \leftrightarrow & CH_3-O-CO-R_1 + & CH-O-CO-R_2 \\
| & & & & | \\
CH_2-O-CO-R_3 & & & & CH_2-O-CO-R_3 \\
\text{Triglyceride} & \text{methanol} & & \text{Methylester} & \text{diglyceride}
\end{array} \quad (11.2a)$$

$$\begin{array}{llcll}
CH_2-OH & & & & CH_2-OH \\
| & & \text{Catalyst} & & | \\
CH-O-CO-R_2 + & CH_3OH & \leftrightarrow & CH_3-O-CO-R_2 & +CH-OH \\
| & & & & | \\
CH_2-O-CO-R_3 & & & & CH_2-O-CO-R_3 \\
\text{Diglyceride} & \text{methanol} & & \text{Methylester} & \text{monoglyceride}
\end{array} \quad (11.2b)$$

$$\begin{array}{llcll}
CH_2-OH & & & & CH_2-OH \\
| & & \text{Catalyst} & & | \\
CH-OH + & CH_3OH & \leftrightarrow & CH_3-O-CO-R_3 + & CH-OH \\
| & & & & | \\
CH_2-O-CO-R_3 & & & & CH_2-OH \\
\text{Monoglyceride} & \text{methanol} & & \text{Methylester} & \text{glycerol}
\end{array} \quad (11.2c)$$

Equation shows the mechanism of transesterification.

The transesterification reaction utilizing basic catalysts is presently the most widely adopted method for biodiesel production. This process necessitates the use of feedstocks, such as vegetable oils or animal fats, that exhibit a low concentration of free fatty acids. When raw materials contain a significant proportion of free fatty acids or moisture, the alkali catalyst may react with the free fatty acids to yield soaps. Furthermore, the presence of water can lead to the hydrolysis of triglycerides into diglycerides, consequently increasing free fatty acids. In this review, the latest effective catalysts for obtaining high FAME yield at low reaction temperatures are discussed below in Table 11.7.

The objective of pre-treating lignocellulosic materials is to improve the accessibility of cellulose for hydrolysis, thereby facilitating its conversion into biofuels through the use of microorganisms. This approach is not only economically viable but also environmentally sustainable and renewable. The pulp and paper industry generates approximately 50% of its energy requirements from biomass residues, while also making extensive use of combined heat and power systems.

Table 11.7 Basic solids as catalyst for transesterification reactions

Catalyst	Feedstock	Optimum reaction condition	Biodiesel FAME yield (%)	References
CaO from waste shells of mollusk and egg	Palm olein oil	T = 60 °C, t = 2 h Methanol/oil = 18:1 Catalyst amount = 10 wt.%	> 90	Viriya-empikul et al. (2010)
CaO from eggshells	Palm oil	T = 65 °C, t = 2 h Methanol/oil = 12:1 Catalyst amount = 1.5 wt.%	98	Cho and Seo (2010)
CaO from natural calcites	Palm olein oil	T = 60 °C, t = 45 min Methanol/oil = 15:1 Catalyst amount = 7 wt.%	95.7	Yoosuk et al. (2010a)
CaO	Sunflower oil	T = 80 °C, t = 5.5 h Methanol/oil = 6:1 Catalyst amount = 1 wt.%	91	Verziu et al. (2011)
CaO	Sunflower oil	T = 75 °C, t = 45 min Methanol/oil = 4:1 Catalyst amount = 1.2 wt.%	80	Vujicic et al. (2010)
CaO (from $CaCO_3$)	Palm olein oil	T = 60 °C, t = 1 h Methanol/oil = 15:1 Catalyst amount = 7 wt.%	93.9	Wen et al. (2010a)
KF/CaO	Chinese tallow seed oil	T = 65 °C, t = 2.5 h Methanol/oil = 12:1 Catalyst amount = 4 wt.%	96.8	Encinar et al. (2010)
KNO_3/CaO	Rape oil	T = 65 °C, t = 3 h Methanol/oil = 6:1 Catalyst amount = 1 wt.%	98	Alba-Rubio et al. (2010)
CaO/ZnO	Ethyl butyrate	T = 60 °C, t = 2 h Methanol/oil = 12:1 Catalyst amount = 1.3 wt.%	> 90	Zabeti et al. (2010)

(continued)

Table 11.7 (continued)

Catalyst	Feedstock	Optimum reaction condition	Biodiesel FAME yield (%)	References
CaO/Al_2O_3	Palm oil	T = 64.29 °C, t = 5 h Methanol/oil = 12.14:1 Catalyst amount = 5.97 wt.%	98.6	Yoosuk et al. (2010b)
CaO/Fe_3O_4	Jatropha curcas oil	T = 70 °C, t = 80 min Methanol/oil = 15:1 Catalyst amount = 2 wt.%	95	Liu et al. (2010)
Li/MgO	Soybean oil	$T = 60$ °C, $t = 2$ h Methanol/oil = 12:1 Catalyst amount = 9 wt.%	93.9	Wen et al. (2010b)
KOH/MgO	Mutton fat	T = 65 °C, t = 20 min Methanol/oil = 22:1 Catalyst amount = 4 wt.%	> 98	Mutreja et al. (2011)

11.13 Chemical Treatment of the Lignocellulosic Materials

11.13.1 Ozonolysis Treatment

Ozonolysis is a cutting-edge pretreatment method that enhances the digestibility of biomass while minimising the generation of chemical waste and preserving the integrity of carbohydrate components. This technique utilises ozone (O_3), a powerful oxidant that decomposes to yield highly reactive hydroxyl ions. Through this process, ozonolysis effectively degrades lignin and a portion of hemicellulose. The method offers several advantages, particularly its capability to efficiently remove lignin, thereby improving the overall accessibility of biomass for subsequent processing; it does not produce toxic residues for the downstream processes, and the reactions are carried out at room temperature and pressure. Furthermore, the decomposability of ozone, which can be achieved through a catalytic bed or by elevating the temperature, facilitates the development of processes designed to reduce environmental pollution.

11.13.2 Acid Pretreatment

Acid pretreatment can be performed under two distinct conditions: at high temperatures with low acid concentrations, known as dilute-acid pretreatment, or at low temperatures with high acid concentrations, referred to as concentrated-acid pretreatment. The application of acid to lignocellulosic materials at elevated temperatures has significantly enhanced enzymatic hydrolysis. Concentrated sulfuric acid is the most widely utilised acid, although hydrochloric acid, phosphoric acid, and nitric acid are employed in this process. Despite their effectiveness in promoting cellulose hydrolysis, concentrated acids present challenges due to their toxicity, corrosiveness, and hazardous nature, requiring reactors that can resist corrosion. Moreover, the recovery of concentrated acid following hydrolysis is critical for making the process economically viable. Dilute acid hydrolysis is arguably the most prevalent method utilised among the various chemical pretreatment techniques. At elevated temperatures ranging from 140 to 190 °C and with low concentrations of acid, specifically between 0.1 and 1% H_2SO_4, applying dilute-acid treatment can facilitate high reaction rates and significantly enhance the hydrolysis of cellulose.

11.13.3 Alkali Treatment

Alkali pretreatment is a process that involves the use of alkaline solutions, such as sodium hydroxide (NaOH), calcium hydroxide ($Ca(OH)_2$), or ammonia (NH_3), to remove lignin and a portion of the hemicellulose effectively. This treatment significantly enhances the accessibility of enzymes to cellulose, thereby increasing the efficiency of subsequent biochemical processes. Sodium, potassium, calcium, and ammonium hydroxides serve as effective agents for alkaline pretreatment processes. Sodium hydroxide (NaOH), in particular, exhibits strong basic activity, which facilitates the removal of lignin and promotes the swelling of cellulose. This base, along with sodium sulfide, is employed in the Kraft paper pulping process. Additionally, calcium hydroxide $Ca(OH)_2$ provides several advantages, including cost-effectiveness and safety. Moreover, this method exhibits similarities to established Kraft paper pulping technologies. Alkali pretreatment processes are characterized by their utilization of lower temperatures and pressures compared to alternative pretreatment technologies.

Oxidative delignification: Lignin biodegradation can be effectively catalysed by the peroxidase enzyme when hydrogen peroxide (H_2O_2) is present. In experiments, nearly 50% of lignin and a significant proportion of hemicellulose were solubilized using a 2% concentration of H_2O_2 at a temperature of 30 °C within an 8-h period. This process resulted in an impressive 95% efficiency in glucose production. The process achieved a remarkable 95% efficiency in glucose production from cellulose during the subsequent saccharification conducted with cellulase at a temperature of 45 °C over a 24-h period.

11.13.4 Organosolv Process

The organosolv process represents an advanced pretreatment strategy that has garnered considerable attention due to its potential applications in the treatment of lignocellulosic materials. This process utilizes a combination of organic or aqueous organic solvents and inorganic acid catalysts, such as hydrochloric acid or sulfuric acid, to effectively disrupt the internal bonds of lignin and hemicellulose. Common solvents employed in the organosolv process include methanol, ethanol, acetone, ethylene glycol, triethylene glycol, and tetrahydrofurfuryl alcohol.

11.13.5 Ionic Liquids (ILs) Pretreatment

The utilization of ionic liquids (ILs) as solvents for the pretreatment of lignocellulosic biomass has recently garnered considerable attention within the scientific community. Ionic liquids are salts that typically consist of large organic cations and small inorganic anions, and they exist as liquids at relatively low temperatures or at room temperature. The properties of solvents can be modified by varying the anions and altering the alkyl groups of the cations. These noteworthy attributes include chemical and thermal stability, non-flammability, low vapor pressures, and the capacity to remain in a liquid state across a broad temperature range. In this process, no toxic or explosive gases are generated, which contributes to the classification of ionic liquids (ILs) as "green" solvents. Carbohydrates and lignin can be effectively dissolved in ILs that demonstrate anionic activity, such as 1-butyl-3-methylimidazolium cations. This phenomenon occurs as ILs establish hydrogen bonds between the non-hydrated chloride ions and the hydroxyl protons of the sugars in a 1:1 stoichiometric ratio. In this study, the ionic liquid 1-ethyl-3-methylimidazolium diethyl phosphate was employed for the pretreatment of wheat straw at a temperature of 130 °C for a duration of 30 min. This procedure yielded reducing sugars amounting to 54.8% following a subsequent 12-h enzymatic hydrolysis (Li et al., 2009). Despite certain existing limitations, advanced research into the potential synthesis of ionic liquids from carbohydrates may play a significant role in reducing production costs. Furthermore, the development of ionic liquid pretreatment methodologies presents considerable promise for enhancing future lignocellulosic biorefinery processes.

11.14 Catalyst Deactivation and Regeneration

Deactivation of the catalyst is a big problem in biofuel production. This can happen for several reasons: coke formation, poisoning, sintering, and the loss of active species. Carbonaceous remains build up on the catalyst surface, blocking active sites and lowering performance. This leads to the formation of coke. When impurities in the material, such as sulphur, nitrogen, chlorine, or phosphorus, bind permanently to active sites, making the catalyst useless, this is called poisoning. High temperatures cause sintering, which groups metal bits together. This makes the catalyst's surface area and activity smaller. The leaching of active metals during liquid-phase reactions reduces catalytic efficiency over time. Deactivation processes can be mitigated through regeneration techniques, including oxidative burn-off for coke removal, chemical treatments to eliminate poisons, thermal reactivation to reverse sintering, and solvent washing to recover leached species. Hydrotreating catalysts, such as CoMo or NiMo on alumina, can be regenerated through a process of oxidation followed by sulfidation to restore their activity. Zeolite-based catalysts, utilized for upgrading pyrolysis vapours, are enhanced by high-temperature oxidative regeneration to eliminate coke. Adopting feedstock pretreatment, optimizing operating conditions, and scheduling regular regeneration cycles are essential to prolong catalyst life, ensuring efficient, cost-effective, and sustainable biofuel production.

11.14.1 Coke Formation (Carbon Deposition)

Coke formation (carbon deposition) is a major cause of catalyst deactivation during biofuel processing, occurring when hydrocarbons decompose or polymerize under high temperatures on the catalyst surface. This process stops active sites and reduces catalytic efficiency. Coke can form through different mechanisms, including dehydrogenation ($C_2H_4 \rightarrow 2C + 2H_2$), thermal cracking of long-chain hydrocarbons ($C_{10}H_{22} \rightarrow C + H_2 + CH_4$), and aromatic condensation ($C_7H_8 \rightarrow C + H_2 +$ light gases). Conditions increasing coke formation include high temperatures, high pressures, acidic catalysts (e.g., zeolites), and feedstocks rich in heavy hydrocarbons or oxygenates. Coke formation leads to a loss of active surface area and shortened catalyst life. To counteract this, oxidative regeneration ($C + O_2 \rightarrow CO_2$) is frequently used to burn off the carbon deposits and restore activity. Preventive measures such as feedstock pretreatment, optimizing reaction conditions, and designing catalysts with reduced coke propensity are also important for maintaining long-term catalyst performance in biofuel production.

11.14.2 Poisoning

Impurities like sulphur, nitrogen, chlorine, and phosphorus in biofeedstocks can greatly deactivate catalysts by binding to their active sites, forming stable and often irreversible compounds. This poisoning disrupts catalytic function, especially in processes such as hydrotreating and gasification.

11.14.2.1 Sulphur Poisoning

Compounds containing sulphur react with the active metal sites to generate stable metal sulphides, which is the cause of sulphur poisoning. For example, sulphur combines with nickel (Ni) or molybdenum (Mo) to generate sulphides in hydrotreating catalysts such as $NiMo/Al_2O_3$ or $CoMo/Al_2O_3$:

$$Ni + H_2S \rightarrow NiS + H_2$$

$$Mo + 2H_2S \rightarrow MoS_2 + 2H_2$$

These metal sulphides are inactive or less active, which hinders the effectiveness of hydrogenation processes.

11.14.2.2 Nitrogen Poisoning

Compounds in feedstocks that contain nitrogen have the potential to create metal nitrides, which take up active sites and lower catalyst efficiency. Nitrogen may react with metal sites such as nickel or molybdenum during hydrotreatment:

$$Mo + NH_3 \rightarrow Mo_2N + H_2$$

Catalytic performance can be reduced by metal nitrides blocking hydrogenation sites.

11.14.2.3 Chlorine Poisoning

Stable metal chlorides, which are normally volatile or inactive under reaction circumstances, can be created when chlorine from bio-derived feedstocks reacts with active metal sites:

$$Ni + Cl_2 \rightarrow NiCl_2$$

This deactivation is often seen in gasification or hydro-processing environments with chlorine contaminants.

11.14.2.4 Phosphorus Poisoning

Deactivation can result from phosphorus compounds forming metal phosphates, which cling firmly to the catalyst surface.

$$Ni + H_3PO_4 \rightarrow Ni_3(PO_4)_2 + H_2$$

These phosphates are difficult to eliminate during routine regeneration because they are thermally stable.

11.14.3 Sintering (Thermal Deactivation)

In high-temperature biofuel production processes including gasification, pyrolysis vapor upgrading, and steam reforming, sintering is a major contributor to catalyst deactivation. When exposed to high temperatures for an extended period, active metal particles (such as Ni, Co, or Fe) migrate and combine to create larger particles with fewer active sites and a smaller surface area. Ni catalysts that are finely disseminated are necessary for the reaction ($C_6H_6 + 6H_2O \rightarrow 6CO + 9H_2$) in steam reforming bio-oil vapours to create hydrogen. Agglomeration of nickel particles during sintering lowers their surface area and hydrogen production. Pyrolysis vapor upgrading, in which oxygenated molecules are transformed into hydrocarbons ($C_6H_8O_2 \rightarrow C_6H_6 + 2H_2O$), reduces deoxygenation efficiency when metal-supported catalysts (such as Ni/ZSM-5) are sintered. Biofuel manufacturing systems use techniques like employing high-surface-area supports (like AlO_3 and SiO_2), adding promoters (like MgO or CeO_2), and operating at controlled temperatures to preserve catalytic performance and longevity to reduce sintering.

11.14.4 Leaching of Active Species

In liquid-phase biofuel production techniques such as aqueous-phase reforming and catalytic transesterification, active metals, or metal oxides dissolve into the reaction media, causing catalyst degradation and leaching of active species. For example, the active species may disso lve in the methanol and glycerol phases during transesterification utilizing alkaline earth metal oxides (such as CaO), decreasing the catalyst's efficiency. Like this, metal ions like Pt^{2+} or Ni^{2+} can seep into the corrosive or acidic aqueous medium during the aqueous-phase reformation of oxygenates produced from biomass (for example, by employing Pt or Ni catalysts). The lifetime and efficiency of the catalyst are shortened by this ongoing loss of active species, which lowers the number of available active sites. Using more durable supports (such as ceramics or mesoporous silica), functionalizing the catalyst surface to enhance metal anchoring, and operating at ideal pH values to reduce solubility are some methods to lessen leaching.

11.14.5 Support Degradation

Support degradation is a major problem in biofuel production processes, especially in hydrothermal environments or in acidic/alkaline conditions. When exposed to hot, wet, or corrosive environments, catalyst supports such zeolites, alumina (AlO_3), and silica (SiO_2) may experience structural collapse or phase transitions. Boehmite ($Al_2O_3 + H_2O \rightarrow 2AlOOH$), for instance, can hydrate from alumina supports during hydrothermal processing, weakening the support structure and causing surface area loss. The crystalline structure and catalytic efficiency of zeolite supports can also be compromised by dealumination in acidic settings, where framework aluminium is leached away. These modifications degrade the catalyst's overall structure, which lowers its stability and efficacy in aqueous-phase reforming or hydrothermal upgrading. Long-term catalyst effectiveness depends on techniques like providing protective coatings, employing hydrothermally stable supports (such zirconia (ZrO_2) or titanium dioxide (TiO_2), and adjusting reaction conditions to reduce support degradation).

11.14.6 Different Methods for Catalyst Regeneration

11.14.6.1 Thermal Regeneration

In the presence of oxygen, the catalyst is heated to burn off carbon deposits or deposited coke. Usually, this procedure takes place at 400–600 °C. used for catalytic cracking, pyrolysis vapor upgrading, and hydro processing catalyst regeneration. To prevent sintering or harming the catalyst structure, controlled temperatures are necessary.

11.14.6.2 Chemical Regeneration

Chemical agents (reducing or oxidizing) are used to treat catalysts to eliminate impurities such as nitrogen, sulphur, or chlorine. For instance, hydrogen gas can be used to desorb sulphur from sulphur-poisoned catalysts. Frequently used in gasification and hydrotreatment procedures. Ideal for catalysts that are susceptible to oxidation at high temperatures.

11.14.6.3 Solvent Washing

To dissolve and eliminate toxins or leached metals, the catalyst can be cleaned using solvents such acids, bases, or organic solvents. Acid leaching, for instance, eliminates metal or phosphorus impurities. Beneficial for liquid-phase procedures such as biomass liquefaction and biodiesel transesterification. Waste should be disposed properly and dangerous solvents be handled carefully.

11.14.6.4 Hydrothermal Treatment

Adsorbed impurities can be removed or support materials can be renewed by subjecting catalysts to hot, pressured water or steam. Regenerating catalysts with deactivated supports, like zeolites in pyrolysis vapor upgrading, is an efficient application. Optimizes temperature and pressure to stop more support degradation.

11.14.6.5 Redispersion of Active Metals

Applying processes such as plasma treatment or hydrogen reduction to break up clumped metal particles into smaller, more active ones. Utilized for gasification catalysts that contain sintered metal particles, as Ni/Al_2O_3. For the best redispersion, exact control is necessary.

11.14.7 Mechanical Regeneration

To separate catalyst agglomerates or eliminate physically inactive particles, use grinding or sieving. Utilized in fixed-bed reactors for bulk catalysts or pellets. Prolongs catalyst life and reduces material loss.

11.14.7.1 Catalyst Recycling

By removing the active ingredients and repurposing them into new catalysts, spent catalysts can be recovered and processed again. Re-impregnating metals onto a fresh or regenerated support may be necessary for this. Perfect for biofuel synthesis using precious metal catalysts (such as Pt and Pd). Economical and environmentally friendly for pricey catalysts.

11.15 Future Directions in Catalysis for Biomass Conversion

Utilising biomass resources and energy in an efficient and sustainable manner can be accomplished through chemical conversion. It is crucial to convey the high-value conversion and replacement of fossil fuels with biomass resources, which is essential for addressing energy and fossil resource limitations, rising environmental pollution, and climate change in many ways. Catalysts have improved the yield and selectivity of reaction products, optimized reaction conditions, and decreased the activation energy of chemical reactions-all of which are significant factors in the effective conversion of biomass. However, obstacles to effective and environmentally friendly biomass conversion have emerged from a lack of knowledge about the mechanisms underlying catalyst-catalysed biomass conversion, improving catalyst-specific target product selection, achieving high conversion efficiency, deactivating and regenerating catalysts, and separating and purifying catalytic products. As a result, in order to carry out more practical and thorough research on the use of catalysts in biomass conversion, we must embrace more sophisticated and scientific approaches in future procedures. Among these are the following:

- Careful examination of catalyst kinetics and reaction processes in biomass conversion. In order to better inform practical applications with theory, people should embrace cutting-edge scientific research methodologies and technologies to further methodically deepen and clarify the fundamental theoretical research of catalysts in each stage of biomass conversion processes, particularly in the study of effective catalytic mechanisms for particular biomass conversion products and high-value chemical components.
- Concentrating on the biomass conversion process's catalytic efficiency, deactivation, regeneration, and recycling of catalysts. implementing innovative ideas and technologies to create high-performance green catalysts that are inexpensive, environmentally friendly, have a long service life, great selectivity, and good stability.
- Optimizing and building catalyst production procedures and reaction conditions for biomass conversion. constructing safe, stable, cost-effective, efficient, and environmentally friendly biomass conversion processes using scientific and methodical techniques; sensibly modifying and optimizing various reaction conditions; and attaining their synergistic effect on the catalytic conversion of biomass.
- Investigating the use of catalysts in biomass conversion by integrating structure–reactivity connections based on machine learning and molecular simulation. We can scientifically integrate experimental and computational data using new technologies like molecular simulation and machine learning. When combined with structure–reactivity relationships, these technologies allow us to directly screen and specifically design efficient catalysts and evaluate their technical and economic performance.

11.15.1 Emerging Trends in Catalyst Design

The efficiency, selectivity, and durability of catalytic processes for biomass conversion are the goals of recent developments in catalyst design. Among the major new trends are:

11.15.1.1 Nanostructured Catalysts

To convert renewable biomass into sustainable fine chemicals and advanced fuels, heterogeneous catalysis plays an important role. Porosity and nanostructure of solid catalysts play crucial roles in determining the accessibility of active sites, influencing reaction pathways, and enhancing selectivity.

Mesoporous materials are recognized for their high BET surface areas, large pore volumes, and adjustable pore diameters. They are widely used in applications such as gas adsorption, catalysis, optical technologies, and sensing. The incorporation of organic functional groups like –SO_3H and –COOH, along with a uniform pore size distribution, improves their performance in these areas. Among mesoporous silica materials, SBA-15 is often preferred over MCM-41 due to its larger pore size and superior hydrothermal and mechanical stability. The high surface area and tunable pore structure of ordered mesoporous silicas, including MCM-41 and SBA-15, make them highly effective for heterogeneous catalysis. Additionally, their high density of silanol groups further enhances their functionality.

The catalyst must be bifunctional, combining acidic and basic characteristics, to facilitate serial cascade reactions in biomass conversion processes. High thermal stability makes functionalized materials like carbon, zirconia, and titania indispensable for biomass conversion applications. Peng et al. (2012) demonstrated that mesoporous silica, modified with both acid and base sites, achieved an HMF yield of roughly 0.54 mol per mol of catalyst per hour at 393 K when used in an ionic liquid. Similarly, Jiang et al. (Sun et al., 2016) developed a bifunctional acid–base catalyst based on heteropoly acids, which effectively catalysed the cascade isomerization and dehydration of glucose to HMF.

Mesopores can be introduced on the surface of metal oxide materials using soft templates, such as aspartic acid (Dutta et al., 2011) and salicylic acid (De et al., 2011). The resulting porous metal oxides have efficiently catalysed the conversion of glucose and fructose to HMF. Since metal oxides with surface hydroxyl groups exhibit limited Brønsted acidity, their Brønsted acid strength can be enhanced through sulfonation. During this process, sulfuric acid groups with S=O bonds form strong interactions with metal atoms, leading to the formation of S=O bonds with hydroxyl groups on the metal oxide surface. Additionally, biopolymers can be used as templates to design porous TiO_2 materials, which have shown excellent catalytic performance in converting underused hemicellulose sugars, such as galactose and lactose, into HMF (Li et al., 2011). Another report shows a mesoporous spherical TiO_2 nanoparticulate material, which serves as a solid catalyst for microwave-assisted conversion of carbohydrates into HMF. The catalyst achieved HMF yields of 34.3% in water and 54.1% in DMSO (Chang et al., 2013). Sulfonated carbons outperform resin- and oxide-based solid acid catalysts due to their superior textural properties. To optimize cellulose hydrolysis, the surface acidity and hydrophobicity/hydrophilicity of these materials are finely tuned through controlled sulfonation. This approach allows nanoporous carbon-supported acidic sites or metal nanoparticles to act as highly effective catalysts for biomass conversion. Recently, Zhang et al. used sulfonated mesoporous carbon for cellulose hydrolysis, achieving a high glucose yield of 75%. Similarly,

Chang et al. (2013) proposed ordered mesoporous carbon was first treated with H_2O_2 to introduce hydroxyl and carboxyl groups. This was followed by sulfonation with concentrated sulfuric acid, resulting in OMC-H_2O_2–SO_3H, which exhibited a significantly higher acid concentration (2.09 mmol/g) than the conventional sulfonated samples (OMC–SO_3H). Furthermore, Hara (2010) conducted extensive research on sulfonated activated carbon, demonstrating its capability to fully hydrolyse cellulose into β-1,4-glucans at 100 °C.

Zeolite crystals have been the primary heterogeneous catalysts in the petrochemical and fine chemical industries for more than forty years. These materials are highly regarded for their combination of strong Brønsted and Lewis acid sites, ordered microporous structures, and outstanding thermal stability, all of which arise from their unique crystalline inorganic framework. The shape-selectivity of zeolites allows them to catalyse specific reactions based on the size and shape of the reactant molecules, making them particularly effective as solid acid catalysts in a wide range of industrial applications (Chen & Sachtler, 1998; Farneth & Gorte, 1995; Tanabe & Hölderich, 1999; West et al., 2010). Zeolites, which are crystalline microporous aluminosilicates, have a structure that includes polar Si–O bonds and ionic charges. These characteristics generate a strong electric field during chemical reactions, enhancing their ability to facilitate reactions by stabilizing reactants or intermediates, thus improving catalytic efficiency. Lercher and co-workers (Zhao et al., 2012) show significant progress in oil pyrolysis has been made using highly active and recyclable Ni- or Pd-based catalysts supported on aluminosilicate zeolites, including ZSM-5 and beta zeolite. Specifically, small microporous zeolites (4.0–5.9 Å) such as ZSM-5, ZSM-11, ZSM-23, TNU-9, MCM-22, and IM-5 have shown to be particularly effective in the first-stage pyrolysis of glucose. These zeolites enhance aromatic yields due to their ability to selectively promote specific reaction pathways, facilitating the formation of aromatic compounds while improving the overall efficiency of the process (Jae et al., 2011).

Metal phosphates form a distinctive group of catalysts that have shown great potential in a variety of dehydration reactions. Hara et al. effectively utilized the Lewis acidity of phosphate-immobilized anatase TiO_2 (phosphate/TiO_2) for the selective conversion of glucose to HMF. When NbPO, AlPO, TiPO, and ZrPO were applied for glucose dehydration to HMF in the aqueous phase, they demonstrated high catalytic activity, with the performance primarily dependent on the number of strong acid sites in the catalyst (Ordomsky et al., 2013). Zhang et al. (2016) proposed an alternative approach for fructose dehydration, achieving a 71% HMF yield using tin phosphate in a water-DMSO mixed solvent at 135 °C for 1 h. In addition, zirconium and titanium phosphates have proven effective in catalysing reactions such as alcohol dehydration and olefin isomerization.

11.15.1.2 Metal–Organic Frameworks (MOFs)

MOFs have attracted considerable interest in heterogeneous catalysis due to their extensive BET surface area, versatile microarchitecture, and adjustable pore sizes.

MOFs have garnered significant attention in heterogeneous catalysis due to their high BET surface area, customizable microstructure, and tunable pore size. This has led to the exploration of MOF-derived solid acids as efficient catalysts for the conversion of carbohydrates into HMF. Through post-synthetic modifications, MOFs can incorporate functional groups into their organic ligands, providing advantages over conventional carbon materials and inorganic solids (Corma et al., 2010; Taylor et al., 2013). Additionally, the highly porous and organized structure of MOFs facilitates improved substrate transfer, boosting catalytic efficiency. Zhang et al. (2011) demonstrated the selective dehydration of fructose to HMF using a MOF-based catalyst, specifically phosphotungstic acid (PTA)-encapsulated MIL-101(Cr), confirming its effectiveness as a solid acid catalyst.

A recent study highlighted the one-pot conversion of cellobiose and cellulose into sorbitol using a bifunctional acid-metal catalyst, [Ru-PTA/MIL-100(Cr)]. The catalyst features ruthenium and phosphotungstic acid (PTA) as the active species, while MIL-100(Cr) functions as both the support and the matrix for encapsulation (Chen et al., 2013). The conversion of fructose to HMF can be efficiently accomplished using sulfonic acid-functionalized metal–organic frameworks (MOF-SO_3H), such as MIL-101(Cr)-SO_3H, UiO-66(Zr)-SO_3H, and MIL-53(Al)-SO_3H. Among them, MIL-101(Cr)-SO_3H stands out, achieving full fructose conversion and a high HMF yield of 90% (Chen et al., 2014). Cirujano et al. (2015) investigated the esterification of levulinic acid with different alcohols, such as ethanol and butanol, revealing that zirconium-based UiO-66 and UiO-66–NH_2 serve as highly efficient catalysts for the reaction. Additionally, Bromberg et al. (2014) presented a straightforward method for synthesizing novel functional composite materials by integrating MOFs with polymer networks. These materials have demonstrated excellent efficiency in producing key chemical intermediates through the dehydration of fructose from renewable sources.

11.15.1.3 Photocatalysts

Biomass, a rich and renewable source of natural carbon, includes a variety of functional categories. As a promising alternative to fossil fuels, it can be converted into chemicals and fuels, helping to reduce carbon emissions and mitigate the energy crisis. However, the complex structure of biomass, its high content of oxygen-containing functional groups, and strong C–C bonds require harsh reaction conditions for transformation. These challenging conditions often lead to unwanted side reactions, producing complex polymers and reducing selectivity for the desired products. To improve biomass utilization, lower energy consumption, and achieve a true carbon–neutral process, there is a need for milder, more efficient, and energy-saving reaction methods. Photocatalysis, which operates under mild conditions, is capable of breaking high-energy bonds in certain chemical compounds. By harnessing natural sunlight or light from renewable electricity, photocatalysis can facilitate one-step conversions that would otherwise require multiple steps. It also offers high atom economy, making it a promising approach for the selective conversion of biomass. Integrating biomass conversion with photocatalysis could provide solutions to current environmental and energy challenges. Significant progress has been made in photocatalytic biomass conversion, including a better understanding of reaction mechanisms, the development of photocatalyst structural modifications, and the design of active sites. Studies have shown the potential of photocatalysis to generate chemicals, fuels, and hydrogen from biomass and its derivatives. These processes could contribute to increasing the share of renewable energy in the global energy mix while complementing conventional petrochemical methods. However, to enhance the apparent quantum efficiency of photocatalytic biomass conversion and boost product yields, further research is needed to develop more efficient catalysts. Although sunlight is a vast energy source, its intensity fluctuates with time and weather, and it is not always located near areas with high energy demand. Typically, regions with abundant sunlight also have abundant biomass resources.

Hydrodeoxygenation plays a key role in upgrading biomass-derived pyrolysis oil into biofuel (Bharath, 2020). Vanillin (4-hydroxy-3-methoxybenzaldehyde) is a widely occurring component in the lignin fraction of pyrolysis oil (Gao et al., 2022). AgPd@g-C_3N_4, produced through the hydrogenation of vanillin, has emerged as a promising candidate for future biofuel applications (Xu et al., 2012). Graphitic carbon nitride (AgPd@g-C_3N_4) co-decorated with Pd and Ag is highly effective in the hydrodeoxygenation of vanillin under visible light exposure (Verma et al., 2016). In this process, formic acid is converted into hydrogen, eliminating the need for high-pressure hydrogen, which is key for the sustainable and safe upgrading of biofuels. AgPd@g-C_3N_4 shows excellent performance in both formic acid decomposition to hydrogen and the hydrodeoxygenation of vanillin, driven by the synergistic effect of Ag and Pd with the C_3N_4 support. Furthermore, AgPd@g-C_3N_4 exhibits remarkable stability, achieving complete vanillin conversion by the 10th cycle. This stability is likely attributed to the noncovalent interactions between the metal nanoparticles and the C_3N_4 support.

Due to their structural similarity to diesel-type hydrocarbons, long-chain bio-derived carboxylic acids serve as promising feedstocks for producing long-chain alkanes (Zhao et al., 2013). Typically containing 12–22 carbon atoms, these carboxylic acids are widely produced as low-value byproducts in industries such as fat and oil processing and pulp manufacturing (Abdul Kapor et al., 2017). TiO_2-based photocatalysts have been shown to effectively facilitate photocatalytic decarboxylation, enabling the conversion of bio-derived fatty acids into alkanes under mild conditions (Huang et al., 2020). Photogenerated holes initiate the process by directly facilitating the decarboxylation of bio-derived fatty acids, resulting in the formation of alkyl radicals (Zhang et al., 2018). A major challenge in photocatalytic decarboxylation is ensuring that radical intermediates preferentially terminate with hydrogen. The conversion efficiency of carboxylic acids is notably enhanced under an H_2 atmosphere (Du et al., 2022). For example, Pt/TiO_2 catalysts have been utilized in solar-driven decarboxylation of fatty acids, resulting in a high yield of $\geq 90\%$ long-chain alkanes. This exceptional efficiency is likely due to the hydrogen-rich surface formed by the interaction between the catalyst and H_2, which facilitates the rapid termination of alkyl radicals with surface hydrogen species. The potential mechanism for photocatalytic reforming of carboxylic acids over Pt/TiO_2 in the presence of H_2 centres around this radical termination process.

Various modified TiO_2 catalysts have been studied for solar-driven decarboxylation of octanoic acid under H_2 atmospheres, resulting in n-heptane and tetradecane (Du et al., 2022). The conversion of octanoic acid under UV–visible light irradiation is enhanced by adding nickel (Ni) or copper (Cu) to TiO_2. Furthermore, the oxygen vacancy (O_v) in TiO_2 plays a crucial role in photocatalytic decarboxylation (Wang, 2021). When NaBH4 is used as a reduction agent, the concentration of Ov in TiO_2 increases proportionally with the amount of $NaBH_4$. As O_v concentrations in TiO_2 increase, the photocatalytic conversion efficiency initially improves but begins to decline beyond a certain threshold. Meanwhile, the selectivity for tetradecane keeps rising with higher O_v concentrations, indicating a potential approach to optimizing longer-chain alkane production.

11.15.1.4 Enzyme Catalysts

For the manufacture of bio-ethanol, enzymatic hydrolysis of pre-treated lignocellulosic biomass is a perfect substitute for acid hydrolysis; the main drawbacks are pre-treatment specifications and financial concerns related to the expenses of producing enzymes and certain operations. Years of bio-prospecting, strain optimization through genetic engineering, enzyme characterisation for both simple and complicated lignocellulosic feedstock, and the creation of pre-treatment techniques to lessen inhibitory effects have all been spurred by the search for better and more affordable enzymes. A potential sign of maturity in this field of inquiry is the recent move to systematic characterizations of de novo mixtures of purified proteins, which has aided in the advancement of feedstock assay-based fast enzyme mixture optimization. To facilitate meaningful comparisons between these investigations and the creation of a database of enzyme activities and kinetics-aspects of which are examined here-international standards must be established. Membrane-confinement exhibits great promise despite the related technological obstacles, and complementary attempts to increase the economic sustainability of enzymatic hydrolysis through process integration and reactor design are also taken into consideration. As systematic research in enzyme activity aligns with that of traditional reaction engineering, significant breakthroughs in enzyme technology towards the economic conversion of lignocellulosic biomass should be anticipated in the next years (Table 11.8).

11.16 Conclusions

Catalysis has proven to be a cornerstone for the efficient and sustainable conversion of biomass into biofuels, addressing key challenges such as energy efficiency, product selectivity, and economic feasibility. Recent advancements in catalyst design-such

Table 11.8 Enzyme-based catalysts in biomass conversion

Enzyme expressed from Host Organism	Purpose/novelty	References
β-glucosidase/β-xylosidase; from *Reticulitermes santonensis*	Determine the activity of enzymes from termite gut suspensions	Vujicic et al. (2010)
Ferulic acid esterase; from *Cellvibrio japonicus*	Confirm the presence of Fee 1B enzyme and substrate specificity	McClendon et al. (2011)
Endoglucanase GH5-1; from *Neurospora crassa*	Endocellulase characterization and validation of *N. crassa* as a host for homologous expression of secreted enzymes	Sun et al. (2011)
Bacillus flexus; from green seaweed (*Ulva lactuca*)	Alkali-halotolerant cellulase production	Trivedi et al. (2011)
Streptomyces sp. *S38* xylanase; in *Pichia pastoris*	Demonstrate expression method and enzyme activity	Fu et al. (2011)

as bifunctional and multifunctional systems, nanostructured materials, zeolites, and metal–organic frameworks (MOFs) have enabled the transformation of diverse feedstocks, including lignocellulosic biomass, algae, and agricultural residues into high-value biofuels. These innovations, combined with green technologies like electrocatalysis, microwave-assisted processes, and supercritical fluids, have significantly improved reaction efficiency while minimizing environmental impacts.

However, significant challenges remain. Feedstock variability, catalyst deactivation, and the high energy demands of specific processes continue to hinder industrial-scale adoption. Additionally, scaling laboratory-scale catalytic systems to commercial operations requires substantial optimization, including advancements in reactor design and process intensification. Addressing these issues requires a multidisciplinary approach integrating materials science, chemical engineering, and environmental research to develop robust, cost-effective, and environmentally sustainable catalytic systems.

The sustainability of biomass conversion could be further enhanced by integrating catalytic processes with renewable energy sources such as solar and wind power. Additionally, developing adaptive catalysts capable of accommodating diverse feedstock compositions and operating conditions will be crucial. Equally important are investments in research and development and policies that promote the adoption of green technologies to drive progress in this field.

Biomass conversion technologies have the potential to significantly reduce greenhouse gas emissions, decrease reliance on fossil fuels, and foster a circular economy by leveraging advanced catalysts, innovative process designs, and robust regulatory frameworks. These efforts will be crucial in creating a sustainable energy future and achieving global environmental and energy goals.

Acknowledgements Authors are thankful to the Institute for providing necessary research support.

References

Abdul Kapor, N. Z., Maniam, G. P., Rahim, M. H. A., & Yusoff, M. M. (2017). Palm fatty acid distillate as a potential source for biodiesel production-a review. Elsevier Ltd. https://doi.org/10.1016/j.jclepro.2016.12.163

Aberilla, J. M., Gallego-Schmid, A., & Azapagic, A. (2019). Environmental sustainability of small-scale biomass power technologies for agricultural communities in developing countries. *Renewable Energy, 141*, 493–506. https://doi.org/10.1016/j.renene.2019.04.036

Adrio, J. L., & Demain, A. L. (2014). Microbial enzymes: Tools for biotechnological processes. *Biomolecules, 4*(1), 117–139. https://doi.org/10.3390/biom4010117

Alba-Rubio, A. C., et al. (2010). Heterogeneous transesterification processes by using CaO supported on zinc oxide as basic catalysts. *Catalysis Today, 149*(3–4), 281–287. https://doi.org/10.1016/j.cattod.2009.06.024

Alvarez, P. J. (2006). Nanotechnology in the Environment—The Good, the Bad. *Journal of Environmental Engineering, 132*, 2006. https://doi.org/10.1061/(ASCE)0733-9372(2006)132

Astrup, T. F., Tonini, D., Turconi, R., & Boldrin, A. (2015). Life cycle assessment of thermal Waste-to-Energy technologies: Review and recommendations. *Waste Management, 37*, 104–115. https://doi.org/10.1016/j.wasman.2014.06.011

Ayodeji, A., & Oluwole, A. (2018). Impact of monetary policy on economic growth in Nigeria. *Oalib, 05*(02), 1–12. https://doi.org/10.4236/oalib.1104320

Balajii, M., & Niju, S. (2019). Biochar-derived heterogeneous catalysts for biodiesel production. *Environmental Chemistry Letters, 17*(4), 1447–1469. https://doi.org/10.1007/s10311-019-00885-x

Basosi, R., Bonciani, R., Frosali, D., Manfrida, G., Parisi, M. L., & Sansone, F. (2020). Life cycle analysis of a geothermal power plant: Comparison of the environmental performance with other renewable energy systems. *Sustainability (Switzerland), 12*(7), 1–29. https://doi.org/10.3390/su12072786

Ben-Iwo, J., Manovic, V., & Longhurst, P. (2016). Biomass resources and biofuels potential for the production of transportation fuels in Nigeria. *Renewable and Sustainable Energy Reviews, 63*, 172–192. https://doi.org/10.1016/j.rser.2016.05.050

Bharath, G. et al. (2020). Catalytic hydrodeoxygenation of biomass-derived pyrolysis oil over alloyed bimetallic Ni_3Fe nanocatalyst for high-grade biofuel production. *Energy Conversion and Management, 213*. https://doi.org/10.1016/j.enconman.2020.112859

Bromberg, L., Su, X., & Hatton, T. A. (2014). Functional networks of organic and coordination polymers: Catalysis of fructose conversion. *Chemistry of Materials, 26*(21), 6257–6264. https://doi.org/10.1021/cm503098p

Brown, R. C. (2007). Hybrid thermochemical/biological processing: Putting the cart before the horse? *Applied Biochemistry and Biotechnology, 137–140*(1–12), 947–956. https://doi.org/10.1007/s12010-007-9110-y

Buckner, C. A., et al. (2016). We are IntechOpen, the world's leading publisher of Open Access books built by scientists, for scientists TOP 1 %. *INTECH, 11*, 13.

Chamkalani, A., Zendehboudi, S., Rezaei, N., & Hawboldt, K. (2020). A critical review on life cycle analysis of algae biodiesel: Current challenges and future prospects. *Renewable and Sustainable Energy Reviews, 134*, 110143. https://doi.org/10.1016/j.rser.2020.110143

Chang, B., Fu, J., Tian, Y., & Dong, X. (2013). Multifunctionalized ordered mesoporous carbon as an efficient and stable solid acid catalyst for biodiesel preparation. *Journal of Physical Chemistry C, 117*(12), 6252–6258. https://doi.org/10.1021/jp312820g

Chapman, J., Ismail, A. E., & Dinu, C. Z. (2018). Industrial applications of enzymes: Recent advances, techniques, and outlooks. *Catalysts, 8*(6), 20–29. https://doi.org/10.3390/catal8060238

Chen, H. Y., & Sachtler, W. M. H. (1998). Activity and durability of Fe/ZSM-5 catalysts for lean burn NOx reduction in the presence of water vapor. *Catalysis Today, 42*(1–2), 73–83. https://doi.org/10.1016/S0920-5861(98)00078-9

Chen, J., Wang, S., Huang, J., Chen, L., Ma, L., & Huang, X. (2013). Conversion of cellulose and cellobiose into sorbitol catalyzed by ruthenium supported on a polyoxometalate/metal-organic framework hybrid. *Chemsuschem, 6*(8), 1545–1555. https://doi.org/10.1002/cssc.201200914

Chen, J., Li, K., Chen, L., Liu, R., Huang, X., & Ye, D. (2014). Conversion of fructose into 5-hydroxymethylfurfural catalyzed by recyclable sulfonic acid-functionalized metal-organic frameworks. *Green Chemistry, 16*(5), 2490–2499. https://doi.org/10.1039/c3gc42414f

Cho, Y. B., & Seo, G. (2010). High activity of acid-treated quail eggshell catalysts in the transesterification of palm oil with methanol. *Bioresource Technology, 101*(22), 8515–8519. https://doi.org/10.1016/j.biortech.2010.06.082

Cirujano, F. G., Corma, A., & Llabrés i Xamena, F. X. (2015). Conversion of levulinic acid into chemicals: Synthesis of biomass derived levulinate esters over Zr-containing MOFs. *Chemical Engineering Science, 124*, 52–60. https://doi.org/10.1016/j.ces.2014.09.047

Collotta, M., Champagne, P., Tomasoni, G., Alberti, M., Busi, L., & Mabee, W. (2019). Critical indicators of sustainability for biofuels: An analysis through a life cycle sustainabilty assessment perspective. *Renewable and Sustainable Energy Reviews, 115*, 109358. https://doi.org/10.1016/j.rser.2019.109358

Corma, A., García, H., & Llabrés I Xamena, F. X. (2010). Engineering metal organic frameworks for heterogeneous catalysis. *Chemical Reviews, 110*(8), 4606–4655. https://doi.org/10.1021/cr9003924

De, S., Dutta, S., Patra, A. K., Bhaumik, A., & Saha, B. (2011). Self-assembly of mesoporous TiO_2 nanospheres via aspartic acid templating pathway and its catalytic application for 5-hydroxymethyl-furfural synthesis. *Journal of Materials Chemistry, 21*(43), 17505–17510. https://doi.org/10.1039/c1jm13229f

Degfie, T. A., Mamo, T. T., & Mekonnen, Y. S. (2019). Optimized biodiesel production from Waste Cooking Oil (WCO) using Calcium Oxide (CaO) nano-catalyst. *Science and Reports, 9*(1), 1–8. https://doi.org/10.1038/s41598-019-55403-4

Demain, A. L., & Adrio, J. L. (2008). Contributions of microorganisms to industrial biology. *Molecular Biotechnology, 38*(1), 41–55. https://doi.org/10.1007/s12033-007-0035-z

Demirbas, T., & Demirbas, A. H. (2010). Bioenergy, green energy. Biomass and biofuels. *Energy Sources, Part a: Recovery, Utilization and Environmental Effects, 32*(12), 1067–1075. https://doi.org/10.1080/15567030903058600

Dey, P., Pal, P., Kevin, J. D., & Das, D. B. (2020). Lignocellulosic bioethanol production: Prospects of emerging membrane technologies to improve the process—A critical review. *Reviews in Chemical Engineering, 36*(3), 333–367. https://doi.org/10.1515/revce-2018-0014

Du, X., Peng, Y., Albero, J., Li, D., Hu, C., & García, H. (2022). Synthetic fuels from biomass: Photocatalytic hydrodecarboxylation of octanoic acid by ni nanoparticles deposited on TiO_2. *ChemSusChem, 15*(2). https://doi.org/10.1002/cssc.202102107

Dutta, S., De, S., Patra, A. K., Sasidharan, M., Bhaumik, A., & Saha, B. (2011). Microwave assisted rapid conversion of carbohydrates into 5-hydroxymethylfurfural catalyzed by mesoporous TiO_2 nanoparticles. *Applied Catalysis, a: General, 409–410*, 133–139. https://doi.org/10.1016/j.apcata.2011.09.037

Encinar, J. M., González, J. F., Rodríguez, J. J., & Tejedor, A. (2002). Biodiesel fuels from vegetable oils: Transesterification of *Cynara cardunculus* L. Oils with ethanol. *Energy and Fuels, 16*(2), 443–450. https://doi.org/10.1021/ef010174h

Encinar, J. M., González, J. F., Pardal, A., & Martínez, G. (2010). Rape oil transesterification over heterogeneous catalysts. *Fuel Processing Technology, 91*(11), 1530–1536. https://doi.org/10.1016/j.fuproc.2010.05.034

Esmaeili, H., Yeganeh, G., & Esmaeilzadeh, F. (2019). Optimization of biodiesel production from *Moringa oleifera* seeds oil in the presence of nano-MgO using Taguchi method. *International Nano Letters, 9*(3), 257–263. https://doi.org/10.1007/s40089-019-0278-2

Fallah Kelarijani, A., Gholipour Zanjani, N., & Kamran Pirzaman, A. (2020). Ultrasonic assisted transesterification of rapeseed oil to biodiesel using nano magnetic catalysts. *Waste Biomass Valorization, 11*(6), 2613–2621 (2020). https://doi.org/10.1007/s12649-019-00593-1.

Farneth, W. E., & Gorte, R. J. (1995). Methods for characterizing zeolite acidity. *Chemical Reviews, 95*(3), 615–635. https://doi.org/10.1021/cr00035a007

Foroutan, R., Esmaeili, H., Mousavi, S. M., Hashemi, S. A., & Yeganeh, G. (2019). The physical properties of biodiesel-diesel fuel produced via transesterification process from different oil sources. *Physical Chemistry Research, 7*(2), 415–424. https://doi.org/10.22036/pcr.2019.173224.1600

Fu, X. Y., Zhao, W., Xiong, A. S., Tian, Y. S., & Peng, R. H. (2011). High expression of recombinant Streptomyces sp. S38 xylanase in Pichia pastoris by codon optimization and analysis of its biochemical properties. *Molecular Biology Reports, 38*(8), 4991–4997. https://doi.org/10.1007/s11033-010-0644-7

Gao, N., Śliz, M., Quan, C., Bieniek, A., & Magdziarz, A. (2021). Biomass CO_2 gasification with CaO looping for syngas production in a fixed-bed reactor. *Renewable Energy, 167*, 652–661. https://doi.org/10.1016/j.renene.2020.11.134

Gao, J., Cao, Y., Luo, G., Fan, J., Clark, J. H., & Zhang, S. (2022). High-efficiency catalytic hydrodeoxygenation of lignin-derived vanillin with nickel-supported metal phosphate catalysts. *Chemical Engineering Journal, 448* (2022). https://doi.org/10.1016/j.cej.2022.137723

Gómez-López, P., et al. (2020). Nanomaterials and catalysis for green chemistry. *Current Opinion in Green and Sustainable Chemistry, 24*, 48–55. https://doi.org/10.1016/j.cogsc.2020.03.001

Gong, J., Darling, S. B., & You, F. (2015). Perovskite photovoltaics: Life-cycle assessment of energy and environmental impacts. *Energy & Environmental Science, 8*(7), 1953–1968. https://doi.org/10.1039/c5ee00615e

Guerra, F. D., Attia, M. F., Whitehead, D. C., & Alexis, F. (2018). Nanotechnology for environmental remediation: Materials and applications. *Molecules, 23*(7), 1–23. https://doi.org/10.3390/molecules23071760

Guldhe, A., Singh, B., Mutanda, T., Permaul, K., & Bux, F. (2015). Advances in synthesis of biodiesel via enzyme catalysis: Novel and sustainable approaches. *Renewable and Sustainable Energy Reviews, 41*, 1447–1464. https://doi.org/10.1016/j.rser.2014.09.035

Gunarathne, V., Ashiq, A., Ramanayaka, S., Wijekoon, P., & Vithanage, M. (2019). Biochar from municipal solid waste for resource recovery and pollution remediation. *Environmental Chemistry Letters, 17*(3), 1225–1235. https://doi.org/10.1007/s10311-019-00866-0

Haarlemmer, G., Guizani, C., Anouti, S., Déniel, M., Roubaud, A., & Valin, S. (2016). Analysis and comparison of bio-oils obtained by hydrothermal liquefaction and fast pyrolysis of beech wood. *Fuel, 174*, 180–188. https://doi.org/10.1016/j.fuel.2016.01.082

Hamza, M., et al. (2021). A review on the waste biomass derived catalysts for biodiesel production. *Environmental Technology and Innovation, 21*, 101200. https://doi.org/10.1016/j.eti.2020.101200

Hara, M. (2010). Biomass conversion by a solid acid catalyst. *Energy & Environmental Science, 3*(5), 601–607. https://doi.org/10.1039/b922917e

Hasan, N. H., & Jazie, A. A. (2021). Hydrothermal liquefaction conversion of lignocelluloses and waste biomass using zeolite catalyst. *IOP Conference Series: Earth and Environmental Science, 790*(1). https://doi.org/10.1088/1755-1315/790/1/012036

Heeres, A., et al. (2018). Synthesis of bio-aromatics from black liquors using catalytic pyrolysis. *ACS Sustainable Chemistry & Engineering, 6*(3), 3472–3480. https://doi.org/10.1021/acssuschemeng.7b03728

Hu, S., Guan, Y., Wang, Y., & Han, H. (2011). Nano-magnetic catalyst KF/CaO-Fe_3O_4 for biodiesel production. *Applied Energy, 88*(8), 2685–2690. https://doi.org/10.1016/j.apenergy.2011.02.012

Huang, Z., et al. (2020). Enhanced photocatalytic alkane production from fatty acid decarboxylation via inhibition of radical oligomerization. *Nature Catalysis, 3*(2), 170–178. https://doi.org/10.1038/s41929-020-0423-3

Inayat, M., Sulaiman, S. A., Bhayo, B. A., & Shahbaz, M. (2020). Application of response surface methodology in catalytic co-gasification of palm wastes for bioenergy conversion using mineral catalysts. *Biomass and Bioenergy, 132*, Article 105418. https://doi.org/10.1016/j.biombioe.2019.105418

Jae, J., et al. (2011). Investigation into the shape selectivity of zeolite catalysts for biomass conversion. *Journal of Catalysis, 279*(2), 257–268. https://doi.org/10.1016/j.jcat.2011.01.019

Jiao, W., et al. (2020). Influencing factors and reaction mechanism for catalytic CO_2 gasification of sawdust char using K-modified transition metal composite catalysts: Experimental and DFT studies. *Energy Conversion and Management, 208*. https://doi.org/10.1016/j.enconman.2020.112522.

Kaloudas, D., Pavlova, N., & Penchovsky, R. (2021). Lignocellulose, algal biomass, biofuels and biohydrogen: A review. *Environmental Chemistry Letters, 19*(4), 2809–2824. https://doi.org/10.1007/s10311-021-01213-y

Kargbo, H., Harris, J. S., & Phan, A. N. (2021). 'Drop-in' fuel production from biomass: Critical review on techno-economic feasibility and sustainability. *Renewable and Sustainable Energy Reviews, 135*, 110168. https://doi.org/10.1016/j.rser.2020.110168

Kashimalla, M., Suraboyina, S., Dubbaka, V., & Polumati, A. (2023). Optimisation of a catalytic hydrothermal liquefaction process using central composite design for yield improvement of bio-oil. *Biomass Conversion and Biorefinery, 13*(5), 3751–3763. https://doi.org/10.1007/s13399-021-01451-8

Kumar, M., Sun, Y., Rathour, R., Pandey, A., Thakur, I. S., & Tsang, D. C. W. (2020). Algae as potential feedstock for the production of biofuels and value-added products: Opportunities and challenges. *Science of the Total Environment, 716*, 137116. https://doi.org/10.1016/j.scitotenv.2020.137116

Li, Q., et al. (2009). Improving enzymatic hydrolysis of wheat straw using ionic liquid 1-ethyl-3-methyl imidazolium diethyl phosphate pretreatment. *Bioresource Technology, 100*(14), 3570–3575. https://doi.org/10.1016/j.biortech.2009.02.040

Li, W., Ma, F., Su, F., Ma, L., Zhang, S., & Guo, Y. (2011). One-step preparation of efficient and reusable SO_{42}-/ZrO_2-based hybrid solid catalysts functionalized by alkyl-bridged organosilica moieties for biodiesel production. *Chemsuschem, 4*(6), 744–756. https://doi.org/10.1002/cssc.201100022

Li, H., & Ma, Y. (2020). Magnetically separable Fe_3O_4-supported Ru–Ni bimetallic catalysts for diformyltricyclodecanes hydrogenation to value-added fine chemicals. *Progress in Reaction Kinetics and Mechanism, 45*. https://doi.org/10.1177/1468678319825732.

Liu, C., Lv, P., Yuan, Z., Yan, F., & Luo, W. (2010). The nanometer magnetic solid base catalyst for production of biodiesel. *Renewable Energy, 35*(7), 1531–1536. https://doi.org/10.1016/j.renene.2009.10.009

Liu, Q., Wang, J., Zhou, J., Yu, Z., & Wang, K. (2020). Promotion of monocyclic aromatics by catalytic fast pyrolysis of biomass with modified HZSM-5. *Journal of Analytical and Applied Pyrolysis, 153*, 1–8. https://doi.org/10.1016/j.jaap.2020.104964

Lu, H., Wang, J., Stoller, M., Wang, T., Bao, Y., & Hao, H. (2016). An overview of nanomaterials for water and wastewater treatment. *Advances in Materials Science and Engineering, 2016*. https://doi.org/10.1155/2016/4964828

Mahmud, M. A. P., Huda, N., Farjana, S. H., & Lang, C. (2019). A strategic impact assessment of hydropower plants in alpine and non-alpine areas of Europe. *Applied Energy, 250*, 198–214. https://doi.org/10.1016/j.apenergy.2019.05.007

Majekodunmi, S. O. (2021). A review on Capa in the pharmaceutical industry. *International Journal of Pharmaceutical Research, 13*(3), 67–78. https://doi.org/10.31838/ijpr/2021.13.03.240

McClendon, S. D., Shin, H. D., & Chen, R. R. (2011). Novel bacterial ferulic acid esterase from Cellvibrio japonicus and its application in ferulic acid release and xylan hydrolysis. *Biotechnology Letters, 33*(1), 47–54. https://doi.org/10.1007/s10529-010-0394-6

Molnár, Á., & Papp, A. (2017). Catalyst recycling—A survey of recent progress and current status. *Coordination Chemistry Reviews, 349*, 1–65. https://doi.org/10.1016/j.ccr.2017.08.011

Montag, M., Zhang, J., & Milstein, D. (2012). Aldehyde binding through reversible C-C coupling with the pincer ligand upon alcohol dehydrogenation by a PNP-ruthenium catalyst. *Journal of the American Chemical Society, 134*(25), 10325–10328. https://doi.org/10.1021/ja303121v

Mutreja, V., Singh, S., & Ali, A. (2011). Biodiesel from mutton fat using KOH impregnated MgO as heterogeneous catalysts. *Renewable Energy, 36*(8), 2253–2258. https://doi.org/10.1016/j.renene.2011.01.019

Narnaware, S. L., & Panwar, N. L. (2021). *Catalysts and their role in biomass gasification and tar abetment: A review*, no. 0123456789. Springer, Berlin. https://doi.org/10.1007/s13399-021-01981-1

Nasresfahani, Z., & Kassaee, M. Z. (2021). Nickel−Copper bimetallic mesoporous nanoparticles: As an efficient heterogeneous catalyst for N-alkylation of amines with alcohols. *Applied Organometallic Chemistry, 35*(1), 1–9. https://doi.org/10.1002/aoc.6032

Nasrollahzadeh, M., Sajjadi, M., Iravani, S., & Varma, R. S. (2021). Green-synthesized nanocatalysts and nanomaterials for water treatment: Current challenges and future perspectives. *Journal of Hazardous Materials, 401*, 123401. https://doi.org/10.1016/j.jhazmat.2020.123401

Navalón, S., Ong, W. J., & Duan, X. (2020). Sustainable catalytic processes driven by graphene-based materials. *Processes, 8*(6), 2020. https://doi.org/10.3390/PR8060672

Nurul Suziana, N. M., Asikin-Mijan, N., Zulkarnain, Z., & Taufiq-Yap, Y. H. (2022). Catalytic hydrothermal liquefaction of empty fruit bunch in subcritical water over bimetallic modified zeolite. *Chemical Engineering Research and Design, 183*, 250–262. https://doi.org/10.1016/j.cherd.2021.10.005

Okolie, J. A., Rana, R., Nanda, S., Dalai, A. K., & Kozinski, J. A. (2019). Supercritical water gasification of biomass: A state-of-the-art review of process parameters, reaction mechanisms and catalysis. *Sustain Energy Fuels, 3*(3), 578–598. https://doi.org/10.1039/c8se00565f

Okolie, J. A., Mukherjee, A., Nanda, S., Dalai, A. K., & Kozinski, J. A. (2021). Catalytic supercritical water gasification of soybean straw: Effects of catalyst supports and promoters. *Industrial and Engineering Chemistry Research, 60*(16), 5770–5782. https://doi.org/10.1021/acs.iecr.0c06177

Ordomsky, V. V., Sushkevich, V. L., Schouten, J. C., Van Der Schaaf, J., & Nijhuis, T. A. (2013). Glucose dehydration to 5-hydroxymethylfurfural over phosphate catalysts. *Journal of Catalysis, 300*, 37–46. https://doi.org/10.1016/j.jcat.2012.12.028

Osman, A. I. (2020). Mass spectrometry study of lignocellulosic biomass combustion and pyrolysis with NOx removal. *Renewable Energy, 146*, 484–496. https://doi.org/10.1016/j.renene.2019.06.155

Pang, J., Sun, J., Zheng, M., Li, H., Wang, Y., & Zhang, T. (2019). Transition metal carbide catalysts for biomass conversion: A review. *Applied Catalysis, B: Environmental, 254*, 510–522. https://doi.org/10.1016/j.apcatb.2019.05.034

Peng, W. H., Lee, Y. Y., Wu, C., & Wu, K. C. W. (2012). Acid-base bi-functionalized, large-pored mesoporous silica nanoparticles for cooperative catalysis of one-pot cellulose-to-HMF conversion. *Journal of Materials Chemistry, 22*(43), 23181–23185. https://doi.org/10.1039/c2jm35391a

Prestigiacomo, C., Proietto, F., Laudicina, V. A., Siragusa, A., Scialdone, O., & Galia, A. (2021). Catalytic hydrothermal liquefaction of municipal sludge assisted by formic acid for the production of next-generation fuels. *Energy, 232*, 121086. https://doi.org/10.1016/j.energy.2021.121086

Rezayan, A., & Taghizadeh, M. (2018). Synthesis of magnetic mesoporous nanocrystalline KOH/ZSM-5-Fe_3O_4 for biodiesel production: Process optimization and kinetics study. *Process Safety and Environmental Protection, 117*, 711–721. https://doi.org/10.1016/j.psep.2018.06.020

Riaz, S., et al. (2022). Biobutanol production from sustainable biomass process of anaerobic ABE fermentation for industrial applications. *Archives of Microbiology, 204*(11), 1–9. https://doi.org/10.1007/s00203-022-03284-z

Sadorsky, P. (1999). Oil price shocks and stock market activity. *Energy Economics, 21*(5), 449–469. https://doi.org/10.1016/S0140-9883(99)00020-1

Sandhya, C., Nampoothiri, K. M., & Pandey, A. (2005). Microbial cells and enzymes. *Microbial Enzymes and Biotransformations, 17*(1), 1–27.

Scarsella, M., de Caprariis, B., Damizia, M., & De Filippis, P. (2020). Heterogeneous catalysts for hydrothermal liquefaction of lignocellulosic biomass: A review. *Biomass and Bioenergy, 140*, 105662. https://doi.org/10.1016/j.biombioe.2020.105662

Schlögl, R. (2015). Heterogeneous catalysis. *Angewandte Chemie - International Edition, 54*(11), 3465–3520. https://doi.org/10.1002/anie.201410738

Schuchardt, U., Sercheli, R., & Vargas, R. M. (1998). Transesterification of vegetable oils: a review. *Journal of the Brazilian Chemical Society, 9*(3), 199–210.

Shafiee, S., & Topal, E. (2009). When will fossil fuel reserves be diminished? *Energy Policy, 37*(1), 181–189. https://doi.org/10.1016/j.enpol.2008.08.016

Singh, R., Krishna, B. B., Mishra, G., Kumar, J., & Bhaskar, T. (2016). Strategies for selection of thermo-chemical processes for the valorisation of biomass. *Renewable Energy, 98*, 226–237. https://doi.org/10.1016/j.renene.2016.03.023

Smets, K., Schreurs, S., Carleer, R., & Yperman, J. (2014). Valorization of raspberry seed cake by flash and slow pyrolysis: Product yield and characterization of the liquid and solid fraction. *Journal of Analytical and Applied Pyrolysis, 107*, 289–297. https://doi.org/10.1016/j.jaap.2014.03.014

Song, B., Lin, R., Lam, C. H., Wu, H., Tsui, T. H., & Yu, Y. (2021). Recent advances and challenges of inter-disciplinary biomass valorization by integrating hydrothermal and biological techniques. *Renewable and Sustainable Energy Reviews, 135*, 110370. https://doi.org/10.1016/j.rser.2020.110370

Subramaniam, B., Licence, P., Moores, A., & Allen, D. T. (2021). Shaping effective practices for incorporating sustainability assessment in manuscripts submitted to ACS sustainable chemistry & engineering: An initiative by the editors. *ACS Sustainable Chemistry & Engineering, 9*(11), 3977–3978. https://doi.org/10.1021/acssuschemeng.1c01511

Sun, J., Phillips, C. M., Anderson, C. T., Beeson, W. T., Marletta, M. A., & Glass, N. L. (2011). Expression and characterization of the *Neurospora crassa* endoglucanase GH5-1. *Protein Expression and Purification, 75*(2), 147–154. https://doi.org/10.1016/j.pep.2010.08.016

Sun, Z., Wang, S., Wang, X., & Jiang, Z. (2016). Lysine functional heteropolyacid nanospheres as bifunctional acid-base catalysts for cascade conversion of glucose to levulinic acid. *Fuel, 164*, 262–266. https://doi.org/10.1016/j.fuel.2015.10.014

Tanabe, K., & Hölderich, W. F. (1999). Industrial application of solid acid-base catalysts. *Applied Catalysis, a: General, 181*(2), 399–434. https://doi.org/10.1016/S0926-860X(98)00397-4

Tang, S., Wang, L., Zhang, Y., Li, S., Tian, S., & Wang, B. (2012). Study on preparation of Ca/Al/Fe_3O_4 magnetic composite solid catalyst and its application in biodiesel transesterification. *Fuel Processing Technology, 95*, 84–89. https://doi.org/10.1016/j.fuproc.2011.11.022

Taylor, J. M., Dawson, K. W., & Shimizu, G. K. H. (2013). A water-stable metal-organic framework with highly acidic pores for proton-conducting applications. *Journal of the American Chemical Society, 135*(4), 1193–1196. https://doi.org/10.1021/ja310435e

Thangalazhy-Gopakumar, S., Adhikari, S., & Gupta, R. B. (2012). Catalytic pyrolysis of biomass over H +ZSM-5 under hydrogen pressure. *Energy and Fuels, 26*(8), 5300–5306. https://doi.org/10.1021/ef3008213

Thitsartarn, W., & Kawi, S. (2011). An active and stable CaO–CeO_2 catalyst for transesterification of oil to biodiesel. *Green Chemistry, 13*(12), 3423–3430. https://doi.org/10.1039/c1gc15596b

Trivedi, N., Gupta, V., Kumar, M., Kumari, P., Reddy, C. R. K., & Jha, B. (2011). An alkali-halotolerant cellulase from Bacillus flexus isolated from green seaweed Ulva lactuca. *Carbohydrate Polymers, 83*(2), 891–897. https://doi.org/10.1016/j.carbpol.2010.08.069

Ullah, K., et al. (2015). Assessing the potential of algal biomass opportunities for bioenergy industry: A review. *Fuel, 143*, 414–423. https://doi.org/10.1016/j.fuel.2014.10.064

Védrine, J. C. (2017). Heterogeneous catalysis on metal oxides. *Catalysts, 7*(11). https://doi.org/10.3390/catal7110341

Verma, S., Nasir Baig, R. B., Nadagouda, M. N., & Varma, R. S. (2016). Visible light mediated upgrading of biomass to biofuel. *Green Chemistry, 18*(5), 1327–1331. https://doi.org/10.1039/c5gc02951a

Verziu, M., Coman, S. M., Richards, R., & Parvulescu, V. I. (2011). Transesterification of vegetable oils over CaO catalysts. *Catalysis Today, 167*(1), 64–70. https://doi.org/10.1016/j.cattod.2010.12.031

Viriya-empikul, N., Krasae, P., Puttasawat, B., Yoosuk, B., Chollacoop, N., & Faungnawakij, K. (2010). Waste shells of mollusk and egg as biodiesel production catalysts. *Bioresource Technology, 101*(10), 3765–3767. https://doi.org/10.1016/j.biortech.2009.12.079

Vujicic, D., Comic, D., Zarubica, A., Micic, R., & Boskovic, G. (2010). Kinetics of biodiesel synthesis from sunflower oil over CaO heterogeneous catalyst. *Fuel, 89*(8), 2054–2061. https://doi.org/10.1016/j.fuel.2009.11.043

Wang, L., et al. (2019). Nature of extra capacity in MoS2 electrodes: Molybdenum atoms accommodate with lithium. *Energy Storage Materials, 16*, 37–45. https://doi.org/10.1016/j.ensm.2018.04.025

Wang, C., Zhang, X., Liu, Q., Zhang, Q., Chen, L., & Ma, L. (2020a). A review of conversion of lignocellulose biomass to liquid transport fuels by integrated refining strategies. *Fuel Processing Technology, 208*, 106485. https://doi.org/10.1016/j.fuproc.2020.106485

Wang, A., Sudarsanam, P., Xu, Y., Zhang, H., Li, H., & Yang, S. (2020b). Functionalized magnetic nanosized materials for efficient biodiesel synthesis: Via acid-base/enzyme catalysis. *Green Chemistry, 22*(10), 2977–3012. https://doi.org/10.1039/d0gc00924e

Wang, S., Li, Z., Yi, W., Fu, P., Zhang, A., & Bai, X. (2021). Renewable aromatic hydrocarbons production from catalytic pyrolysis of lignin with Al-SBA-15 and HZSM-5: Synergistic effect and coke behaviour. *Renewable Energy, 163*, 1673–1681. https://doi.org/10.1016/j.renene.2020.10.108

Wang, D., et al. (2021). Catalytic upgrading of lignocellulosic biomass pyrolysis vapors: Insights into physicochemical changes in ZSM-5. *Journal of Analytical and Applied Pyrolysis, 156*. https://doi.org/10.1016/j.jaap.2021.105123.

Weldu, Y. W. (2017). Life cycle human health and ecosystem quality implication of biomass-based strategies to climate change mitigation. *Renewable Energy, 108*, 11–18. https://doi.org/10.1016/j.renene.2017.02.046

Wen, L., Wang, Y., Lu, D., Hu, S., & Han, H. (2010a). Preparation of KF/CaO nanocatalyst and its application in biodiesel production from Chinese tallow seed oil. *Fuel, 89*(9), 2267–2271. https://doi.org/10.1016/j.fuel.2010.01.028

Wen, Z., Yu, X., Tu, S. T., Yan, J., & Dahlquist, E. (2010b). Synthesis of biodiesel from vegetable oil with methanol catalyzed by Li-doped magnesium oxide catalysts. *Applied Energy, 87*(3), 743–748. https://doi.org/10.1016/j.apenergy.2009.09.013

West, R. M., et al. (2010). Zeolite H-USY for the production of lactic acid and methyl lactate from C3-sugars. *Journal of Catalysis, 269*(1), 122–130. https://doi.org/10.1016/j.jcat.2009.10.023

Wiltschi, B., et al. (2020). Enzymes revolutionize the bioproduction of value-added compounds: From enzyme discovery to special applications. *Biotechnology Advances, 40*, Article 107520. https://doi.org/10.1016/j.biotechadv.2020.107520

Xu, X., Li, Y., Gong, Y., Zhang, P., Li, H., & Wang, Y. (2012). Synthesis of palladium nanoparticles supported on mesoporous n-doped carbon and their catalytic ability for biofuel upgrade. *Journal of the American Chemical Society, 134*(41), 16987–16990. https://doi.org/10.1021/ja308139s

Yang, C., et al. (2020). Hydrothermal liquefaction and gasification of biomass and model compounds: A review. *Green Chemistry, 22*(23), 8210–8232. https://doi.org/10.1039/d0gc02802a

Yang, J., et al. (2021). Study on the hydrothermal liquefaction of antibiotic residues with molecular sieve catalysts in the ethanol-water system: Focus on product distribution and characterization. *RSC Advances, 11*(43), 26763–26772. https://doi.org/10.1039/d1ra03860e

Yaşar, F. (2019). Biodiesel production via waste eggshell as a low-cost heterogeneous catalyst: Its effects on some critical fuel properties and comparison with CaO. *Fuel, 255*, 115828.

Yoosuk, B., Udomsap, P., Puttasawat, B., & Krasae, P. (2010a). Modification of calcite by hydration-dehydration method for heterogeneous biodiesel production process: The effects of water on properties and activity. *Chemical Engineering Journal, 162*(1), 135–141. https://doi.org/10.1016/j.cej.2010.05.013

Yoosuk, B., Udomsap, P., Puttasawat, B., & Krasae, P. (2010b). Improving transesterification acitvity of CaO with hydration technique. *Bioresource Technology, 101*(10), 3784–3786. https://doi.org/10.1016/j.biortech.2009.12.114

Zabeti, M., Daud, W. M. A. W., & Aroua, M. K. (2010). Biodiesel production using alumina-supported calcium oxide: An optimization study. *Fuel Processing Technology, 91*(2), 243–248. https://doi.org/10.1016/j.fuproc.2009.10.004

Zhang, Y., Degirmenci, V., Li, C., & Hensen, E. J. M. (2011). Phosphotungstic acid encapsulated in metal-organic framework as catalysts for carbohydrate dehydration to 5-hydroxymethylfurfural. *Chemsuschem, 4*(1), 59–64. https://doi.org/10.1002/cssc.201000284

Zhang, M., Tong, X., Ma, R., & Li, Y. (2016). Catalytic transformation of carbohydrates into 5-hydroxymethyl furfural over tin phosphate in a water-containing system. *Catalysis Today, 264*, 131–135. https://doi.org/10.1016/j.cattod.2015.06.031

Zhang, H., Zhou, P., Ji, H., Ma, W., Chen, C., & Zhao, J. (2018). Enhancement of photocatalytic decarboxylation on TiO_2 by water-induced change in adsorption-mode. *Applied Catalysis, B: Environmental, 224*, 376–382. https://doi.org/10.1016/j.apcatb.2017.10.020

Zhao, C., Kasakov, S., He, J., & Lercher, J. A. (2012). Comparison of kinetics, activity and stability of Ni/HZSM-5 and Ni/Al_2O_3-HZSM-5 for phenol hydrodeoxygenation. *Journal of Catalysis, 296*, 12–23. https://doi.org/10.1016/j.jcat.2012.08.017

Zhao, C., Brück, T., & Lercher, J. A. (2013). Catalytic deoxygenation of microalgae oil to green hydrocarbons. *Royal Society of Chemistry*. https://doi.org/10.1039/c3gc40558c

Chapter 12
Waste to Bioenergy: Biofuel Production from Seafood Waste

Ulaganathan Arisekar, K. Udhayakumaar, Rajendran Shalini, P. Padmavathy, and Balasubramanian Sivaraman

Abstract Global fisheries and aquaculture production reached 232.2 million tonnes, a 4.4% rise from 2020, with Asia contributing over 70%, followed by Europe (9%) and Latin America (9%). Seafood is exported for human consumption, generating waste including internal organs. The seafood industry and domestic market create millions of tonnes of waste from pre-processing, containing notable fat/oil levels. Biofuel production from seafood waste has gained attention due to valorization in line with circular economy approaches and demand for sustainable energy solutions. This chapter explores extraction methods of biofuel/bioethanol from seafood waste, including biochemical approaches like anaerobic digestion, fermentation, and enzymatic hydrolysis, and thermochemical techniques like pyrolysis and gasification. It addresses benefits of biofuel production from waste, environmental benefits, and challenges in extraction, storage, and scale-up. Case studies of biofuel production from waste in various regions are discussed, providing insights into efficient production systems and technological advancements. The chapter emphasizes the potential use of seafood waste in bioenergy within the circular economy model. Further research could provide more efficient biofuel production technology for sustainable energy in the future.

Keywords Seafood waste · Extraction · Conversion methods · Bioethanol · Biobutanol · Environmental impact · Future scopes

U. Arisekar (✉) · R. Shalini · B. Sivaraman
Department of Fish Quality Assurance and Management, Fisheries College and Research Institute, Tamil Nadu Dr. J. Jayalalithaa Fisheries University, Thoothukudi, Tamil Nadu, India
e-mail: arisekar@tnfu.ac.in

K. Udhayakumaar · P. Padmavathy
Department of Aquatic Environment Management, Fisheries College and Research Institute, Tamil Nadu Dr. J. Jayalalithaa Fisheries University, Thoothukudi, Tamil Nadu, India

V. Singh et al. (eds.), *Bioenergy and Value Added Byproducts from Biomass Waste*, Sustainable Environmental Waste Management Strategies,
https://doi.org/10.1007/978-3-032-22849-9_12

12.1 Introduction

The rapid exhaustion of limited non-renewable energy resources has emerged as a critical challenge in the twenty-first century, prompting a substantial shift towards sustainable, cost-effective, and environmentally friendly alternatives. For over a century, the combustion of fossil fuels has released carbon dioxide (CO_2) and other substances into the atmosphere, altering the Earth's climate and increasing ocean acidity (Nisar et al., 2021). Since the Industrial Revolution began, fossil fuel consumption has led to a 43% increase in atmospheric carbon dioxide levels (kumar et al., 2016). Concurrently, our reliance on fossil fuels for transportation and domestic lighting has resulted in adverse effects such as pollution and climate change (Perera, 2018). These pollutants are linked to numerous severe health issues, including respiratory disorders, cardiac arrests, cerebrovascular events, asthma, premature mortality, and increased absenteeism from work and school. A positive trend in the quest for sustainable energy sources is evident, with biofuels currently accounting for approximately 3.4% of global transportation fuel (Usman et al., 2022). Bioenergy derived from biomass fuels contributes to mitigating global warming by reducing carbon dioxide emissions. The carbon–neutral nature of biomass combustion for power generation stems from its absorption of carbon dioxide during photosynthesis. This process creates a closed carbon cycle, supporting sustainability objectives and reducing the overall impact on atmospheric carbon dioxide concentrations (Amoah et al., 2019). Biofuels are primarily derived from biological sources, including bacteria, plants, animals, and waste materials (Berla et al., 2013). Currently, solid and liquid biofuels are extensively utilized for energy production due to their abundance, high energy content, and cost-effectiveness. While both conventional fuels and biofuels originate from biomaterials, the formation of coal and petroleum fossil energy resources requires millions of years, whereas biofuels can be rapidly produced and extracted with significant yields (Zou et al., 2016). The initial generation of biofuels, derived from sugars, starches, and vegetable oils, sometimes incorporating animal fats, relies heavily on food crops such as corn and sugar beet. However, their usage presents several challenges, including potential food price increases due to resource competition, limited greenhouse gas emission reductions, substantial land requirements, and relatively higher costs compared to fossil fuels. Notably, increased corn demand results in shifts in soybean and wheat production, leading to price increases of 40, 20, and 17% for corn, soybeans, and wheat, respectively (Datta et al., 2019). In this context, biofuels from fish waste emerge as a promising alternative to conventional fuels, potentially meeting human energy needs without altering climate patterns (Foster et al., 2017). Our focus is on examining various biofuel extraction methods from seafood waste, assessing their production benefits, to recommending the optimal extraction technique for commercial combustion applications. The discussion then transitions to examining the challenges associated with producing biofuels from seafood waste, concluding with an analysis of ongoing obstacles within seafood waste biorefineries. Figure 12.1 showed the overall process of bioenergy extraction from seafood waste.

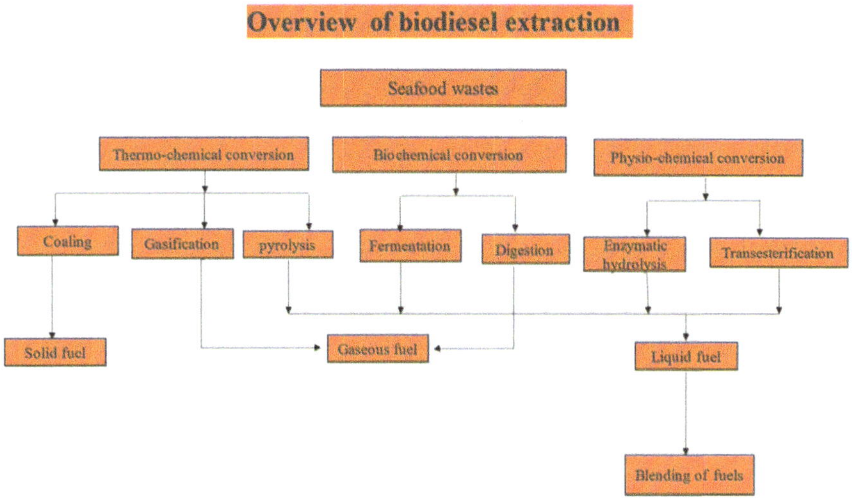

Fig. 12.1 Flow diagram for extraction of bioenergy from seafood waste

12.2 Seafood Waste: A Source of Biofuel

The utilization of seafood waste for biofuel production is gaining traction due to its widespread availability and high fatty acid content. This waste can serve as an excellent alternative to conventional biofuel sources, potentially conserving resources through reduced emissions during engine combustion and promoting environmental sustainability. Seafood waste offers promise for sustainable energy solutions and functions as a natural carbon sink, aiding in carbon sequestration. The biochemical composition of seafood, characterized by low protein and high fatty acid levels, makes it an efficient and sustainable biofuel feedstock. In a world urgently seeking renewable energy sources, seafood exploration presents a viable option, paving the way for an environmentally friendly and sustainable future in the face of climate change and fossil fuel reliance. According to the Food and Agriculture Organization (FAO), global fish production reached 242 MT in 2023. Approximately 75% of this production is intended for direct human consumption, while the remaining 25% is allocated for non-food purposes. The global seafood market is projected to attain a value of 732.87 billion USD with a production of 174 MT. It is estimated that processing plant waste constitutes between 40 and 65% of all processed fish, significantly impacting the surrounding environment. Sardine, mackerel, and tuna are the three most common oily fish used in the canning industry. The processing of sardines, a large fish widely used in commercial markets, generates substantial waste, including the stomach, liver, and head (Khoddami et al., 2009). As the search for alternative energy sources intensifies, the prospect of carbon–neutral emissions from seafood-derived biofuel is gaining traction. Seafood-based biofuels offer a unique advantage over conventional fossil fuels by absorbing and blending, resulting in reduced carbon dioxide emissions into the atmosphere (Handayani et al., 2022). The conversion of seafood waste biomass into biofuel creates a balance through blending, where the carbon released during combustion is offset. This process leads to near-zero net emissions throughout the biofuel's production and usage cycle, establishing it as a carbon–neutral energy source. Fossil fuel consumption in transportation and energy production sectors is a primary contributor to global warming. Biofuels such as biodiesel, bioethanol, biogas, and other syn-gas can help mitigate climate change and global warming by reducing reliance on petroleum-based fossil fuels. Fish processing bio-waste pollutes marine or aquatic ecosystems by decreasing water bodies' pH, temperature, and dissolved oxygen levels. However, fish waste can be repurposed for various applications, including biofuel production, which can enhance energy security, promote socioeconomic stability, and build resilience against climate change-induced environmental risks by reducing greenhouse gas emissions. Biodiesel, a liquid biofuel used in diesel engines, and biogas, a gaseous fuel primarily composed of methane used for cooking, heating, and power generation in internal combustion engines, are two examples of such biofuels. The aquaculture and fishing processing industries generate fish waste that poses environmental challenges and threatens biodiversity when disposed of in terrestrial, aquatic, and marine environments. Consequently, the development of biodiesel and biogas from fish waste and byproducts is crucial for sustainable development and long-term growth in the renewable energy sector.

12.2.1 Physio Chemical Conversion

12.2.1.1 Biodiesel Production

Fatty acid methyl esters (FAME), which are monoalkyl esters derived from long-chain fatty acids, are commonly found in biodiesel. These substances can be extracted from various renewable lipid sources, including sunflower, palm, soybean, rapeseed, canola, jatropha, and cottonseed. However, they tend to be more costly than conventional diesel fuel. There is increasing interest in investigating alternative sources for biodiesel production, such as non-edible vegetable oils, used cooking oil, and oil derived from seafood waste (Aien & Mahdavi, 2020). The conversion of seafood waste into biodiesel involves multiple steps, including oil extraction and transesterification. The production of three moles of methyl esters and one mole of glycerol typically requires three moles of alcohol for each mole of triglyceride (Costa et al., 2022). While this process can utilize various alcohols like ethanol, butanol, methanol, propanol, and amyl alcohol, methanol is preferred due to its cost-effectiveness, favorable physical and chemical properties, and widespread use in commercial production (Mathiyazhagan et al. 2011). The reaction can be facilitated by inorganic catalysts, such as acids or alkalies (Di sierio et al., 2007), or by lipase enzymes (Pinyaphong et al., 2011). Al Azad et al. (2019) noted that fish waste components, including viscera, bones, trimmings, tails, fins, and skin, were traditionally processed into fishmeal for livestock and aquaculture feed. They suggested that fish waste offers a diverse fatty acid profile suitable for biodiesel production. Crude oils were obtained through Soxhlet extraction using petroleum ether, and the extracted fish oil was subsequently methylated to produce fatty acid methyl esters using a reactive apparatus. Ghaly et al. (2010) highlight the substantial amount of fish waste generated by the fish processing industry and its potential as a sustainable raw material for biodiesel production. This underscores the importance of utilizing fish waste as a feedstock to meet the increasing demand for renewable energy sources.

12.2.1.2 Preparation of Seafood Waste

The less desirable edible components of various seafood byproducts, including heads, intestines, and tails, were gathered and transported to the lab within 30–45 min in an insulated cooler maintained at approximately (− 4 °C). Initially, a pre-treatment process involved cleaning with fresh water to eliminate blood residue and contaminants (Karkal & Kudre, 2023). Larger fish waste pieces were cut into smaller, more uniform sizes, as the waste originated from different fish parts (Tayib et al., 2024). The fish waste's fat content was evaluated, and the remaining material was stored in a freezer at (− 18 °C) to prevent spoilage. The seafood waste was then utilized according to their respective extraction methods.

12.2.1.3 Oil Extraction

The extraction of oil from combined and separate fish waste samples was conducted using the wet reduction technique described by Crexi et al. (2009). Frozen samples were defrosted under running water until ice-free at the core. Each extraction batch utilized approximately one kg of fish waste. The measured waste was placed in a metal container covered with aluminum foil and inserted into a horizontal autoclave. Steam cooking was performed at 121 °C for three different durations (25, 35, and 45 min) to determine the optimal extraction time for maximum oil yield by promoting protein coagulation and cell membrane breakdown (Kiran et al., 2021). Following steam cooking, the fish waste was manually pressed to separate the liquid (containing dissolved dry matter and oil) from the solid (press cake). The extracted liquid was then decanted for 30 min in a separatory funnel to isolate the oil (triglycerides) layer. To remove remaining impurities, the obtained oil with residual sediments underwent centrifugation in an Eppendorf 5430 centrifuge at 3000 rpm for 15 min, facilitating fat separation (Girish, 2019).

12.2.1.4 Transesterification of Oils

To convert fish oil into biodiesel, a transesterification process was conducted using a 250 ml flat-bottom flask equipped with a water-cooled reflux condenser and magnetic stirrer. An oil bath provided temperature control, while a thermocouple probe immersed in the reaction mixture allowed continuous temperature monitoring. The experiment utilized specific concentrations of catalyst, methanol, and fish oil as required for each trial. To ensure uniform temperature distribution and catalyst suspension, the mixture was stirred at 800 rpm. Following the designated reaction time, the mixture was either centrifuged for 10 min at 5000 rpm or separated by gravity using a separatory funnel to extract the biodiesel. The extracted biodiesel was then analyzed using GC–MS to determine the FAME yield (Arumugam & Ponnusami, 2017). When fish oil contains high levels of free fatty acids, using a base homogeneous catalyst is not recommended as it may lead to emulsified soap formation during the reaction, hindering biodiesel production (Sharma et al. 2016). To avoid saponification, an acid catalyst can be employed instead. It is important to note that waste-derived catalysts with high concentrations of essential catalytic compounds demonstrate enhanced catalytic efficiency in transesterification, resulting in a more efficient process with faster reaction rates and shorter durations. To address high free fatty acid content in the feedstock, a two-step process was implemented. This involved acid esterification using H_2SO_4 to eliminate free fatty acids, followed by base transesterification using NaOH to produce biodiesel.

12.2.1.5 Enzymatic Hydrolysis

Immobilized lipase enzymes were employed to transform fish oil into biodiesel through enzymatic hydrolysis and transesterification. This process requires maintaining a low alcohol-to-oil molar ratio and a relatively cool reaction temperature. Unlike chemical catalysts, which necessitate multi-stage pretreatment for feedstocks containing free fatty acids (FFA), especially when FFA content exceeds 0.5 weight per cent, enzyme catalysts can initiate biodiesel production without additional preparation and can be reclaimed and reused to minimize processing expenses. Enzyme immobilization enables recovery and allows for multiple reuses of costly enzymes (Tan et al., 2010). The enzyme solution was prepared in a pH 7 phosphate buffer to prevent deactivation during immobilisation. Immobilized lipase was created by combining 0.5 g of activated carbon with 150 ml of enzyme solution. This mixture was agitated at room temperature for 24 h. After equilibration, the immobilized lipase was extracted by filtering the solution through the Whatman filter paper (Lam et al., 2010). The feedstock oil was mixed with methanol and water. As the reaction duration directly influences biodiesel production, an appropriate enzyme, such as 100 mg of immobilized Aspergillus niger lipase, was introduced to the mixture. The reaction mixture was then maintained at a constant temperature (30–50 °C) in an orbital shaker at 200 RPM for 10 h. Typical reaction times for enzymatic transesterification of various oils range from 4 to 72 h. The most crucial process parameters affecting biodiesel production are temperature, water content in the mixture, and methanol-to-oil ratio. The optimal values for these factors depend on the immobilization method, lipase source, and biodiesel feedstock (Arumugam & Ponnusami, 2017). Linder, (2005) utilized three different enzymes—neutrase, alcalase, and flavourzyme—at a concentration of 0.055 to extract fat from salmon heads. Pinyaphong et al. (2011) used fish oil extracted from discarded fish parts with Carica papaya lipase as a catalyst. They found that Carica papaya lipase was suitable for methanolysis of fish oil to produce methyl ester. The maximum methyl ester yield reached 83% under optimal conditions: 1:4 oil-to-methanol molar ratio, 20% lipase (based on oil), initial lipase water activity of 0.23, and 20% tert-butanol (based on oil) at 40 °C after 18 h of reaction time. According to Ward et al. (2011), the production of fatty acid methyl esters from waste fish oil through enzymatic catalysis could offer both economic and environmental benefits by utilizing seafood waste products. Despite the significant potential of enzymatic hydrolysis for biodiesel production, it faces obstacles in effectively competing with petroleum-based fuels. These challenges stem from the necessity of implementing strict sanitation and preservation protocols to safeguard against contamination from external sources.

12.2.2 *Biochemical Conversion*

12.2.2.1 Bioethanol

In contemporary research, numerous investigations have focused on the use of seafood as a substrate for bioethanol production, offering an alternative to conventional crops such as sugarcane, maize, and soybeans. The transformation of seafood into bioethanol involves several critical steps. Initially, the raw seafood undergoes pretreatment, which includes washing and cutting to eliminate debris and stones. Fish waste materials are then placed in a cylinder and heated to 80 °C for five hours in an incubator. After cooling at room temperature, the oil and residue are separated (Hossain et al., 2008). The fish byproduct biomass is collected through filtration, weighed, dried, and stored for bioethanol production. To convert the biomass into sugars, the processed feedstock is combined with suitable fermentators like Saccharomyces cerevisiae, Clostridium acetobutylicum, and Clostridium beijerinckii (Sheldon, 2014). Hydrolysis, a chemical process catalyzed by enzymes, breaks down complex polysaccharides in the raw feedstock into simple sugars. The resulting sugars undergo fermentation using specific bacterial and yeast strains, followed by distillation and dehydration to produce bioethanol (Alam et al., 2015). Fermentation, a series of chemical reactions, converts glucose into ethanol. Various factors, including fermentation conditions and acid content, influence the efficiency of bioethanol synthesis (Adewale et al., 2015). Certain bacterial strains can also ferment seafood hydrolysates to produce butanol. Fish waste biomass typically contains less lignin compared to other biomass sources, potentially facilitating easier conversion of its components into sugars. A significant challenge in using fish wastes for bioethanol production is the scarcity of suitable bacteria capable of efficiently converting monosaccharides from seafood wastes into ethanol. Saccharomyces cerevisiae, the most commonly used yeast in commercial ethanol production, primarily ferments sugars present in seafood hydrolysate. Additionally, bacteria such as Scheffersomyces, Kluyveromyces marxianus, and Zymomonas mobilis are employed for this purpose. The bacterium Zobellia palmae demonstrates a high rate of ethanol production when fermenting fatty fish wastes.

12.2.2.2 Biobutanol

Research has explored biobutanol as a potential substitute for ethanol in gasoline, given its higher energy content and lower vapor pressure. Fish waste polysaccharides show promise as a source for biobutanol production (Rustad et al., 2011). The conversion of seafood waste biomass into biobutanol is facilitated by enzymes such as cellulases and xylanases, along with Clostridium saccharoper butylacetonicum (Koçar et al., 2013). The anaerobic production of biobutanol through ABE fermentation was first documented by Louis Pasteur in 1862. This process initially generates butyric and acetic acids, which are subsequently converted into solvents including ethanol, butanol, and acetone. Additionally, Clostridium strains have been employed to produce biobutanol from fish waste by metabolizing sugars like xylose and glucose.

12.2.2.3 Anaerobic Digestion

Anaerobic digestion represents the most straightforward technique for generating biofuel from marine waste, yielding biogas primarily composed of methane. Initially, fish waste used for biogas production undergoes rinsing with tap water. This preprocessing step is crucial as it ensures higher methane gas production by improving feedstock quality through the removal of foreign matter. The presence of foreign substances on the reactor's fish waste feedstock can impede the anaerobic digestion process. The recommendation to rinse fish waste for sand removal remains valid, as sand can substantially increase the ash content in biomass by introducing large quantities of SiO_2 to the ash composition, thereby reducing its calorific value. Biogas is produced through the anaerobic decomposition of organic molecules found in biodegradable sources like fish waste. This process releases gases including methane, carbon dioxide, and trace amounts of hydrogen sulphide (Ivanovs et al., 2018). The transformation occurs in four primary stages. Initially, certain obligate anaerobes, such as Streptococci and Clostridium, generate enzymes that convert insoluble organic materials and larger molecules (lipids, carbohydrates, and proteins) into soluble organic compounds. The second stage, acidogenesis, involves acidogenic bacteria catalyzing enzymatic processes that transform soluble organic molecules into volatile fatty acids and alcohols. Subsequently, acetogenesis takes place with the assistance of acetogenic bacteria, resulting in the production of hydrogen and acetic acid through various biochemical processes. Finally, methanogens break down these substances to produce carbon dioxide and methane (Alam et al., 2015).

The digestate, containing nitrogen and phosphorus components, can be utilized to create fertilizers derived from seafood waste or biological feedstocks, potentially generating additional revenue streams. Moreover, the use of biogas obtained from seafood waste can significantly reduce greenhouse gas emissions. The chemical composition of the specific fish waste feedstock entirely influences the methane-producing bacteria during anaerobic digestion. The lipid content in feedstock affects its suitability for digestion, with higher lipid levels yielding increased biogas volumes per gram. Anaerobic fermentation is particularly effective for biogas production from seafood waste due to its potentially high lipid content (Milledge et al., 2014). Several factors can impact biogas production, including the carbon to nitrogen (C/N) ratio, particle size, duration, temperature, type of fish waste, and seasonal variations. Additionally, the reaction temperature, heating rate, and oxygen supply during the anaerobic digestion process may influence the quantity and quality of the final biogas (Demirbas & Demirbas 2011). The high ash content in fish waste leads to a lower carbon-to-nitrogen ratio, which hinders the growth of anaerobic microbes and negatively affects biogas production. Employing microorganisms that are tolerant to harsh conditions during the anaerobic digestion of fish waste results in improved biogas quality.

12.2.3 Thermochemical Conversion

12.2.3.1 Pyrolysis

Thermochemical processes are relatively simple, typically requiring only thermal media and appropriate synthetic alcohols for fat separation. A fixed-bed horizontal tube furnace was employed for pyrolysis of fish waste at three specific temperatures: 400, 500, and 600 °C, with a 25 °C/min heating rate and 0.5 L/min nitrogen gas flow. The process continued for an additional hour after reaching the target temperature, with biochar collected from the reactor and bio-oil gathered from the flask (Reza et al., 2022). Rapid pyrolysis of organic materials without oxygen produces charcoal, bio-oil, and gaseous residues, enabling the creation of bio-oil and biochar from marine waste (Fadhil et al., 2017). In comparison, hydrothermal liquefaction involves subjecting biomass to high temperatures and pressures in the presence of water, either in liquid or supercritical state. Using water as a solvent instead of flash pyrolysis eliminates the need for biomass drying and allows reactions to occur at lower temperatures. Hydrothermal liquefaction primarily yields bio-oil or bio-crude, with solid residue, biochar, and water containing soluble organic compounds as main byproducts. Both techniques belong to the thermochemical technology family, which generates bio-oil from organic components in feedstock.

12.2.3.2 Gasification

Syngas is produced through a process called gasification. This method involves the partial oxidation of organic matter at temperatures ranging from 800 to 1000 °C, primarily resulting in syngas, a mixture of combustible gases. The production of syngas occurs in two stages: first, pyrolysis creates char, which is then gasified using a syngas agent such as oxygen (O_2) or water vapor (H_2O) (Hannon et al., 2010). The resulting syngas comprises hydrogen, carbon monoxide, methane, and ethylene, along with nitrogen, carbon dioxide, and water vapor. This versatile gas can be utilized for heat generation through combustion or converted into power and heat via integrated gas turbine systems (Reza et al., 2022). Additionally, syngas derived from gasification can serve as a precursor for producing methanol and hydrogen, which have various applications, including transportation. However, it is important to note that the production of methanol from biomass-derived methane is more expensive than producing it from fossil fuel gas.

12.3 Biofuel and Conventional Diesel vs Environmental Impact

Substantial resources have been allocated to converting seafood byproducts into biofuels, owing to their high oil content, dense growth in fish waste disposal, and rapid fish production rates. Recently, blending fish waste with conventional diesel has become economically viable, reducing dependence on fossil fuels and subsequently decreasing CO_2 emissions. However, optimizing oil extraction methods and ensuring raw material quality posed significant challenges in biorefinery development. Notable obstacles include preserving fish waste and ensuring its availability for biofuel production in suitable locations, as well as the lack of an efficient distribution network. Additional challenges arise from the need for specific fish species selection and the effective hydrolysis, conversion, and ingestion of certain polysaccharides by conventional bacteria. Moreover, ensuring the scalability of biotechnological processes remains a major hurdle in developing seafood waste-based biorefineries. Despite ongoing laboratory research, the near-term practicality of these technologies remains uncertain. In the context of sustainable biofuel production, seafood waste presents unique advantages compared to other feedstock sources. Proper preservation and handling of seafood industry waste can significantly contribute to large-scale biofuel production. It is worth noting that seafood waste has historically been used for poultry feed or improperly discarded on land and in water, with its biofuel potential remaining largely unexplored in recent years. Biofuels derived from seafood waste are carbon–neutral, as they are renewable resources that reduce atmospheric CO_2 emissions. This waste has the potential to mitigate greenhouse gas emissions and slow global warming. Additionally, utilizing seafood waste helps reduce coastal and land pollution, promoting a cleaner environment. The use of seafood waste for biofuel production can lower fuel costs and foster green energy development. However, a significant drawback of seafood waste is the considerable seasonal variation in its fatty acid content and susceptibility to spoilage. If immediate processing is not possible, preservation measures must be implemented to prevent microbial degradation, resulting in additional energy-related expenses and potential transportation costs. Converting seafood waste into biofuels requires improved access to the organic content within the feedstock through physical or chemical extraction methods. Ultimately, producing biofuels and bioenergy from marine waste is more expensive than other biomass sources, primarily due to the numerous pretreatment steps, raw material transportation, and the energy required to operate large bioreactors.

12.4 Biofuel: Current Trends and Future Perspectives

Seafood waste's high fatty acid content makes it an excellent resource for biofuel production. Various pretreatment and extraction methods are utilized to optimize this fish waste's vast potential, resulting in higher-quality biofuel. This approach enhances financial viability by enabling biofuel production and creating opportunities to use seafood waste in other industrial sectors, such as fertilizers and cosmetics. Implementing environmentally sustainable practices is crucial for maintaining ecosystem health and management. Biofuels are considered the most effective and economically viable alternative to conventional fuels. However, there is a trend towards adopting more eco-friendly techniques, including the use of green solvents like water or deep eutectic solvents, combined with efficient heating systems. Developing cost-effective and environmentally friendly methods to break down non-cellulosic polysaccharides in seafood remains a significant challenge. Cellulolytic enzymes can be employed to degrade polysaccharides in seafood waste for bioethanol production. Another promising approach involves exploring enzymes from marine microorganisms, which may lead to the discovery of effective enzymes for biobutanol production. Advancements in biotechnology are anticipated to result in the development of microorganisms capable of fermenting and decomposing these sugars through a single bioprocessing method. This strategy combines genetic modifications with metabolic engineering techniques. Utilizing appropriate pretreatment and bioprocessing methods can significantly accelerate the development of processes that convert seafood waste biomass into biofuels.

12.5 Conclusion

The development of third-generation biofuels from seafood waste presents a highly promising opportunity. The success of this process hinges on various factors that affect the nature of the biomass and the extraction techniques employed. Enhancing methods for extracting oil from seafood byproducts will bolster the accumulation of these waste materials. Fuel production from fish waste has the potential to become environmentally friendly, cost-effective, widely renewable, and feasible. To achieve this, a deeper understanding is required, and research initiatives should be coordinated internationally. Such efforts can help reduce land pollution and minimize harmful emissions during the combustion of biofuel-powered engines.

Acknowledgements Authors are thankful to the Fisheries College and Research Institute, Tamil Nadu for providing necessary research support.

References

Aien, M., & Mahdavi, O. (2020). On the way of policy making to reduce the reliance of fossil fuels: Case study of Iran. *Sustainability, 12*(24), 10606.

Adewale, P., Dumont, M. J., & Ngadi, M. (2015). Recent trends of biodiesel production from animal fat wastes and associated production techniques. *Renewable and Sustainable Energy Reviews, 45*, 574–588.

Alam, F., Mobin, S., & Chowdhury, H. (2015). Third generation biofuel from algae. *Procedia Engineering, 105*, 763–768.

Al Azad, S., Kiu, M. Y., & Lal, M. T. B. M. (2019). Optimization of physical and chemical factors for the production of biomass and lipid from *Chlorella vulgaris. International Journal of Energy and Environmental Engineering, 10*(4), 419–429.

Amoah, J., Kahar, P., Ogino, C., & Kondo, A. (2019). Bioenergy and biorefinery: Feedstock, biotechnological conversion, and products. *Biotechnology Journal, 14*(6), 1800494.

Arumugam, A., & Ponnusami, V. (2017). Production of biodiesel by enzymatic transesterification of waste sardine oil and evaluation of its engine performance. *Heliyon, 3*(12).

Berla, B. M., Saha, R., Immethun, C. M., Maranas, C. D., Moon, T. S., & Pakrasi, H. B. (2013) Synthetic biology of cyanobacteria: Unique challenges and opportunities. *Frontiers in Microbiology, 4*(246), 1–14.

Costa, A. A. F. D., de Oliveira, A. D. N., Esposito, R., Len, C., Luque, R., Noronha, R. C. R., Rocha Filho, G. N. D., & Nascimento, L. A. S. D. (2022). Glycerol and catalysis by waste/low-cost materials—A review. *Catalysts, 12*(5), 570.

Crexi, V. T., Souza-Soares, L. A., & Pinto, L. A. (2009). Carp (Cyprinus carpio) oils obtained by fishmeal and ensilage processes: Characteristics and lipid profiles. *International Journal of Food Science & Technology, 44*(8), 1642–1648.

Datta, A., Hossain, A., & Roy, S. (2019). An overview on biofuels and their advantages and disadvantages. *Asian Journal of Chemistry.*

Demirbas, A., & Demirbas, M. F. (2011). Importance of algae oil as a source of biodiesel. *Energy Conversion and Management, 52*(1), 163–170.

Di Serio, M., Cozzolino, M., Giordano, M., Tesser, R., Patrono, P., & Santacesaria, E. (2007). From homogeneous to heterogeneous catalysts in biodiesel production. *Industrial & Engineering Chemistry Research, 46*(20), 6379–6384.

Fadhil, A. B., Ahmed, A. I., & Salih, H. A. (2017). Production of liquid fuels and activated carbons from fish waste. *Fuel, 187*, 435–445.

Foster, S. R., Reames, T. G., & Stephens, J. C. (2017). Energy for metropolis. *University of Miami Law Review, 71*(2), 313–355.

Ghaly, A. E., Dave, D., Brooks, M. S., & Budge, S. (2010). Production of biodiesel by enzymatic transesterification. *American Journal of Biochemistry and Biotechnology, 6*(2), 54–76.

Girish, C. R. (2019). Catalyst preparation from various naturally available waste Materials for biodiesel production: A review. *International Journal of Mechanical and Production Engineering Research and Development, 9*(6), 117–130.

Handayani, K., Anugraha, A. S., McLellan, B. C., & Nguyen, T. K. (2022). Indonesia's pathway to net-zero emissions: The role of renewable energy and the impact on the power sector. *Energy Strategy Reviews, 44*, 1–14.

Hannon, M., Gimpel, J., Tran, M., Rasala, B., & Mayfield, S. (2010). Biofuels from algae: Challenges and potential. *Biofuels, 1*(5), 763.

Hossain, A. B. M. S., Saleh, A. A., Aishah, S., Boyce, A. N., Chowdhury, P. P., & Naqiuddin, M. (2008). Bioethanol production from agricultural waste biomass as a renewable bioenergy resource in biomaterials. In *4th Kuala Lumpur International Conference on Biomedical Engineering 2008: BIOMED 2008 25–28 June 2008 Kuala Lumpur, Malaysia* (pp. 300–305). Springer.

Ivanovs, K., Spalvins, K., & Blumberga, D. (2018). Approach for modelling anaerobic digestion processes of fish waste. *Energy Procedia, 147*, 390–396.

Karkal, S. S., & Kudre, T. G. (2023). Valorization of marine fish waste biomass and Gallus gallus eggshells as feedstock and catalyst for biodiesel production. *International Journal of Environmental Science and Technology, 20*(7), 7993–8016.

Khoddami, A., Ariffin, A. A., Bakar, J., & Ghazali, H. M. (2009). Fatty acid profile of the oil extracted from fish waste (head, intestine and liver)(Sardinella lemuru). *World Applied Sciences Journal, 7*(1), 127–131.

Kiran, C. H., Ganesh, D. B., Banapurmath, N. R., & Ganachari, S. V. (2021, February). Experimental investigations on high pressure assisted waste mix cooking oil powered biodiesel fuelled CRDI engine and injection parameter studies. In *IOP Conference Series: Materials Science and Engineering* (Vol. 1070, No. 1, p. 012119). IOP Publishing.

Koçar, G., & Civaş, N. (2013). An overview of biofuels from energy crops: Current status and future prospects. *Renewable and Sustainable Energy Reviews, 28*, 900–916.

Kumar, M., Gupta, A., & Thakur, I. S. (2016). Carbon dioxide sequestration by chemolithotrophic oleaginous bacteria for production and optimization of polyhydroxyalkanoate. *Bioresource Technology, 213*, 249–256.

Lam, M. K., Lee, K. T., & Mohamed, A. R. (2010). Homogeneous, heterogeneous and enzymatic catalysis for transesterification of high free fatty acid oil (waste cooking oil) to biodiesel: A review. *Biotechnology Advances, 28*(4), 500–518.

Linder, M. (2005). *Opportunities and barriers for biodiesel and bioethanol in Germany, the United Kingdom and Luxembourg*. Master Thesis. Lund University, Sweden, pp 1–116.

Mathiyazhagan, M., Ganapathi, A., Jaganath, B., Renganayaki, N., & Sasireka, N. (2011). Production of biodiesel from non-edible plant oils having high FFA content. *International Journal of Chemical and Environmental Engineering, 2*(2).

Milledge, J. J., Smith, B., Dyer, P. W., & Harvey, P. (2014). Macroalgae-derived biofuel: A review of methods of energy extraction from seaweed biomass. *Energies, 7*(11), 7194–7222.

Nisar, A., Khan, S., Hameed, M., Nisar, A., Ahmad, H., & Mehmood, S. A. (2021). Bio-conversion of CO_2 into biofuels and other value-added chemicals via metabolic engineering. *Microbiological Research, 251*, 126813.

Perera, F. (2018). Pollution from fossil-fuel combustion is the leading environmental threat to global pediatric health and equity: Solutions exist. *International Journal of Environmental Research and Public Health, 15*(1), 16.

Pinyaphong, P., Sriburi, P., & Phutrakul, S. (2011). Biodiesel fuel production by methanolysis of fish oil derived from the discarded parts of fish catalyzed by Carica papaya lipase. *International Journal of Chemical and Molecular Engineering, 5*(4), 336–342.

Reza, M. S., Azad, A. K., Abu Bakar, M. S., Karim, M. R., Sharifpur, M., & Taweekun, J. (2022). Evaluation of thermochemical characteristics and pyrolysis of fish processing waste for renewable energy feedstock. *Sustainability, 14*(3), 1203.

Rustad, T., Storrø, I., & Slizyte, R. (2011). Possibilities for the utilisation of marine by-products. *International Journal of Food Science & Technology, 46*(10), 2001–2014.

Sharma, M. P. (2016). Selection of potential oils for biodiesel production. *Renewable and Sustainable Energy Reviews, 56*, 1129–1138.

Sheldon, R. A. (2014). Green and sustainable manufacture of chemicals from biomass: State of the art. *Green Chemistry, 16*(3), 950–963.

Tan, H. T., Lee, K. T., & Mohamed, A. R. (2010). Pretreatment of lignocellulosic biomass: A review on recent advances. *Renewable and Sustainable Energy Reviews, 14*(6) 1827–1841.

Tayib, Y. M., Al–Sheikh, F., Shakor, Z. M., & Anderson, W. A. (2024). Biodiesel production from by extraction of carp fish (Cyprinidae) oil and transesterification using CaO derived from limestone and eggshells as heterogeneous catalysts. *Biofuels, 15*(6), 699–709.

Usman, M., Cheng, S., & Cross, J. S. (2022). Biomass feedstocks for liquid biofuels production in Hawaii & Tropical Islands: A review. *International Journal of Renewable Energy Development, 11*(1), 111–132.

Ward, A. J., Lewis, D. M., & Green, F. B. (2011). Biofuels from algae: Challenges and potential. *Proceedings of the Institution of Civil Engineers - Energy, 164*(2), 93–107

Zou, S., Wu, Y., Yang, M., Li, C., Jun, J. (2016). Pyrolysis characteristics and kinetics of microalgae *Dunaliella tertiolecta* under different heating rates. *Journal of Thermal Analysis and Calorimetry, 125*(1), 519–528.

Chapter 13
A Recent Update on Biofuel Production from Microalgae

Aman Kumar, Mansi Rani, and Shubhankar Kumar Singh

Abstract The microalgae due to its ability of high growth rate, high lipid content and ability to survive in different environment make it a suitable and important feed stock for production of biofuel. Due to these advantages, microalgae supersede over other traditional biomass resources. This chapter deals with other important aspect of biofuel production from microalgae. The role of microalgae in meeting global energy challenges by reducing the dependency upon fossil fuel is addressed. The biological aspects of microalgae explaining the requirements for growth and mechanisms for lipid production and carbohydrate accumulation are taken in consideration. A summarized discussion has been done on various types of biofuels derived from the microalgae such as biodiesel, bioethanol, biobutanol, biogas and biohydrogen. The development in cultivation techniques, harvesting and processing system and the technologies such as photobioreactors and genetic engineering also taken into discussion. This chapter deliberates for the increase in yield and economic feasibility. Along with the environmental and economic aspect, the potential of carbon sequestration, life cycle assessment and scalability of microalgae are explored. Further the global policies and market trends for microalgal fuel are also analyzed. In conclusion, a talk on future direction and recent trend like synthetic biology, circular bioeconomy strategies have been done where microalgal biofuel is positioned as a key factor in the transition to realiable renewable energy.

Keywords Microalgae · Biofuel production · Lipid accumulation · Algal biomass · Biorefinery · Sustainable energy

A. Kumar · M. Rani · S. K. Singh (✉)
Department of Microbiology, ICMR-Rajendra Memorial Research Institute of Medical Sciences, Patna, India
e-mail: shubhankark.singh@icmr.gov.in

V. Singh et al. (eds.), *Bioenergy and Value Added Byproducts from Biomass Waste*, Sustainable Environmental Waste Management Strategies,
https://doi.org/10.1007/978-3-032-22849-9_13

13.1 Introduction

The greenhouse gas has devastating effect on the environment. This captured the attention of policy maker for thinking about a sustainable source of energy. The idea for such thinking is also guided with the view to combat the ill effect of greenhouse emission and find out an alternative to the fossil fuel. There are several solutions before the thinkers to combat the ill effect of greenhouse gas. Among the solutions, biofuels have shown its ability to play an important role. The biofuel has shown its efficiency in combating the climatic change effect, which is its renewable nature and availability in the environment. The microalgae — used for preparation of biofuels — is available in nature. The use of these biofuels give a pathway for carbon neutral energy system. The ability of biofuels and its integration with the existing energy infrastructure makes it a better choice for addressing the issue of energy security and environmental sustainability. The recent research and development in the field of biofuel technology has further improved their efficiency and also widen its applicability showing the capacity of biofuels as an essential component in climate change strategy (Jeswani et al., 2020; Robertson et al., 2017).

The microalgae have attracted the attention of researcher towards its use in production of biofuel. Apart from the traditional sources of biofuels, for example, crops and agriculture residue, the microalgae by its character have shown its efficiency for production of biofuels. The microalgae have its characteristics of rapid growth, high lipid content and ability to sustain on non-arable lands using saline or wastewater. Thus, the microalgae have proved its efficacy of avoiding competition with agriculture land used for production of food grains. The microalgae have also shown its efficacy to absorb carbon by utilizing the CO_2 from the industrial emissions. By this way the microalgae performing the function of capturing of carbon and a storage system (Fischer et al., 2009; Lal, 2013). The microalgae have its ability to produce several types of biofuels such as biodiesels, biobutanol, bioethanol, biogas and biohydrogen. This makes the microalgae a versatile and dependable solution for the need of alternative energy to combat the greenhouse gas effect (Fig. 13.1).

In this chapter, a comprehensive and updated development in the field of biofuel production from the microalgae has been dealt with. This chapter has also focused on the technological development, its environmental development and economically viability of such development. The chapter also deals at the latest trend adopted in cultivation techniques, harvesting methods and biomass conversion technologies. Further in this chapter the transformative potential of microalgae in the field of renewable energy system is also highlighted. The objective of this study is to explain the recent advancement and innovations in the field of utilising the microalgae for the production of biofuels. The role of microalgae-based biofuel in addressing the sustainable energy requirement also studied This chapter also emphasises on the biofuel research and climate action which focuses toward the role of microalgae as a super most in the field of renewable energy revolution.

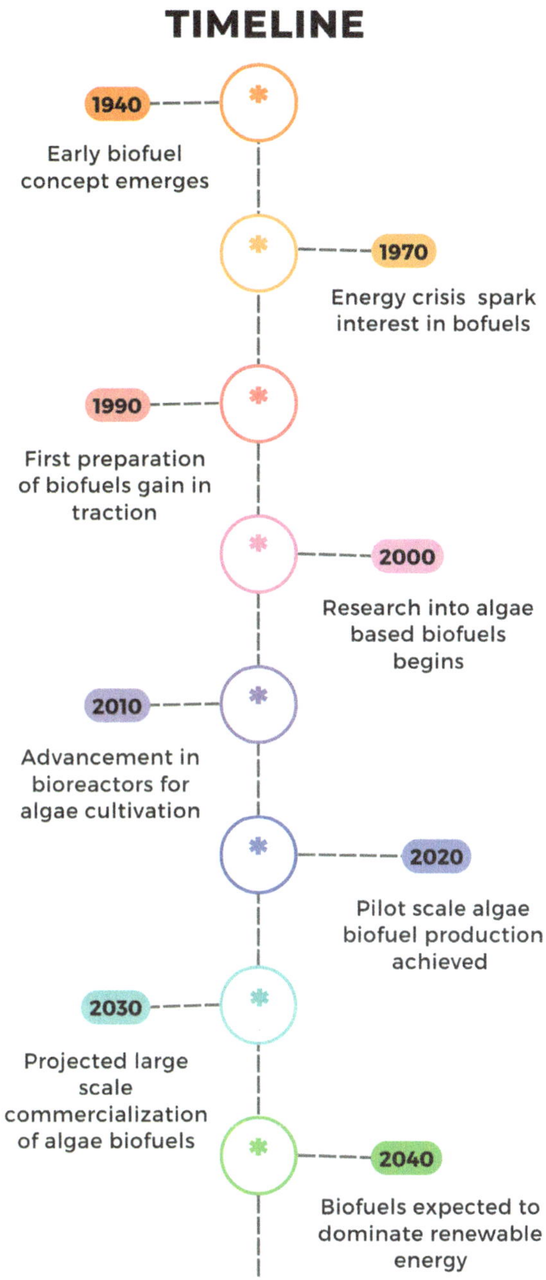

Fig. 13.1 Timeline explaining the rise of biofuel

13.2 Advances in Microalgal Biology

The development in microalgal research related to the algal taxonomy and physiology which revealed its potency to optimise the production of biofuel. Further the studies of physiology of different algal species also reveals its potency. The researchers have selected the most efficient strain of microalgae for production of biofuel. This development has enhanced the overall productivity and sustainability in the field of biofuel production (Brito-Bello, & Lopez-Arredondo, 2023). The recent development in the field of microalgal research has attracted the attention across the globe. It aims to find a sustainable renewable energy source that can address the issue of environmental challenges associated with the fossil fuels.

13.2.1 Recent Discoveries in Algal Taxonomy and Physiology Affecting Biofuel Yields

The use of recent discoveries and advancement in the field of algal taxonomy and physiology, has led in understanding the mechanisms, which has its influence in production of biofuel. All the discoveries are the basis of utilizing the microalgal strain at its optimal level for the increased production of biofuel as well as overall efficiency in production.

The algal diversity has also been explained in recent researches and particularly in relation to their growth condition and biochemical composition. These researches are very useful in identifying the different algal species with their growth rates and superior lipid profiles. Thus, it helps the scientist to use the selective cultivation of most promising strains of algae for the production of biofuel. The algal taxonomy plays very important role in biofuel production. The different species show unique metabolic system which influence their capacity to accumulate lipids, carbohydrate and proteins which are necessary for production of biofuel (Falfushynska, 2024).

The recent discoveries explained the physiological responses of microalgae to their respective environments. These studies have further explained about their manipulation and growth and the factors for their optimum use in biofuel production The influencing factors like light intensity, temperature, availability of nutrients and salinity — impacts the growth of algae and its lipid synthesis (Kim et al., 2023). For example, studies have revealed that the stress condition (like nutrient starvation or salt stress) can significantly increase the lipid accumulation in specific algal species (Maltsev et al., 2023). This discovery emboldens researcher to customize cultivation strategy for maximising the biomass and lipid production from microalgae.

13.2.2 Genetic Engineering Breakthroughs for Enhancing Lipid and Biomass Productivity

The genetic engineering has been playing a very important role in the field of increasing the productivity of lipid and biomass in microalgae. The microalgae have become a reliable source for biofuel production due to use of genetic engineering. Recent development in the genetic engineering has provided a range of methodologies for increasing and understanding the metabolic system involved in lipid synthesis and strain improvement.

The recent breakthrough utilises the advanced genetic engineering techniques like CRISPR/Cas9, TALENs, ZFNs. By using these techniques, the modification of genomes in algae made out to which has boosted the lipid productivity and biomass (Ahmad Kamal et al., 2024). These genetic engineering methods are targeted to select specific gene which associated with the lipid metabolisms and production. For example, over expression of Acyl-ACP synthetase has shown to increase lipid production by 5.4% in *Synechocystis* sp. PCC 6803 (Ahmad Kamal et al., 2024).

Researcher have applied the metabolic and genetic engineering methods to maximise the lipid accumulation system. The research has shown that by suppression of particular metabolic pathway has led to improve lipid content. As the down regulation of phospholipase A2 has led to enhance lipid content up to 64.25% in engineered strains. The multi-omics strategy helps in describing the transcription factors and regulatory network which influence the lipid accumulation. This type of advanced strategies also offers internal information for more targeted engineering efforts (Shin et al., 2019).

By applying the CRISPR/Cas9 technology, researchers have disrupted the metabolic pathway in microalgae. This technology reduces the competitive gene which leads to degradation of lipid and allow to retain more lipid within cell with lipid biosynthesis methods. Researches on *Chlamydomonas reinhardtii* has shown that loss of certain gene which related to lipid catabolism results in increased in overall lipid content in algae and shifts in fatty acid composition to improve the production of biofuels (Dhokane et al., 2023).

The technique of CRISPR/Cas9 provides capability of targeting multiple genes. This approach is used to adjust suitably the metabolic networks which underlie in biosynthesis process of lipid. By using this CRISPR/Cas9 technique the researchers knock out or edit several genes which involved in lipid associated systems. For instance, the simultaneous modification of genes such as acyl-CoA diacylglycerol acyltransferase (*DGAT*) and phospholipid diacylglycerol acyltransferase (*PDAT*) which has exhibited to produce strain of algae with significantly improved lipid content (Dhokane et al., 2023).

CRISPR/Cas9 also facilitate introduction of heterologous gene that encode enzymes in algae which are involved in lipid biosynthesis (Nymark et al., 2016). Using this technology of expressing specific genes for increasing lipid biosynthesis, the researcher has created and engineered strains with superior lipid productivity.

This method of introduction of such developed genes has led to increased Triacyl glycerol production level in comparison to wild type strain.

The researches have shown the impact of genetic modification on algal strain by using of CRISPR/Cas9 and related technologies. These technologies have led to increase metabolic productivity and other applications. In *Chlamydomonas reinhardtii* the strain CC-4349 has shown remarkable result. The method of knocking out of *CpFTSY* gene and zeaxanthin epoxidase (*ZEP*) genes by using Cas9 technique have increased 13-fold higher zeaxanthin production and improved productivity in photosynthesis with increased biomass under high light conditions. Similarly, the mutants under *ZEP* knock out method has displayed 56 times increase in zeaxanthin content and 47 times improvement in productivity without reduction in lutein (Baek et al., 2018). The knocking out process of phospholipase A2 (*PLA2*) and *ELT* genes increased lipid productivity by 64.25% and also increased fatty acid accumulation for biofuel application (Nguyen et al., 2020; Shin et al., 2019). In double knockout method of *ZEP* and *LCYE*, the yield production of zeaxanthin has shown a 60% increase and in simultaneous knock out methods of *ADP-glucose pyrophosphorylase* (*AGP*) and *ZEP* increased the oil proactivity by 81% (Song et al., 2022). In other advancement system using CRISPRi mediated downregulation of *CrPEPC1* in strain CC-400 has shown a result of increasing lipid content by 28.5% t (Kao & Ng, 2017). Using of bacterial phytase gene in strain UVM11 by knocking out method, has highlighted the potency for recombining the protein production (Zadabbas Shahabadi et al., 2023).

The genetic interventions have yielded similar success for *Nannochoropsis spp.* The knock out of *ZnCys* gene in *N. gaditana* has doubled the productivity of lipid as well as retaining CO_2 fixation capacity (Ajjawi et al., 2017). The knockout of *Aco1* in *N. gaditana* has resulted in two-fold enhancement in lipid productivity (Verruto et al., 2018). In *N. salina, cesA1* knockout mutant has shown reduced cellulose and chrysolaminarin content but has increased lipid extractability by 1.7-fold in the region of thinner cell walls (Jeong et al., 2020). The knock out of *beta-glucan synthase* and *transglycosylase* genes have significantly enhanced the level of fatty acid whereas targeted knocking of *delta-12-fatty acid desaturase (FAD12)* has increased linoleic acid production by four folds (Ryu et al., 2021). The *NobZIP77* knockout in *N. oceannica* increased production of tri-acyl glycerol under nitrogen deprivation condition (Zhang et al., 2022).

The genetic modification in *Chlorella spp.* have shown diverse results. The knockout of *fad3* gene in *C. vulgaris* has boosted the lipid by 46% (Lin & Ng., 2020). The gene regulation in *C. sorokiniana* by using CRISPRa/i technology has enhanced the protein and lipid production significantly (Lin et al., 2022). Under the dark light cycle when knockout of plastidic ATP/ADP translocase gene is used in *Parachlorella kessleri*, the productivity of lipid increased by more than 30% (Kasai et al., 2023).

The studies over the other algal species have shown a better result. The knock out of *ADP-glucose pyrophosphorylase* (*AGP*) gene used in *Tertraselmis spp* has exhibithed increased in production of lipid upto 2.3 to 2.7-times and has also shown increase in fatty acid by 314% (Chang et al., 2020). By knockout of *CHS1* in *Porhyridium purpureum* has significantly enhance the production of phycoerythrin (Jeon et al., 2021). The above noted has shown the potency of crisper technologies in increasing productivity of microalgae and in diversification of their application in production of biofuels, nutraceuticals and recombinant for proteins.

13.2.3 Advances in Understanding Metabolic Pathways for Biofuel Precursors

The application of genetic engineering techniques has significantly enhanced the efficiency and yield of biofuel production from various biological materials. The advancement in technology has led to understand the metabolic pathways for biofuel precursor. The advancements in technique and methodology have been playing an important role in optimising the conversion of biomass into sustainable biofuels.

The recent advancement in the field of sustainable biofuel, has focused on utilizing novel strains and organisms including the exploration of microorganisms of non-conventional category such as cyanobacteria and archaea. These organisms have unique metabolic systems which can be engineered to produce biofuels in comparison to traditional yeast or bacterial strain with more efficiency. Researches, on this method, has shown that the genetically engineered cyanobacteria have capacity to convert CO_2 directly to biofuel precursor. This activity results in to the minimum energy input required for biomass conversion into biofuel (Savakis and Hellingwerf, 2015).

The researchers have studied the pathway like glycolysis and pentose phosphate pathway. These pathways show the lipid biosynthesis by reductive power and various precursor (Sarkar & Shimizu, 2015). Further analysis of these metabolic pathway in photosynthetic microorganisms is used for short chain fatty acid production as biofuel precursor (Cheon et al., 2016). The studies have also explained about the role of microalgae for production of long chain hydrocarbons by using lipid metabolic pathways. This makes algae as a dependable feed stock for production of biodiesel (Hu et al., 2008).

The synthetic biology has emerged as a very crucial field for the researchers. It combines the biological science and principle of engineering to design and construct new biological apparatus and systems. This emerging field biology has become a very integral part of biofuel production from the biomass. The approaches of combining the principle of engineering in biological science has revolutionized the production of biofuel precursor. The synthetic biology enables the researchers to assemble a new synthetic pathway in those organisms which naturally do not have biofuel production capacity. These innovative researches have led to higher yield and more efficient biofuel production process developed for microbial metabolisms synthesis (Rock et al., 2021).

The studies have shown the integration of biological synthetic tools to modify key pathways in microalgae for increasing biofuel production through precise genetic engineering (Abdullah et al., 2024)). Further the studies have also revealed that engineered triacylglycerol (TAG) profiles a vital biodiesel precursor by manipulating the acetyl-CoA flux and other critical enzymes (Shahid et al., 2020). Studies of proteomic and metabolomic tools has identified the several bottlenecks in metabolic flux and has given a pathway for maximising the lipid production in microalgae (Chakdar et al., 2021).

The development in omics technologies has played an important role in uncovering noble metabolic pathway and regulating the mechanisms in microalgae. The study over the enzymes has shown its involvement in lipid biosynthesis like acetyl-CoA carboxylase (De Bhowmick et al., 2015). This technique is now used for increasing lipid accumulation in microalgae through genetic modifications. In a study in which mapping by applied system biology used in metabolic pathway in algae has revealed a favourable condition for biofuel precursors like TAG and fatty acid (Jinkerson et al., 2011). Further in another study it revealed the capability of environmental manipulation for increasing the lipid metabolisms for increasing the production of biofuel (Srivstava et al., 2018).

13.3 Latest Developments in Cultivation Techniques

The cultivation techniques play very important role in biofuel production. Recent advancement has been made in cultivation technique of microalgae for biofuel production to increase yield with efficiency and sustainability. The integration of advanced technology and emphasis on environmental issues has shown a significant change toward more sustainable bioenergy solution to combat the greenhouse gas effect.

13.3.1 Innovative Growth Systems

Among the innovative cultivation technique, the high efficiency photobioreactors (PBR) are at the front position of microalgae cultivation technologies. Advanced light delivery system like spectral filtration and luminescent materials are used to increase photosynthetic efficiency for cultivation. For increasing the biomass yield and lipid content, improved light transmission and its distribution within the bioreactors have significant role which are very important for production of biofuel (Nwoba et al., 2019). The recent studies have shown the technique in which membrane PBR used for separating the microalgae. In this method the microalgae separated from the growth medium for increasing the nutrient uptake. The increased in nutrient uptake is done by forming biofilms which promotes metabolic activity and increases overall productivity of biofuels (Penloglou et al., 2024). The modular and hybrid system of cultivation has also emerged as innovative cultivation technique for optimising the biomass production and lipid accumulation phases. The combined cultivation system

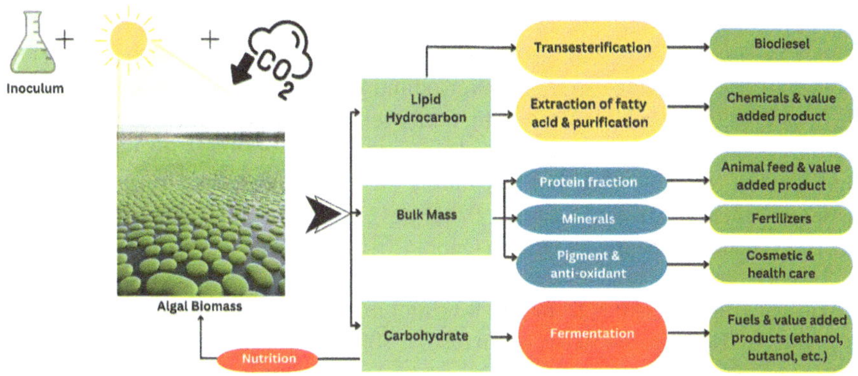

Fig. 13.2 A conceptual diagram of photosynthesis performed and products derived from microalgae biomass

(modular and hybrid) is beneficial to use in open pond cultivation. This modular cultivation system is used in nutrient rich phases. In nutrient rich phase, the controlled environment of PBR is used during lipid induction period for maximising products of biomass. And in the hybrid system, the contamination risk is also minimised which is related to open pond cultivation. This integrated system of cultivation of biomass allows more efficient and sustainable production system which is capable for responding the fluctuating environmental conditions (Fig. 13.2).

The impact of light intensity and photoperiods on the biomass productivity of various algal species have been studied extensively over a range of bioreactor design. *Arthospira (Spirulina) platensis* has shown an increased biomass productivity when subjected to a 10 L bioreactor with a diameter of 22 cm and height of 29 cm. The productivity of *A. platensis* has shown 620 mg L^{-1} day^{-1} under white LED light intensities of 635, 980, 1300 and 2300 μmol m^{-2} s^{-1} during a 12 h:12 h light–dark cycle (Chaiklahan et al., 2022). A flat plate bioreactor with a volume of 4 L (33 cm length, 22 cm diameter and 10 cm height) used for *Oscillatoria sp.* It has shown the biomass productivity of 75, 80 and 92 mg L^{-1} day^{-1} with fluorescent lamp at 160 μE m^{-2} s^{-1} depending upon frequencies of supply (one, twice and thrice daily) in identical ligh-dark cycles (Nithiya et al., 2017).

A 30 L photobioreactor when used for mixed algal cultures has exhibited varied productivity which depends upon the species and condition. Using red, blue and white LED. light at 110 μmol m^{-2} s^{-1} under a 16 h:8 h light–dark cycle, productivity of biomass increased. It was observed that the productivity of biomass increased by 47.75, 50.04, 54.87 and 54.5 mg L^{-1} day^{-1} from UQ-Lake A, UQ-Lake F, Storm Water A, and Storm Water F, respectively and all at 10% CO_2 (Aslam et al., 2017). In a research with sintered disc glass bubble columned bioreactor used for *Spirulina platensis,* the growth shows a maximum productivity of 457.5 mg L^{-1} day^{-1} under white LED light with a light–dark cycle of 14 h:10 h, by supplementing with CO_2 and NaOH in the culture medium Z(20)3 (Kumari et al., 2014).

An experiment with vertical tubular bioreactor with a volume of 2.5 L used for *Chlorella vulgaris (BA 002)* has shown productivity of 27.08 ± 7.80 and 24.21 ± 8.89 mg L^{-1} day^{-1} under high pressure sodium and LED light, in a photoperiod of 12 h:12 h, 18 h:6 h and 24 h:00 h light–dark cycle (Ratomski & Hawrot-Paw, 2021). Further studies of *Scenedesmus abundans* cultivated under white tube light having light intensities of 27, 40.5, and 54 μmol m^{-2} s^{-1} has shown productivity of 47.4 ± 0.002, 70.23 ± 0.001, and 52.46 ± 0.002 mg L^{-1} day^{-1} using BG-11, Fogg's, and CHU-13 media, respectively (Rai & Gupta, 2017).

In a research, in which 40 L bioreactor used for growing *Chlorella vulgaris* using LED light with intensities from 0 to 80 μmol m^{-2} s^{-1}, shows varying productivities (Geiman, 2021). A sequential column reactor 300 ml working volume (56 mm diameter and 160 mm height) used, it achieves 950 mg L^{-1} day^{-1} for *Chlorella PY-ZU1* under white and plain lights with intensities of 4500–6000 lux for 10 min of empty bed residence time (Cheng et al., 2013). The above studies highlight the important role of light intensity, photoperiods and bioreactor configurations in maximising the productivity of algal biomass.

13.3.2 Integration of Smart Technologies

The recent studies and researches have integrated the Artificial intelligence (AI) and machine learning in monitoring and maximising the microalgae cultivation condition. By using these technologies, the accuracy of data and predictive capabilities for biomass production has increased and become more reliable. Analysis of historical data and real-time variables through the AI automated system helps in optimising the growth. The study on the variables and parameters like light intensities, CO_2 concentration and nutrient labels is controlled for improving production. These techniques lead to improvement in biomass production and also reduced operational costs (Imamoglu, 2024; Teng et al., 2020).

For the collection of real time data by deployment of Internet of Things (IoT) device has played an important role for setting up the large-scale setups. The IoT sensor has capacity to monitor the critical parameters which tract remotely to provide timely intervention for increasing cultivation efficiency with sustainability. The integration of such smart technologies proves a transformative step toward precision agricultural technique in microalgal cultivation.

The computational modelling has emerged as valuable tool in increasing production of biofuel from algae. The computational model is used to predict the performance of key enzymes and pathways for maximising the biofuel precursors in oleaginous microalgae (Tibocha-Bonilla et al., 2018). The use of engineered algal strain with optimise acetyl-CoA flux has shown improved precursor ability and hydrocarbon products (Mishra & Joshi, 2022). This result exhibits the ability of integrated modelling and engineering approach in meeting the challenges of biofuel production.

13.3.3 Sustainable Approaches

The recent development in use of waste water and CO_2 emission for the production of microalgae as growth media, represents an important sustainability milestone. The microalgae show an exceptional capability to absorb nutrients from several types of waste water like effluents of municipal and industrial waste. This ability is useful in minimising the environmental pollution in the process of producing biomass for biofuel production (Srimongkol et al., 2022). The microalgae have also capacity of utilising the CO_2 emission comes out from industrial process. The microalgae for its growth utilising the CO_2 emission and thus reducing the greenhouse gas in the course of producing biofuel (Ezhumalai et al., 2024; He et al., 2023).

The concept of circular bioeconomy is getting momentum. In this concept, the microalgal production integrates with environmental remediation practices for converting the waste materials into valuable resources (Kuppan et al., 2023). The concept of circular bioeconomy provides a sustainable model for leveraging renewable biological resources for production of biofuels.

13.4 Cutting-Edge Harvesting and Processing Techniques

For the production of microalgae, the harvesting and processing techniques play very important role. The advancement in harvesting and processing techniques have improved the production efficiency of microalgae for production of biofuel with sustainability. The assimilation of energy efficient method and innovative conversion processes signifies a transformative approach toward production of microalgal biofuels.

13.4.1 Energy-Efficient Harvesting

In the harvesting process of microalgae, the innovative techniques of flocculation and magnetic separation have made significant steps forward in the efficient harvesting of microalgae. The flocculation process including the chemical and biological methods allows microalgal cells to accumulate and concentrate enabling easier recovery process (Matter et al., 2019). The bio-flocculation in which naturally flocculating strain has been used, has emerged as an ecofriendly alternative to chemical flocculants and thus minimising the operational costs and environmental effect (Salim et al., 2010).

The magnetic separation is another highly improved technique which is used for the harvesting of microalgae. The magnetic nanoparticles are used to facilitate the capture of algal biomass. This harvesting method is more resource efficient and rapid approach in comparison to traditional method like centrifugation. In research it is found that the use of magnetic separation method has reduced the energy input requirement. This method of harvesting has enhanced the economic feasibility for large scale microalgal biofuel production (Wang et al., 2015).

For reducing the operational costs of microalgal production, automated harvesting system have also been in use. Development in automation technologies have enabled the real time monitoring and control of the harvesting process. This development has resulted in enhanced efficiency and reduced labour costs. The process of automation includes several methods like sedimentation and centrifugation into automated work flow. This helps in stream lining the operation and minimising human intervention.

13.4.2 Advances in Biomass Conversion

For the conversion of microalgal biomass into biofuels, the thermochemical process including the hydrothermal liquefaction (HTL) represents an important development. The HTL techniques uses the system of subjecting the wet microalgae to high temperature and pressure for conversion of biomass into biocrude oil. This process does not use energy intensive drying technique which is specifically required in conventional processes. Thus, the advancement in HTL technique has highlighted the cost-effective technique for high quality biofuel yield. This HTL technique is also helpful in reducing the moisture content present in biomass and addressing the issue relating to feedstock variability (Deepika et al., 2024).

In recent time, the development and innovation in enzymes and catalyst for transesterification and fermentation process have contributed toward increasing the biofuel production efficiency. The enzymes specially lipases are being used for transesterification of algal oil into biodiesel which offers advantages like milder reaction condition and reduced biproduct formation. New coming up technologies are focusing on utilization of immobilized lipases and optimization of reaction parameter to increase products and also minimise the cost associated with production of biodiesel (Spanou et al., 2024).

Apart from the above techniques, noble catalyst for fermentation process is being developed to improve the conversion of microalgal biomass into biofuel. All these researches and advancements aims to stream line the processes by using non edible feed stocks and integrating biorefinery processes (Ghedini et al., 2021). All these above noted advancements and innovation in the field of biomass conversion technologies represent a promising pathway for the improvement of sustainable and accessible biofuels from the microalgae.

13.4.3 Coupled Systems

The coupled systems recognise the process of combining, harvesting and conversion processes for efficient microalgal biofuel production. This technique facilitates biomass recovery and also initiate conversion process at the same time which offers a more holistic step for the production of biofuels (Zheng et al., 2024). In this system, harvesting and conversion process works simultaneously thereby reducing the intermediate processing steps which lead to minimises the losses. Thus, it enhances the overall efficiency in production of biofuel from microalgae.

Bio-flocculation is another important method. In this method, the harvested microalgae is directly linked with conversion process to create a continuous transition from harvesting to processing step (Salim et al., 2010). This technique is very important in meeting the challenges related with current harvesting methods. This technique has reduced the high energy consumption and costs related with downstream processes of third-party separation techniques. The researcher by developing a system wherein these techniques have simultaneously shown a higher recovery rate. Researches also shown the overall reduction in energy expenditures which contribute to economic sustainability of microalgal biofuels. The developments in the field of harvesting and conversion system marks an important innovation in development of sustainable solution for energy.

13.5 New Horizons in Biofuel Types from Microalgae

The microalgae have emerged as an important source for various biofuels such as biodiesel, bioethanol, biobutanol, biogas and biohydrogen. The development in the process of extraction and production technology with innovative researches has been shaping the future of biofuels obtained from microalgae.

13.5.1 Biodiesel

Biodiesel is one of the biofuels produced from microalgae by increasing lipid extraction method and this method has gained importance in the recent years. To improve the efficiency and reduce environmental effect, solvent free techniques have been developed (Jung et al., 2024). Further the researches have also developed an enzymatic extraction method for biofuel production for better yield and in maintaining the quality of extracted lipids. The transesterification processes for the production of biodiesel are also considered as a feasible, efficient, and ecofriendly method. In this method, the waste residue and non-toxic solvents are used, thus the overall carbon foot prints related to the production of biodiesel is reduced (Pérez, 2016).

13.5.2 Bioethanol and Butanol

Advancement in to the fermentation technology for fermentation of carbohydrate rich strains of microalgae is important for production of bioethanol and biobutanol. The studies published in recent times have highlighted the potency of using the specific microalgal strains. These strains have high carbohydrate content and used for fermenting to produce these biofuels i.e., bioethanol and biobutanol. For instance, the strains like *Scenedesmus* and *Chlorella* has shown high carbohydrate yields suitable for fermentation processes (Onay et al., 2018). These developments have significantly contributed to use the microalgae as a source for renewable energy. The microalgae have shown its benefit of reduction in the waste material as well as reducing the reliance on fossil fuel (Upreti & Ambwani, 2016). By using genetic engineering and metabolic systems simultaneously the microalgae increase their carbohydrate yield. This makes the microalgae suitable substance for production of bioethanol and biobutanol (Lakhawat et al., 2022).

13.5.3 Biogas

The advanced process of anaerobic digestion is the key factor for increasing the biogas production from microalgae. The co-digestion strategies in which microalgae are combined with the other organic waste material have shown an improved efficiency for the production of methane. (Elsayed, 2017). This method facilitates nutrient balance and also facilitates the synergistic breakdown of materials which leads to higher biogas production. The development in recent time in the field of digestor and operational strategies have resulted in sustainable use of microalgae with more efficiency. This is very important for increasing the biogas production.

13.5.4 Biohydrogen

Development in biohydrogen production has been achieved through using the engineered microalgae in photobiological processes. By using the genetic engineering technology, the microalgae are suitably modified to increase their hydrogen production capability under a specific light condition (Burgel et al., 2022). The photosynthetic efficiency of microalgae is optimised for better result and improvement in biohydrogen production which makes it suitable alternative source of energy. The research in the field of exploring the metabolic systems has led to more efficient production of hydrogen with a benefit of minimising the cost and maximising the production.

13.6 Integration of Microalgal Biofuels in Modern Energy Systems

Use of microalgal biofuels in modern energy system is increasing rapidly and it is need of time for growth of sustainable renewable energy sources to combat the greenhouse gas effect. The process of integration of microalgal biofuel into the energy system involves the innovative production technique. The blending techniques with traditional fuel is also part of integration of microalgal biofuel with traditional biofuel and also development in biorefineries. In the process of integration, different co-product from the microalgae is also produced.

13.6.1 Co-products

The new development in the field of microalgal biofuel production have focused on valorising the residual biomass from the microalgae which produces valuable co-products like pigments and proteins. The remaining biomass after the lipid extraction processed will be used to extract carotenoids which is used in food and cosmetic industry due to their properties of antioxidant. The protein extracted from microalgal biomass are rich with essential amino acid which is used as food supplement or animal feed. Thus, these co-products increase the overall economic viability of microalgal biofuel production (Kuo et al., 2021). The valorisation improves the economics of biofuel production and has also contributed to a circular economy by maximising the usefulness of each cultivated microalgal crops.

13.6.2 Blending

Blending is a process of mixing microalgal biofuel with traditional fuels in the energy system and this is next important step in integrating this renewable energy source into the existing energy system. Researches on blending of biofuels such as B20 and E10 has shown their ability to maintain performance of engine in comparison to the traditional fuels. The review on blending of biodiesels with fossil fuels, like B20 has achieved similar torque and power output. The result has shown similar output in standard condition in comparison to conventional diesel in which minimum engine modification is required. Researches on blend of ethanol and gasoline like E10, confirm comparable power output and efficiency similar to as traditional gasoline provides (Manikandan & Walle, 2013). The data gathered from engine testing experiment provides the information that B20 blends perform optimally in diesel engines. This also confirmed in comprehensive review which shows that both B20 and E10 provides similar engine performance and also reduce emissions. This ability of microalgal biofuel is feasible option for reducing fully dependence on fossil fuels and it is a move towards more sustainable energy system (Adhikesavan & Rajadurai, 2014; Adib et al., 2024).

13.6.3 Advances in Biorefineries

Researchers have developed a specially designed biorefineries for microalgal based production. It is an important step in increasing the efficiency and scalability of production of biofuels. The biorefineries are devised to integrate multiple process. These modular biorefineries provided the simultaneous extraction of lipids for production of biofuels, and proteins for food applications, and other high value bioproducts (Rathinavel et al., 2024). The development in modular bioreactor design, the improved cultivation technique, and the integrated processing systems allow the production with flexibility and efficiency. These are used for various algal strain and different operational conditions. This state of art approach maximises the production and minimises the waste and operational costs. This makes microalgae as a more competitive alternative to the traditional biofuels' sources.

13.7 Environmental and Economic Impacts

The ability of carbon absorption *i.e.*, carbon sequestration by microalgal system presents the environmental benefits. This ability of microalgae also has economic benefits at different stages such as cultivation, harvesting and processing. The capability of microalgal system has improved with the development of technological advancement which emphasises efficiency in production and cost effectiveness.

13.7.1 Carbon Sequestration Potential of Microalgal Systems

The microalgae are considered as a highly effective tools in the hands of researchers for carbon capture in the present biological system due to its rapids growth rate and high mass productivity. Microalgae uses atmospheric CO_2 and fuel gas emissions to increase its growth and also captures the carbon from atmosphere (Soreanu et al., 2022). The studies have indicated that the algal system has a capacity to achieve a higher carbon sequestering efficiency in open pond. This highlights the important role in reducing carbon footprints. Furthermore, in term of production of biomass, the microalgae has the capacity to produce higher biomass which is more than the terrestrial biomass production system (Sayre, 2010). This ability of microalgae contributes to reduce the greenhouse gas emission and a developed effort in sustainable biomass production for various applications.

13.7.2 Life Cycle Analysis of Microalgal Production Technologies

The study of life cycle of microalgae and its analysis is one of the important steps to understand the environmental effect related to various stages like cultivation harvesting and processing. The analysis of microalgal lifecycle reveal that the microalgae have a lower carbon footprint in comparison to the traditional agriculture practices. This analysis has taken into account of energy inputs and the emission (Cheah et al., 2015). The studies also reveal that using of the waste stream like waste water and CO_2 from industrial process for the cultivation of microalgae, results in the significant reductions of water pollution and also recycles nutrients. Thus, creates a more sustainable production cycle. The development in closed photobioreactor systems have increased the efficiency of mass transfer and productivity. This is very important for maximising the environmental benefits of microalgal cultivation (Moberg et al., 2010).

13.7.3 Economic Viability and Cost Reduction Strategies

Studies in the recent time have discussed the economic viability of microalgal systems. These studies have emphasised on important inventions in the production technologies which has the ability to reduce the cost significantly. Development in the field of efficient cultivation technique and improved photobioreactor designs has reduced the operational cost and expenses (Moberg et al., 2010). The studies have revealed different techniques for improving the cultivation parameter and in selecting the high yield microalgae strain. These studies concluded that these methods are very important for getting maximum result for increasing the biomass production as well as in reducing the input costs. Assimilation of microalgae in the existing agriculture and industrial setup offers a dual benefit. The benefits derived by assimilation of microalgae is generating economic return through use of biomass and reducing the carbon emission at the same time (Cheah et al., 2015).

13.8 Global Policies and Market Trends

The policies adopted in recent times relating to the microalgal biofuel production supports the research and commercialization of biofuel. The policy also supports rapid development of the same and recognises the importance of microalgae as a source of renewable energy.

13.8.1 Global Policies Supporting Microalgal Biofuel Research

The governments across the world are adopting policies that are meant for favouring the development of renewable energy sources including the energy from microalgal biofuel (Echendu et al., 2023). Such policies focus on reduction of greenhouse gas emissions, increasing energy security and encouraging sustainable economic growth. Several countries have adopted the policies of financial incentive for research and development in the field of renewable energy technology to encourage the innovation and development of biofuel production from microalgae.

Adopting such type of policy in which carbon pricing mechanisms are used, is very important. It is so because this creates a financial incentive for the industries to adopt cleaner source of energy including microalgal biofuel for reducing the greenhouse gas effect. For instance, the policies of granting subsidies for production of biofuel from microalgae. The research grants for increasing the progress in microalgae technologies making it more competitive against the fossil fuels. All the developed policies aim to achieve a broader environmental objective like achieving carbon neutrality with promotion of viable economic growth.

13.8.2 Regional Case Studies

The regional case studies of several countries highlight the researches in use and benefits of microalgal biofuel products. The projects involved in studies use the developed techniques and highlights the economic and environment benefit associated with the microalgal biofuel.

In the Unites States of America, to support the microalgal fuel-based project derived from the algae, several initiatives have been issued specially for those states which has ample sunlight and water resources. The result of these product has shown the increased carbon sequestration, reduced waste and significant ability for production of biofuel. This led to the job creation for the local people and energy independence of that area. (Abreu et al., 2023).

In India the government has also undertaken initiative by launching policy. Under the policy the government has initiated project funding programme for development and implementation of several renewable energy programmes. This is for reducing the carbon foot print and increase the sustainable energy ecosystem with a view to provide energy to the remote areas (National Policy on Biofuels, Govt of India).

The Qatar university has undertaken a study on the use of microalgae as a bioenergy feed stock. The research based on selection of appropriate strains of microalgae for development of biomass and use of wet microalgal biomass for production of biofuel. This study also highlighted the ability of microalgae to produce different type of biofuels like biodiesel, bioethanol, biomethane, and biocrude oil (Khan et al., 2023).

13.8.3 Recent Partnerships Between Academia, Industry, and Government

The researches in the present time have developed a new era of collaboration among the academy, industry and the government for the research and development of microalgal biofuel. The academic institutions with its talent are playing important role in the field of microalgal biology. The academic institutions which develop the technologies and system under the initiatives and incentives from the Government with joint initiatives of the industries. Thus, the gap between the theoretical research and practical application are fully bridged out and the researches are put to production (Soyombo et al., 2024).

There are several examples of the collaborative efforts. In India, where the university and the government have initiated partnership in collaboration with the biotech companies. This collaboration has increased the researches on techniques of microalgae cultivation. This type of partnership encourages sharing of resources. Such type of partnership has led to invention of techniques to revitalise and encourage the commercialization of microalgal biofuel production. These are necessary for meeting the demand of energy in future.

13.9 Future Prospects and Research Directions

The future of energy ecosystem is to provide greenhouse gas free energy ecosystem. The objective for renewable energy which comes out from the biofuel from microalgae plays important role. The direction of research in the field of microalgae is to provide a sustainable source of energy. The main area of focus which requires to be taken into consideration is exploration of improved genetic technologies for strain development. The development of such techniques is used in large scale integration of biofuel with the other renewable energy sources. The future prospect and research are also directed for the long-term sustainability in the production of biofuel from microalgae with increased production capacity. Since the world is moving towards achieving the net carbon zero target, the microalgae are emerging as one of the important solutions for renewable energy sources and carbon capture.

13.9.1 Improved Genetic Tools for Strain Development

The development in genetic engineering is very important for increasing the productivity of micro algae strain for the production of biofuel. With the development of advance genetic tools researchers have taken effort for manipulation of metabolic pathway of microalgae. This genetic tool will enhance the lipid accumulation in cell which is a crucial factor for biofuel production. The advance techniques like CRISPR/

Cas 9 and genome editing are proving more effective techniques to the scientists. These techniques have used for creation of high yield strains which tolerate various environmental stress and pressure (Li-Beisson & Peltier, 2013).

It is necessary to establish a strong platform for selection and transformation of algal strain which will be used in accelerating research and development efforts. Such strong platforms will facilitate the introduction of beneficial traits like better and improved photosynthesis efficiency and resistance to pathogen. By keeping continuous focus on genetic diversity among microalgal species, the noble path can be identified. Those noble pathways will be used to increase biofuel production and feasible agriculture practices.

Investment in the field of biotechnical development will result in development and expansion of tool for manipulation of gene in microalgae. This step includes development of synthetic biology. The synthetic biology will help in developing the design for algae strains made for specific environment for the production of biofuel. The improved method for screening of genetically engineered strains will also play a very important role in selecting the best strains and techniques for its commercial application.

13.9.2 Large-Scale Integration with Other Renewable Energy Systems

Addition of biofuel with other renewable energy sources requires a holistic approach for achieving the energy sustainability. This can be done by using the ability of carbon sequestration by microalgae and its efficiency for producing biomass. The researchers can take steps for creating a coordinated approach to integrate it with solar, wind and bioenergy system (Jaumard et al., 2021). This type of combination increases the overall efficiency and reliability of energy production system.

The microalgae cultivation system when used for the treatment of waste—provide a dual benefit of production of biofuel and waste management. The algae have capacity to use nutrients from the wastewater and thus reducing the environmental pollutant and generates energy. The collaboration between biofuel production and renewable energy systems provides sustainable energy solution in a circular economic practice.

It is necessary to access the technoeconomic feasibility for large scale addition of biofuel with the existing energy system. It is necessary to have a study on economic implication like capital and operational expenditures in comparison to the other renewable source of energy (Jaumard et al., 2021). The pilot projects are required to be undertaken for test on the large scale for obtaining valuable data and in-depth knowledge for future development.

13.9.3 Scalability in Microalgae Biofuel Production

For the successful implementation of microalgae production as a source for production of biofuel is wholly dependent on its scalability. For the setting up large scale production on the basis of techniques developed in laboratories, the strategy is to be developed for maximise the growth condition with bioprocessing techniques. To achieve this target, the pilot projects which utilises both open pond system and photobioreactors are setup to calculate its efficiency in large scale application in industries (Muthukrishnan, 2022).

It is necessary to focus on identifying the most suitable strains which shows high growth rate and lipid production under different environmental conditions. The selection of strain is an important effort. The genetic engineering leads the development for such strong and healthy strain that can sustain in different climatic conditions. Thus, such approach increases the feasibility of large-scale cultivation (Pahariya et al., 2024).

It is necessary to develop an effective harvesting and processing system for transition of biomass production to conversion into biofuel. The developed techniques like centrifugation, flocculation and extraction in which green technologies are used for reduction of cost. The energy requirement related with biofuel production are also reduced which makes it more economically feasible.

13.9.4 Long-Term Sustainability of Microalgae-Based Biofuels

It is very important to have a long-term sustainability for establishment of microalgae as a reliable source of biofuel. To reduce the environmental effect, it is necessary to undertake a sustainable practice in the field of cultivation, harvesting and processing of microalgae at priority basis. An advanced and automated integrated system be also developed. The waste like agriculture runoff and waste water should be used as a source for nutrient of microalgae to promote a circular economy which contribute toward resources recovery system.

The ecological effect of large-scale microalgae cultivation is essentially assessed. It is necessary to study the effect of ecological impact of large scale microalgae cultivation. The research can also be conducted on the effect of nutrient flow, light, pollution, and change in land use. The study on these parameters will also be taken into consideration for protecting the ecosystem. Regulation of algae farming and the productivity of algae farming is also considered in terms of protecting ecosystem. The life cycle of microalgae biofuel can also be accessed for its sustainability. It is necessary to evaluate carbon footprint energy balance and environmental consequences during the entire production process of microalgal biofuel.

13.9.5 Microalgae's Role in Achieving Global Net-Zero Targets

The microalgae have ability to play a crucial role in achieving the net zero emission in the world. The capability of carbon sequestration, the microalgae can absorb significant amount of CO_2 in the process of photosynthesis. This makes it a valuable source for carbon capture and storage strategies. Use of microalgae with fossil fuel in combustion process have ability to lower the greenhouse gas emission from the industry.

Use of microalgal biofuel into the existing energy and industrial system has led to transitioning the renewable energy source. For example, the microalgal biofuel has capacity to be used as a complement to fossil fuel in the transition phase. This biofuel offers a cleaner alternative till the infrastructure for a fully renewable system is developed.

The use of microalgae is very versatile and it extends beyond biofuel. It is also used for extraction of high value co-products like Omega3 fatty acid, pigments and bioplastic. These co-products have its own additional revenue chain which increase the economic feasibility of microalgal biofuel products (Kuo et al., 2021). The multi-facet use microalgal supports in a reduction of emissions and promotes economic activity and awareness in the rural and coastal communities.

13.10 Conclusion

Development in the field of microalgal research have put emphasis on its growing ability in addressing the energy and environmental challenges. The cultivation technique of microalgae with its development has increased the economic feasibility of microalgae biofuel production with reduction in input cost and promotion of sustainability. For example, the cultivation of microalgae in waste water system, the new invention such as advanced photobioreactor design and integration of microalgae biofuel into other renewable energy sources like solar power has increased the efficiency and scalability of production of microalgae. The advance technologies like CRISPR, genetic engineering have provided sufficient techniques for enhancement in lipid production and optimising the biofuel production from algae and also meeting the issue of commercialization of biofuels (Kuppan et al., 2023; Xue et al., 2021).

Apart from the biofuels, the role of microalgae in carbon sequestration adds another layer of benefit for redressal of greenhouse effect in the environment. The ability to absorb CO_2 in large amount in the course of photosynthesis is in consonance with global strategies of reducing the climate effect. Researches had revealed that the cultivation of microalgae at industrial scale level can reduce the industrial emission up to a considerable level. This makes the microalgae as an integral factor for achieving the carbon neutral system. The co-products of the microalgae like nutraceutical and pharmaceuticals have shown their multi-facet ability and economic versatility. The applications of microalgae co-products from antioxidant to Omega3 fatty acid which contributes to health and wellness sector of industries (Zieliński et al., 2023; Kuppan et al., 2023).

The future of large-scale production of microalgae will depends on its ability of adaptability, multiproduct potential and low environmental effect to meet the scalability and economic challenges. The development in the field of artificial intelligence and auto-monitoring support system will play an important role in making feasible of large-scale cultivation. The government policies are also required to support the large-scale cultivation by adopting the allocation of subsidies in the form of carbon credit. The world is transitioning toward the sustainable renewable energy economy. The microalgae have ability to play crucial role by offering the solution to be integrated with renewable energy production, environmental restoration and economic development as well. To harness the ability of microalgal at its maximum level a regular interdisciplinary research and global cooperation are necessary to achieve a greener future.

Acknowledgements Authors are thankful to ICMR-RMRIMS, Patna for providing necessary support.

References

Abdullah, M., Ali, Z., & Yasin, M. T. (2024). Advancements in sustainable production of biofuel by microalgae: Recent insights and future directions. *Environmental Research*. https://www.sciencedirect.com/science/article/pii/S0013935124018073

Abreu, A. P., Martins, R., & Nunes, J. (2023). Emerging applications of Chlorella sp. and Spirulina (Arthrospira) sp. *Bioengineering, 10*(8), 955. https://doi.org/10.3390/bioengineering10080955

Adhikesavan, C., & Rajadurai, J. S. (2014). Influence of B20 biodiesel blends on diesel engine combustion. *Journal of the Japan Institute of Energy, 93*(7), 675–683. https://doi.org/10.3775/jie.93.675

Adib, A. R., Rahman, Md. M., Hassan, T., Ahmed, M., & Rifat, A. A. (2024). Novel biofuel blends for diesel engines: Optimizing engine performance and emissions with *C. cohnii* microalgae biodiesel and algae-derived renewable diesel blends. *Energy Conversion and Management: X*. https://doi.org/10.1016/j.ecmx.2024.100688

Ahmad Kamal, A. H., Mohd Hamidi, N. F., Zakaria, M. F., Ahmad, A., Harun, M. R., Chandra Segaran, T., & Jusoh, M. (2024). Genetically engineered microalgae for enhanced bioactive compounds. *Discover Applied Sciences*. https://doi.org/10.1007/s42452-024-06116-5

Ajjawi, I., Verruto, J., Aqui, M., Soriaga, L. B., Coppersmith, J., Kwok, K., et al. (2017). Lipid production in *Nannochloropsis gaditana* is doubled by decreasing expression of a single transcriptional regulator. *Nature Biotechnology, 35*, 647–652. https://doi.org/10.1038/nbt.386510.1038/nbt.3865

Aslam, A., Thomas-Hall, S. R., Mughal, T. A., & Schenk, P. M. (2017). Selection and adaptation of microalgae to growth in 100% unfiltered coal-fired flue gas. *Bioresource Technology, 233*, 271–283. https://doi.org/10.1016/j.biortech.2017.02.111

Baek, K., Yu, J., Jeong, J., Sim, S. J., Bae, S., & Jin, E. (2018). Photoautotrophic production of macular pigment in a Chlamydomonas reinhardtii strain generated by using DNA-free CRISPR-Cas9 RNP-mediated mutagenesis. *Biotechnology and Bioengineering, 115*, 719–728. https://doi.org/10.1002/bit.26499

Brito-Bello, A. A., & Lopez-Arredondo, D. (2023). Bioactive compounds with pesticide activities derived from aged cultures of green microalgae. *Biology, 12*(8), 1149. https://doi.org/10.3390/biology12081149

Burgel, G., Ribas, P. G., Ferreira, P. C., Passos, M. F., Santos, B., Savi, D. C., Ludwig, T. A. V., Vargas, J. V. C., Galli-Terasawa, L. V., Kava, V. M., & Kava, V. M. (2022). Morphology, molecular phylogeny and biomass evaluation of *Desmodesmus abundans* (Scenedesmaceae-Chlorophyceae) from Brazil. *Brazilian Journal of Biology, 82*, e265235. https://doi.org/10.1590/1519-6984.265235

Chaiklahan, R., Chirasuwan, N., Srinorasing, T., Attasat, S., Nopharatana, A., & Bunnag, B. (2022). Enhanced biomass and phycocyanin production of Arthrospira (Spirulina) platensis by a cultivation management strategy: Light intensity and cell concentration. *Bioresource Technology, 343*, 126077. https://doi.org/10.1016/j.biortech.2021.126077

Chakdar, H., Hasan, M., & Pabbi, S. (2021). High-throughput proteomics and metabolomic studies guide re-engineering of metabolic pathways in eukaryotic microalgae. *Bioresource Technology*. https://www.sciencedirect.com/science/article/pii/S0960852420317697

Chang K. S., Kim J., Park H., Hong S.-J., Lee C.-G., Jin E. (2020). Enhanced lipid productivity in *AGP* knockout marine microalga *Tetraselmis sp.* using a DNA-free CRISPR-Cas9 RNP method. *Bioresource Technology 303*, 122932. https://doi.org/10.1016/j.biortech.2020.122932

Cheah, W. Y., Show, P. L., Chang, J. S., Ling, T. C., & Juan, J. C. (2015). Biosequestration of atmospheric CO_2 and flue gas-containing CO_2 by microalgae. *Bioresource Technology, 184*, 190–201. https://doi.org/10.1016/j.biortech.2014.11.026

Cheng, J., Huang, Y., Feng, J., Sun, J., Zhou, J., & Cen, K. (2013). Improving CO2 fixation efficiency by optimizing Chlorella PY-ZU1 culture conditions in sequential bioreactors. *Bioresource Technology, 144*, 321–327. https://doi.org/10.1016/j.biortech.2013.06.122

Cheon, S., Kim, H. M., Gustavsson, M., & Lee, S. Y. (2016). Recent trends in metabolic engineering of microorganisms for the production of advanced biofuels. *Current Opinion in Chemical Biology, 35*, 10–21. https://doi.org/10.1016/j.cbpa.2016.08.003

De Bhowmick, G., Koduru, L., & Sen, R. (2015). Metabolic pathway engineering towards enhancing microalgal lipid biosynthesis for biofuel application. *Renewable and Sustainable Energy Reviews*. https://www.sciencedirect.com/science/article/pii/S1364032115004013

Deepika, C., Mrinal, Pohrmen, C. B., Jaiswal, K. S., Sangmesh, B., Jaiswal, K. K., Ramasamy, A. P., & Jaiswal, A. K. (2024). Hydrothermal liquefaction of wet microalgal biomass for biofuels and platform chemicals: advances and future prospects. *Discover Applied Sciences*. https://doi.org/10.1007/s42452-024-05911-4

Dhokane, D., Shaikh, A., Yadav, A., Giri, N., Bandyopadhyay, A., Dasgupta, S., & Bhadra, B. (2023). CRISPR-based bioengineering in microalgae for production of industrially important biomolecules. *Frontiers in Bioengineering and Biotechnology, 11*, 1267826. https://doi.org/10.3389/fbioe.2023.1267826

Echendu, A., Hampo, C. C., Olatunde, D., Obasih, J. I., Oni, O., Ojo, D., Mathew, M., Adedeji, M., & Oladipo, S. (2023). Biofuels as a key renewable energy source: A review of life cycle assessment studies in South Africa. *Biofuels*. https://doi.org/10.1080/17597269.2023.2264037

Elsayed, K. (2017). *Production of biofuels from microalgae* (Doctoral dissertation, IRC-Library, Information Resource Center der Jacobs University Bremen).

Ezhumalai, G., Arun, M., Manavalan, A., Rajkumar, R., & Heese, K. (2024). A holistic approach to circular bioeconomy through the sustainable utilization of microalgal biomass for biofuel and other value-added products. *Microbial Ecology., 87*(1), 61. https://doi.org/10.1007/s00248-024-02376-1

Falfushynska, H. (2024). Advancements and prospects in algal biofuel production: a comprehensive review. *Phycology*. https://doi.org/10.3390/phycology4040030

Fischer, G., Hizsnyik, E., Prieler, S., Shah, M., & Van Velthuizen, H. T. (2009). Biofuels and food security. https://pure.iiasa.ac.at/8969

Geiman, C. (2021). Improving algae photobioreactor energy efficiency through active irradiance control for dynamic carbon dioxide fixation. http://hdl.handle.net/1773/46957

Ghedini, E., Taghavi, S., Menegazzo, F., & Signoretto, M. (2021). A review on the efficient catalysts for algae transesterification to biodiesel. *Sustainability*. https://doi.org/10.3390/su131810479

He, S., Barati, B., Hu, X., & Wang, S. (2023). Carbon migration of microalgae from cultivation towards biofuel production by hydrothermal technology: A review. *Fuel Processing Technology, 240*, 107563. https://doi.org/10.1016/j.fuproc.2022.107563

Hu, Q., Sommerfeld, M., & Jarvis, E. (2008). Microalgal triacylglycerols as feedstocks for biofuel production. *The Plant Journal, 54*(4), 621–639. https://doi.org/10.1111/j.1365-313X.2008.03492.x

Imamoglu, E. (2024). Artificial intelligence and/or machine learning algorithms in microalgae bioprocesses. *Bioengineering, 11*(11), 1143. https://doi.org/10.3390/bioengineering11111143

Jaumard, B., Chen, L., & Devaux, F. (2021). Techno-economic assessment and life cycle assessment of CO_2 to biofuel via micro-algae process. SSRN *Electronic Journal*.

Jeon, M. S., Han, S.-I., Jeon, M., & Choi, Y.-E. (2021). Enhancement of phycoerythrin productivity in *Porphyridium purpureum* using the clustered regularly interspaced short palindromic repeats/CRISPR-associated protein 9 ribonucleoprotein system. *Bioresource Technology, 330*, 124974. https://doi.org/10.1016/j.biortech.2021.124974

Jeong, S. W., HwangBo, K., Lim, J. M., Nam, S. W., Lee, B. S., Jeong, B., et al. (2020). Genetic impairment of cellulose biosynthesis increases Cell Wall fragility and improves lipid extractability from oleaginous alga *Nannochloropsis salina*. *Microorganisms, 8*, 1195. https://doi.org/10.3390/microorganisms8081195

Jeswani, H. K., Chilvers, A., & Azapagic, A. (2020). Environmental sustainability of biofuels: A review. *Proceedings of the Royal Society A, 476*(2243), 20200351. https://doi.org/10.1098/rspa.2020.0351

Jinkerson, R. E., Subramanian, V., & Posewitz, M. C. (2011). Improving biofuel production in phototrophic microorganisms with systems biology. *Biofuels*. https://www.tandfonline.com/doi/abs/10.4155/bfs.11.7

Jung, P., Briegel-Williams, L., Werner, L., Jost, E., Schultz, M., Nürnberg, D. J., Grube, M., & Lakatos, M. (2024). A direct PCR approach with low-biomass insert opens new horizons for molecular sciences on cryptogam communities. *Applied and Environmental Microbiology*. https://doi.org/10.1128/aem.00024-24

Kao, P.-H., & Ng, I.-S. (2017). CRISPRi mediated phosphoenolpyruvate carboxylase regulation to enhance the production of lipid in *Chlamydomonas reinhardtii*. *Bioresource Technology, 245*, 1527–1537. https://doi.org/10.1016/j.biortech.2017.04.111

Kasai, Y., Takagi, S., Ota, S., Ishii, K., Takeshita, T., Kawano, S., & Harayama, S. (2023). Development of efficient genetic-transformation-and genome-editing systems, and the isolation of a CRISPR/Cas9-mediated high-oil mutant in the unicellular green alga Parachlorella kessleri strain NIES-2152. https://doi.org/10.21203/rs.3.rs-3287503/v1

Khan, S., Das, P., Quadir, M. A., Thaher, M. I., Mahata, C., Sayadi, S., & Al-Jabri, H. (2023). Microalgal feedstock for biofuel production: Recent advances, challenges, and future perspective. *Fermentation, 9*(3), 281. https://doi.org/10.3390/fermentation9030281

Kim, S., Kim, M., Chang, Y. K., & Kim, D. (2023). Lipid production under a nutrient-sufficient condition outperforms starvation conditions due to a natural polarization of lipid content in algal biofilm. *Fuel, 339*, 126902. https://doi.org/10.1016/j.fuel.2022.126902

Kumari, A., Kumar, A., Pathak, A. K., & Guria, C. (2014). Carbon dioxide assisted Spirulina platensis cultivation using NPK-10: 26: 26 complex fertilizer in sintered disk chromatographic glass bubble column. *Journal of CO_2 Utilization, 8*, 49–59. https://doi.org/10.1016/j.jcou.2014.07.001

Kuo, C., Sun, Y., Lin, C.-H., Lin, C., Wu, H.-T., & Lin, C.-S. (2021). Cultivation and biorefinery of microalgae (Chlorella sp.) for producing biofuels and other byproducts: A review. https://doi.org/10.3390/su132313480

Kuppan, P., Sudharsanam, A., Venkateswarlu, K., & Megharaj, M. (2023). Solar technology-closed loop synergy facilitates low-carbon circular bioeconomy in microalgal wastewater treatment. *NPJ Clean Water, 6*(1), 43. https://doi.org/10.1038/s41545-023-00256-8

Lakhawat, S. S., Malik, N., Kumar, V., Kumar, S., & Sharma, P. K. (2022). Implications of CRISPR-Cas9 in developing next generation biofuel: A mini-review. *Current Protein and Peptide Science, 23*(9), 574–584. https://doi.org/10.2174/1389203723666220907110310

Lal, R. (2013). Soil carbon management and climate change. *Carbon Management, 4*(4), 439–462. https://doi.org/10.4155/cmt.13.31

Li-Beisson, Y., & Peltier, G. (2013). Third-generation biofuels: current and future research on microalgal lipid biotechnology. https://doi.org/10.1051/ocl/2013031

Lin, W.-R., & Ng, I.-S. (2020). Development of CRISPR/Cas9 system in *Chlorella vulgaris* FSP-E to enhance lipid accumulation. *Enyzme and Microbial Technology, 133*, 109458. https://doi.org/10.1016/j.enzmictec.2019.109458

Lin, J. Y., Lin, W. R., & Ng, I. S. (2022). CRISPRa/i with A daptive S ingle G uide A ssisted R egulation D NA (ASGARD) mediated control of Chlorella sorokiniana to enhance lipid and protein production. *Biotechnology Journal, 17*(10), 2100514. https://doi.org/10.1002/biot.202100514

Maltsev, Y., Kulikovskiy, M., & Maltseva, S. (2023). Nitrogen and phosphorus stress as a tool to induce lipid production in microalgae. *Microbial Cell Factories, 22*, 239. https://doi.org/10.1186/s12934-023-02244-6

Manikandan, K., & Walle, M. (2013). The effect of gasoline–ethanol blends and compression ratio on SI engine performance and exhaust emissions. *International Research Journal of Engineering and Technology, 2*, 3142–3154.

Matter, I. A., Bui, V. K. H., Jung, M., Seo, J. Y., Kim, Y.-E., Lee, Y.-C., & Oh, Y.-K. (2019). Flocculation harvesting techniques for microalgae: A review. *Applied Sciences, 9*(15), 3069. https://doi.org/10.3390/app9153069

Mishra, S. D., & Joshi, S. (2022). Recent advances in biofuel production through metabolic engineering. *Bioresource Technology*. https://www.sciencedirect.com/science/article/pii/S0960852422003662

Moberg, A., Ellem, G., Jameson, G., & Herbertson, J. (2010). Fluid dynamical considerations in the design of a microalgae photobioreactor. In *Chemeca 2010: Engineering at the Edge*, 26–29 Sept 2010, Hilton Adelaide, South Australia. Engineers Australia. https://doi.org/10.3316/informit.968412396350788

Muthukrishnan, L. (2022). Bio-engineering of microalgae: Challenges and future prospects toward industrial and environmental applications. *Journal of Basic Microbiology, 54*(4), 621–639. https://doi.org/10.1002/jobm.202100417

National Policy on Biofuels. Government of India Ministry of New & Renewable Energy, hareda-808646251

Nguyen, T. H. T., Park, S., Jeong, J., Shin, Y. S., Sim, S. J., & Jin, E. (2020). Enhancing lipid productivity by modulating lipid catabolism using the CRISPR-Cas9 system in *Chlamydomonas*. *Journal of Applied Phycology, 32*, 2829–2840. https://doi.org/10.1007/s10811-020-02172-7

Nithiya, E. M., Tamilmani, J., Vasumathi, K. K., & Premalatha, M. (2017). Improved CO_2 fixation with Oscillatoria sp. in response to various supply frequencies of CO_2 supply. *Journal of CO2 Utilization, 18*, 198–205. https://doi.org/10.1016/j.jcou.2017.01.025

Nwoba, E. G., Parlevliet, D. A., Laird, D. W., Alameh, K., & Moheimani, N. R. (2019). Light management technologies for increasing algal photobioreactor efficiency. *Algal Research, 39*, 101433. https://doi.org/10.1016/j.algal.2019.101433

Nymark, M., Sharma, A. K., Sparstad, T., Bones, A. M., & Winge, P. (2016). A CRISPR/Cas9 system adapted for gene editing in marine algae. *Scientific reports, 6*(1), 24951.https://doi.org/10.1038/srep24951

Onay, M. (2018). Investigation of biobutanol efficiency of chlorella sp. Cultivated in municipal wastewater. *Journal of Geoscience and Environment Protection*. https://doi.org/10.4236/gep.2018.610003

Pahariya, R., Chauhan, A., Ranjan, A., Thakur, S. K., Tuli, H., Ramniwas, S., Shahwan, M., & Jindal, T. (2024). Impact of nanoparticles on microalgae and the prospects for biofuel production:

Current advancements and future outlook. *Journal of Applied Biology & Biotechnology., 12*(3), 17–25. https://doi.org/10.7324/JABB.2024.160045

Penloglou, G., Pavlou, A., & Kiparissides, C. (2024). Recent advancements in photo-bioreactors for microalgae cultivation: A brief overview. *Processes, 12*(6), 1104. https://doi.org/10.3390/pr12061104

Pérez, L. (2016). Biofuels from microalgae, A promising alternative. https://www.semanticscholar.org/paper/f2983084b5c7d5426e9b28cb34b995f4c73b3e85

Rai, M. P., & Gupta, S. (2017). Effect of media composition and light supply on biomass, lipid content and FAME profile for quality biofuel production from *Scenedesmus abundans*. *Energy Conversion and Management, 141*, 85–92. https://doi.org/10.1016/j.enconman.2016.05.018

Rathinavel, L., Singh, S., Rai, P. K., Chandra, N., Jothinathan, D., Gaffar, I., Pandey, A. K., Choure, K., Waoo, A. A., Joo, J. C., & Pandey, A. (2024). Sustainable microalgal biomass for efficient and scalable green energy solutions: Fueling tomorrow. *Fuels, 5*(4), 868–894. https://doi.org/10.3390/fuels5040049

Ratomski, P., & Hawrot-Paw, M. (2021). Production of *Chlorella vulgaris* biomass in tubular photobioreactors during different culture conditions. *Applied Sciences, 11*(7), 3106. https://doi.org/10.3390/app11073106

Robertson, G. P., Hamilton, S. K., Barham, B. L., Dale, B. E., Izaurralde, R. C., Jackson, R. D., Jackson, R.D., Landis, D.A., Swinton, S.M., Thelen, K.D. & Tiedje, J. M. (2017). Cellulosic biofuel contributions to a sustainable energy future: Choices and outcomes. *Science, 356*(6345), eaal2324. https://doi.org/10.1126/science.aal2324

Rock, A., Novoveská, L., & Green, D. (2021). Synthetic biology is essential to unlock commercial biofuel production through hyper lipid-producing microalgae: A review. *Applied Phycology, 2*(1), 41–59. https://doi.org/10.1080/26388081.2021.1886872

Ryu, A. J., Jeong, B., Kang, N. K., Jeon, S., Sohn, M. G., Yun, H. J., et al. (2021). *Bioresour. Technol., 340*, 125676. https://doi.org/10.1016/j.biortech.2021.125676

Salim, S., Bosma, R., Vermuë, M. H., & Wijffels, R. H. (2010). Harvesting of microalgae by bioflocculation. *Journal of Applied Phycology, 23*(5), 849–855. https://doi.org/10.1007/s10811-010-9591-x

Sarkar, D., & Shimizu, K. (2015). An overview on biofuel and biochemical production by photosynthetic microorganisms with understanding of the metabolism. *Bioresources and Bioprocessing, 2*(1), 17. https://doi.org/10.1186/s40643-015-0045-9

Savakis, P., & Hellingwerf, K. J. (2015). Engineering cyanobacteria for direct biofuel production from CO_2. *Current Opinion in Biotechnology, 33*, 8–14. https://doi.org/10.1016/j.copbio.2014.09.007

Sayre, R. (2010). Microalgae: The potential for carbon capture. https://doi.org/10.1525/bio.2010.60.9.9

Shahid, A., Rehman, A., & Usman, M. (2020). Engineering the metabolic pathways of lipid biosynthesis to develop robust microalgal strains for biodiesel production. *Biotechnology and Bioengineering*. https://iubmb.onlinelibrary.wiley.com/doi/abs/10.1002/bab.1812

Shin, Y. S., Jeong, J., Nguyen, T. H. T., Kim, J. Y. H., Jin, E., & Sim, S. J. (2019). Targeted knockout of phospholipase A2 to increase lipid productivity in *Chlamydomonas reinhardtii* for biodiesel production. *Bioresource Technology, 271*, 368–374. https://doi.org/10.1016/j.biortech.2018.09.121

Song, I., Kim, S., Kim, J., Oh, H., Jang, J., Jeong, S. J., Baek, K., Shin, W. S., Sim, S. J., & Jin, E. (2022). Macular pigment-enriched oil production from genome-edited microalgae. *Microbial Cell Factories, 21*(1), 27. https://doi.org/10.1186/s12934-021-01736-7

Soreanu, G., Cretescu, I., Lutic, D., Harja, M., & Maier, S. S. (2022). Study of microalgae influence on carbon capture from gaseous streams within the biotrickling filtration process. *International Multidisciplinary Scientific GeoConference: SGEM, 22*(4.1), 391–398. https://doi.org/10.5593/sgem2022/4.1/s19.50

Soyombo, O. T., Odunaiya, O. G., Abioye, K. M., & Adeleke, A. G. (2024). SME collaboration models for accelerating clean energy innovations: Insights from emerging market economies.

International Journal of Management & Entrepreneurship Research. https://doi.org/10.51594/ijmer.v6i9.1584

Spanou, A., Moschona, A., Theodosiou, E., Patsios, S. I., & Pavlidis, I. V. (2024). Novel concepts for the biocatalytic synthesis of second-generation biodiesel. *Frontiers in Catalysis, 4*, 1360702. https://doi.org/10.3389/fctls.2024.1360702

Srimongkol, P., Sangtanoo, P., Songserm, P., Watsuntorn, W., & Karnchanatat, A. (2022). Microalgae-based wastewater treatment for developing economic and environmental sustainability: Current status and future prospects. *Frontiers in Bioengineering and Biotechnology, 10*, 904046. https://doi.org/10.3389/fbioe.2022.904046

Srivstava, R., Sharma, P. K., & Saharia, M. (2018). Tailoring microalgae for efficient biofuel production. *Frontiers in Marine Science*. https://www.frontiersin.org/articles/10.3389/fmars.2018.00382/full

Teng, S. Y., Yew, G. Y., Sukačová, K., Show, P. L., Máša, V., & Chang, J. S. (2020). Microalgae with artificial intelligence: A digitalized perspective on genetics, systems and products. *Biotechnology Advances, 44*, 107631. https://doi.org/10.1016/j.biotechadv.2020.107631

Tibocha-Bonilla, J. D., Zuñiga, C., & Godoy-Silva, R. D. (2018). Advances in metabolic modeling of oleaginous microalgae. *Biotechnology for Biofuels*. https://link.springer.com/article/10.1186/s13068-018-1244-3

Upreti, P., & Ambwani, S. (2016). A review of third-generation biofuels from microalgae. *Journal of Biofuels, 7*(1), 14–19. https://doi.org/10.5958/0976-4763.2016.00003.9

Verruto, J., Francis, K., Wang, Y., Low, M. C., Greiner, J., Tacke, S., et al. (2018). Unrestrained markerless trait stacking in *Nannochloropsis gaditana* through combined genome editing and marker recycling technologies. *Proceedings of the National Academy of Sciences, 115*, E7015–E7022. https://doi.org/10.1073/pnas.1718193115

Wang, S. K., Stiles, A. R., Guo, C., & Liu, C. Z. (2015). Harvesting microalgae by magnetic separation: A review. *Algal Research, 9*, 178–185. https://doi.org/10.1016/j.algal.2015.03.005

Xue, J., Balamurugan, S., Li, T., Cai, J. X., Chen, T. T., Wang, X., Yang, W. D., & Li, H. Y. (2021). Biotechnological approaches to enhance biofuel producing potential of microalgae. *Fuel, 302*, 121169. https://doi.org/10.1016/j.fuel.2021.121169

Zadabbas, S. H., Akbarzadeh, A., Ofoghi, H., & Kadkhodaei, S. (2023). Site-specific gene knock-in and bacterial phytase gene expression in *Chlamydomonas reinhardtii* via Cas9 RNP-mediated HDR. *Frontiers in Plant Science, 14*, 1150436. https://doi.org/10.3389/fpls.2023.1150436

Zhang, P., Xin, Y., He, Y., Tang, X., Shen, C., Wang, Q., Lv, N., Li, Y., Hu, Q., & Xu, J. (2022). Exploring a blue-light-sensing transcription factor to double the peak productivity of oil in *Nannochloropsis oceanica*. *Nature Communications, 13*(1), 1664. https://doi.org/10.1038/s41467-022-29337-x

Zheng, Y., Liu, X., Du, M., Lu, Q., Li, Y., Zhou, X., & Wang, D. (2024). One-step strategy streamlining microalgae biofuel production: integration of microalgae harvesting and methane production by dual-functional magnetic nanoparticles. *ACS ES&T Water, 4*(12), 5834–5843. https://doi.org/10.1021/acsestwater.4c00816

Zieliński, M., Dębowski, M., Kazimierowicz, J., & Świca, I. (2023). Microalgal carbon dioxide (CO_2) capture and utilization from the European Union perspective. *Energies, 16*(3), 1446. https://doi.org/10.3390/en16031446

Chapter 14
Environmental Sustainability of Second-Generation Biofuels for Climate Mitigation and Achieving Circular Economy Objectives

Tarun Sharma and Bharat Bhushan Nagar

Abstract Global energy demand and related climate crisis, characterized with global warming and environmental pollution. Biofuels have caught global attention worldwide as a mitigation option for this problem. Second generation biofuels have been found to be more circular and sustainable in terms of materiality and resource consumption. Multiple sources of second-generation biofuels exist, which are primarily derived from agri-residues biomass; agri-residue derivatives; source segregated organic fraction of MSW; agri-processing industries byproducts and other related streams. This chapter lists all above-mentioned second-generation biofuel feedstocks and their potential for global attainment of fossil fuel reduction. Feedstock availability potential across regional and international scenario are mentioned and their derivatives conversion potential, utilizing various engineering processes and related performances are presented. The chapter analyses the impact of biochemical composition of feedstocks, requirement of pre-processing and other biochemical treatment requirements for optimization of resource utilization and environmental performance of these feedstocks. The chapter describes environmental performance of second-generation biofuels and share its circular materiality aspect, which are supported from life cycle assessment studies. The chapter cites environmental performance; circular aspect of materiality and climate neutrality as the key to unlocking the wider potential of such feedstocks to fulfill energy demands in a sustainable manner.

Keywords Bioenergy · Biofuels · Biomass · Lignin · Sustainability · Feedstock

T. Sharma (✉) · B. B. Nagar
Amity Institute of Organic Agriculture, Amity University, Noida 201313, Uttar Pradesh, India
e-mail: tarunagronomos@gmail.com

V. Singh et al. (eds.), *Bioenergy and Value Added Byproducts from Biomass Waste*, Sustainable Environmental Waste Management Strategies,
https://doi.org/10.1007/978-3-032-22849-9_14

14.1 Introduction

The world's population reached a new milestone of eight billion on 15th November 2022. Energy demand has been on rising due to the mushrooming population and urbanization (Anonymous, 2022). Today's global economies and societies depend on continuous energy availability because it touches every aspect of human life. The major source of energy that we currently use is non-renewable fossil fuels-based products like petrol, diesel, and coal, which deplete with time. As of 2019, 84% of global energy demands were being met using non-renewable fossil fuels (Khan et al., 2021). These fossil fuels are the biggest contributor of Green House Gas (GHG) emissions, resulting in global warming.

We need sources of energy that are sustainable, affordable, accessible and renewable (Rokicki et al., 2022). These energy resources comprise of wind, solar, thermal, and biomass-derived energy resources (biofuels). Biofuels are the most environmentally friendly alternatives to conventional fossil fuels. Biofuels are like fossil fuels based physical form and exist in liquid, solid, or gaseous states. They are derived from biomass, such as maize, sugarcane, woodchips, charcoal, and in the form of biogas produced from anaerobic digestion process.

Biohydrogen, bioethanol and biodiesel are currently used for biofuels worldwide (Devi et al., 2022). Biofuels are also of diverse types, divided on the basis of the level of technology and generation of products and feedstocks (Neupane, 2022). Classification of biofuels as per feedstock and technology utilized is given in Table 14.1.

Table 14.1 Classification of biofuels as per feedstocks and technology utilized

First generation biofuels	Feedstocks	Edible crops such as Sugarcane, mustard, Wheat, corn, and rice grains
	Technology utilized	Fermentation, Transesterification
	Limitations	Competition for food resources, high global warming potential
Second generation biofuels	Feedstocks	Lignocellulosic biomass (Forest or grass-based), agricultural residues, industrial and household organic waste
	Technology utilized	Pretreatment, Enzymatic hydrolysis, Fermentation/ Transesterification
	Limitations	Low yield, wide variations in feedstocks
Third generation biofuels	Feedstocks	Algae
	Technology utilized	Harvesting, drying, Oil extraction, Transesterification, Pretreatment, Saccharification, and fermentation
	Limitations	Low biomass production, Energy intensive
Fourth generation biofuels	Feedstocks	Genetically modified algae
	Technology utilized	Genetically modified feedstocks and genetically modified microbes
	Limitations	

Among the classified biofuels, second-generation biofuels altogether avoid land use changes brought about by the cultivation of energy crops (first generation biofuels). Since they are derived from residual biomass processing of agri-crops responsible for food supply, they do not pose any competition/threat to agri-crops in terms of water and land consumption (Kowalski et al., 2022; Naik et al., 2010). Second generation biofuels hold potential to reduce global fuel related GHG emissions significantly. They are poised to replace conventional fossil fuel-based energy generation and consumption to an extent of 40% by the year 2040 (Prasad & Dhanya, 2011). Waste or agri residue derived biofuels follow the concept of circular economy wherein by-products or a process or product waste is converted into fuels, thus maximizing its utility with minimized environmental footprint (Devi et al., 2022). Parallel to this approach, lignocellulosic biomass keeps releasing carbon dioxide in a circular loop by absorbing the carbon dioxide during photosynthesis (Tonini et al., 2016). The term circular economy was given by KE Boulding in 1966.

A circular economy is a restorative, regenerative system that balances the economic, social, and environmental goals to ensure that all its parts and outputs are at their greatest levels of utility and value throughout the system (Korhonen et al., 2018). Deriving biofuels from waste or residues in the circular economy can cope with the upcoming energy crises with declining fossil fuel reservoirs (de Carvalho Freitas et al., 2022).

Second-generation biofuels are fuels produced from non-food lignocellulosic-rich biomass (Rai et al., 2022). The biomass here encompasses agricultural residues, organic wastes, by-products of agri-processing industries and energy crops, represented by fast-growing woody mass. A detailed description of biomass or feedstocks used in second-generation biofuels is given in Fig. 14.1.

Biopolymers of cellulose, hemicelluloses, and lignin are the main compositions of these biomasses. Second-generation biofuels are more sustainable than other fuels because of their advantages such as the usage of waste residues and barren land (Piccolo & Bezzo, 2009; Rai et al., 2022). In addition to existing advantages for second-generation biofuels, there are also some dis-advantages, which are associated with lack of mature technology; impact of seasonal availability; competing utilization of bio-plastics production and un-economical production, as they are post-harvest biomass form of agriculture crops (Demirbas, 2009; Kowalski et al., 2022).

Two approaches are used for the production of biofuels. These are biochemical and thermo- chemical processes and involve various steps under them to produce biofuel from raw materials. In a bio-chemical process involving the conversion of hemicelluloses and cellulose of biomass into fermentable reducing sugar is achieved with support of enzymes or acid hydrolysis, followed by ethanol production from sugars by micro-organisms. These microorganisms and enzymes act as biocatalysts in this process. Various microorganisms used for this process include *Rhodospirillum rubrum, Citrobacter spp., Rubrivivax gelatinosus, Clostridium aceticum, Butyribacterium methylotrophicum,* etc. (Abrini et al., 1994; Grady et al., 1998; Love, 2022, Maness Weaver, 2002). In the case of waste streams with high water contents, the process of conversion leads to formation of methane (CH_4), carbon-dioxide (CO_2) and mercaptans such as hydrogen sulphide (H_2S). Methane is separated from H_2S

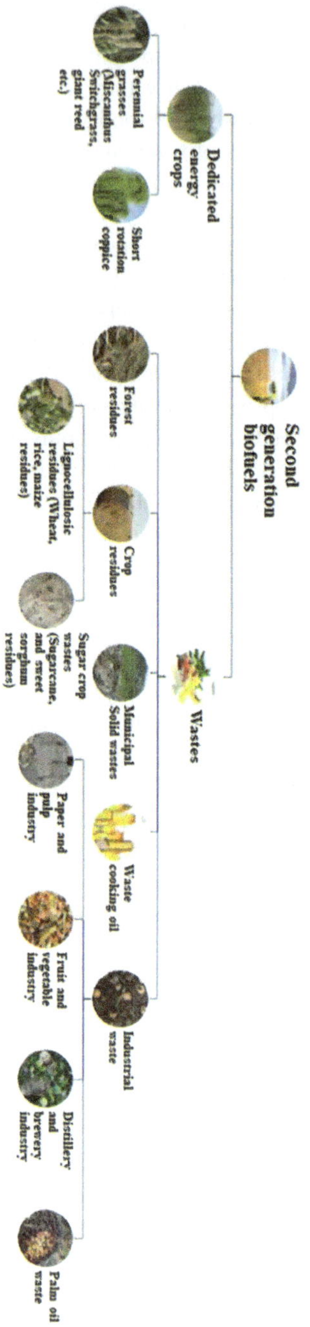

Fig. 14.1 Classification of biofuels

and CO_2 and is used for power generation. It may further be refined and compressed to produce bio-methane and used for energizing vehicles for transportation purposes or domestic cooking.

On the other hand, the thermo-chemical process involves pyrolysis and gasification technology for conversion of the lignocellulosic structure of biomass into liquid or gas at elevated temperature in limiting presence of oxygen. Each of these processes has its own advantages and disadvantages. The biochemical process requires a complimentary pre-treatment process for changing the structure of biomass to exhibit the cellulose and hemicelluloses for enzymatic hydrolysis. This process also requires specific microorganisms which makes the process costly (Baruah et al., 2018; Foust et al., 2009). In the thermo-chemical process, efficient and robust technologies are available for processing of lignocellulose feedstock, but the cost of collection, transportation, and stand-alone processing of biomass is a major issue.

Advanced biofuels are being promoted under bioenergy policies across the globe such as the National Biofuel Policy of India, the Energy Independence and Security act of the United States of America, and the Renewable Energy Directive of the European Union (Anonymous, 2018; EISA, 2007). In this chapter, we were going to discuss second-generation biofuels in a circular economy perspective, which are derived from non-edible feedstocks, wastes, and residues for a sustainable environment.

14.2 Method and Materials

14.2.1 Global Outlook for Second Generation Biofuels

Globally, bioenergy is the largest source of renewable energy, which accounts for 55% of total renewable energy and contributes to 6% of global energy supply. Due to prevailing global Net Zero Emission targets promulgated globally, bioenergy holds significant potential to displace fossil fuels by 2030. Use of modern bioenergy has increased on average by about 7% per year between 2010 and 2021 and is on an upward trend. Global trends of bioenergy are shown in Fig. 14.2.

More efforts are needed to accelerate modern bioenergy deployment to get on track with the Net Zero Scenario, which sees deployment increase by 10% per year between 2021 and 2030, while simultaneously ensuring that bioenergy production does not incur negative social and environmental consequences.

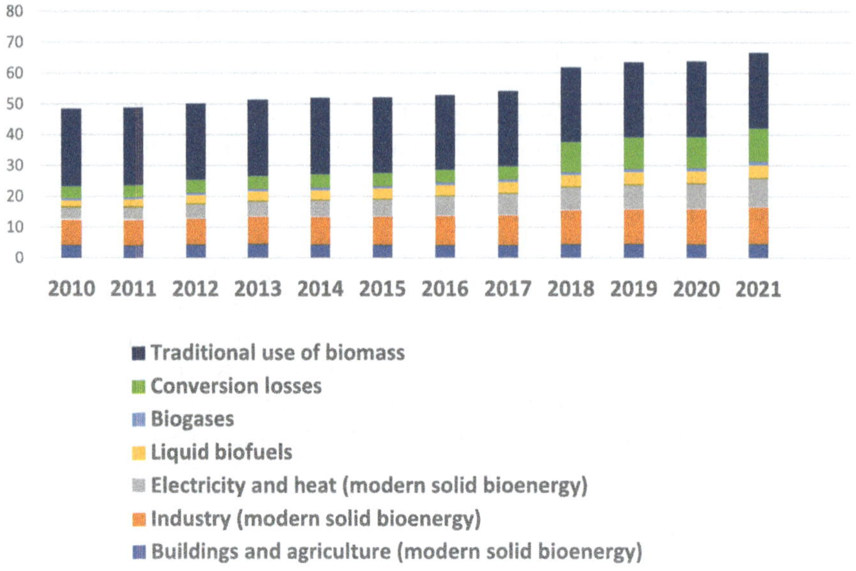

Fig. 14.2 Global growth trends of bioenergy

14.2.2 Status of Wastes and Residues Available for Biofuel Production

Organic waste-based biofuel feedstocks are available in plenty across the globe in the form of crop residues, Agro-industrial wastes such as palm oil wastes, fruit and vegetable industry waste, distillery and brewery waste, source separated organics from municipal solid and paper and pulp industry discarded material. Organic wastes especially rich in sugars are highly productive for biofuel generation. Details of the various feedstocks available for advanced biofuel generation are mentioned in subsequent sections.

14.2.3 Sustainable Biomass

Sustainable biomass production using dedicated bioenergy crops especially on marginal lands can be a viable alternative to food vs fuel conflict as well as potential candidate for land rehabilitation (Mehmood et al., 2017). Switchgrass, elephant grass (Napier grass), Giant reed and miscanthus grass are the major perennial dedicated bioenergy crops being utilized for bio-fuel production in the world (Zabed et al., 2017). Lignocellulosic biomass is considered as the most abundant renewable candidate for bioenergy production globally. Lignocellulosic biomass is available in plenty, with average global quantity of 1.3 billion tons per annum (Kowalski et al., 2022). Cellulosic and hemicellulose can be exposed to hydrolysis from such biomass to release fermentable sugars, which are prime precursors for production of biofuels (Resch & Emme, 2022). Cellulosic ethanol derived from lignocellulosic biomass is a significant contributor to presently commercialized second generation biofuels (Ayodele et al., 2020). European and North American countries in bio-fuel technology have significantly promoted lignocellulosic biomass-based biofuels as environmentally sustainable energy sources. For the year of 2022, Brazil produced around fifty-five million liters of cellulosic ethanol as against forty million liters for the previous year i.e., 2021 (Broda et al., 2022). On the Biomass Energy sector, the India government committed to increasing the share of non-fossils fuel in total capacity to 40% by 2030. At present, India produces about 450–500 million tonnes of biomass per year.

14.2.3.1 Crop Residues

India produces 118.43 Mt of grain and an estimated 165.8 Metric Tons (Mt) of straw from its 43.8 million Hectares (Mha) of rice-growing land (Kaur et al., 2022). As the second largest worldwide crop producer (Food and Agricultural Organization of the United Nations), India generates ~ 500 million metric tonnes (MT) of crop residue annually, of which 100 MT is burned (Fig. 14.3).

Burning has become an increasingly popular method of managing rice crop wastes due to the volume of rice residue with no practical value; brief 17 days window of next sowing and delayed degradation of rice residues in prevailing cold weather in northern parts of India. Every year, during the months of October and November,

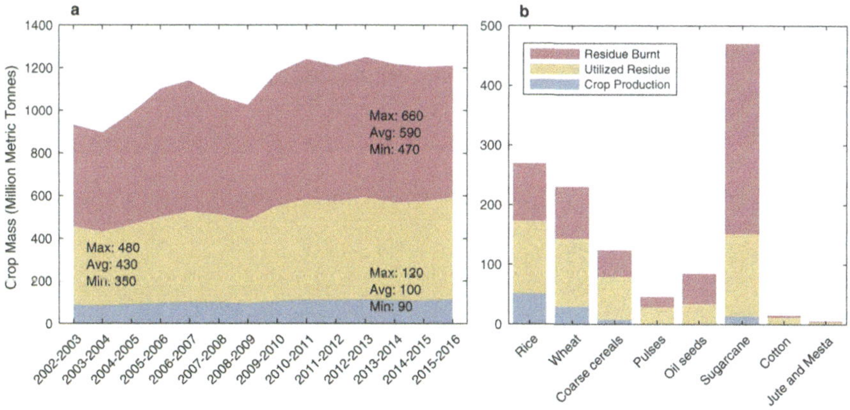

Fig. 14.3 Data on annual crop production and agri residue burnt in India. **a** Annual total crop production (blue) utilized residue (yellow), and residue burned (red) in million metric tonnes in India from 2003 to 2016; **b** Annual crop production (blue) including crop for food and fiber, utilized residue (yellow) and residue burned (red) by crop type in India, averaged from 2003 to 2016 in million metric tonnes (Pandey et al., 2021)

around 50 Mt of rice straw is burned. Burning residue is a significant source of air pollution, releasing about 1.5 Mt of particulate matter, 150 Mt of carbon dioxide, as well as other greenhouse gases and volatile organic compounds (e.g., NO_2, SO_2, CO, CH_4, NH_3) (Kaur et al., 2022). These emissions cause a variety of respiratory issues in people, reduce soil nutrient, and carbon inputs, and disrupt soil microbial activity. Each year, Asia burns around 730 Mt of biomass, mostly burning forests (45%), agricultural residues (34%), and grasslands (20%). Retention and integration of rice leftovers in soil might result in a 150 Mt/year reduction in CO_2 emissions and a 240 Mt/year increase in soil organic carbon (SOC) sequestration (Chen et al., 2019).

Crop residues pose an opportunity as a cleaner and greener energy source with their bioconversion into biofuels such as bioethanol, biodiesel, biohydrogen, biomethane and biobutanol (Meramo et al., 2022). Bio-butanol can be produced from waste biomass (by ABE fermentation using microbes. Biomass Atlas published in 2015 by Dhar & Purohit shows the total biofuel Potential from Net Availability of Agricultural Residues in India (Fig. 14.4).

Crop residues such as wheat straw, barley straw, rice straw, maize stover, sugarcane baggase and sweet sorghum bagasse can be utilized as feedstocks for biobutanol production.

Crop Residue	Agricultural residue used for fodder, fuel, and other purposes (%)			Net agri-residue availability for biofuels (Mt)			Net Ethanol Availability (BL)		
	Fodder	Fuel	Other	2010/11	2020/21	2030/31	2010/11	2020/21	2030/31
Rice straw and husk	80.8	11.1	8	13.8	15.8	17.8	3	4	5.3
Wheat straw	86.4	0	13.6	0	0	0	0	0	0
Jowar stalk	100	0	0	0	0	0	0	0	0
Bajra straw	89.8	0	10.2	0	0	0	0	0	0
Maize stalk and cobs	81	19	0	7.4	8.5	9.7	1.6	2.1	2.9
Other cereals stalk	100	0	0	0	0	0	0	0	0
Gram waste	0	100	0	9.5	9.7	10	2	2.4	3
Tur shell and waste	3.5	48.5	48	2.9	3.1	3.3	0.6	0.8	1
Lentil shell and waste	3.5	48.5	48	1	1.3	1.4	0.2	0.3	0.4
Other pulses - shell and waste	3.5	48.5	48	6.3	6.4	6.9	1.3	1.6	2.1
Groundnut waste	0	13.2	86.8	1.8	1.9	2.1	0.4	0.5	0.6
Rape and Mustard Waste	0	100	0	11.8	13.9	15.9	2.5	3.5	4.8
Other oilseed waste	0	100	0	23.1	27.8	32.2	4.9	7	9.7
Cotton seeds and waste	0	100	0	14.1	15.3	16.2	3	3.8	4.9
Cotton gin and trash	0	100	0	0.3	0.3	0.4	0.1	0.1	0.1
Jute and Mesta waste	0	100	0	2.2	2.6	2.8	0.5	0.7	0.9
Sugarcane bagasse and leaves	11.8	41	47.2	35.9	42.6	48.2	7.7	10.7	4.5
Total				130.2	149.3	166.8	27.9	37.3	50.1

Fig. 14.4 Biofuel potential from net availability of agricultural residues in India. *Source* Purohit and Dhar (2015)

Biohydrogen is another form of promising bioenergy source produced by dark fermentation of lignocellulosic biomass or waste materials (crop residues, fruit and vegetable waste, municipal solid waste). Apart from biohydrogen, dark fermentation also results in co-products such as acetic acid, ethanol, or propionic acid. Dark fermentation is the acidogenic bacteria-based fermentation of feedstocks rich in carbohydrates in absence of light and oxygen (Łukajtis et al., 2018). It has been estimated that 1.23 billion tons of waste feedstocks (rice straw, wheat straw, barley straw, corn stover, sugarcane bagasse etc.) can be collected for biohydrogen production having a potential to generate 10.56 million tons of biohydrogen. The potential of such feedstocks can be further enhanced by pre-treatment of crop residues or feedstocks.

Biomethane, an anaerobic digestion based renewable energy source from organic wastes (plant or animal wastes) can act as a partial substitute for heat, power generation and as a transport fuel (Assandri et al., 2022). India produces around 500–550 million tons of crop residues annually of which substantial amount is burned. With such a huge availability of organic wastes, India poses a potential of around 125 billion cubic metres of biomethane generation. It is found that more than 125 billion m3 of biomethane could be produced annually from waste resources in India, which would be able to supply around 47% of India's present electricity generation (Akashdeep et al., 2022). The corresponding avoided GHG emissions from the displacement of fossil fuels and the prevention of crop residue field burning and municipal waste dumping are between 284 and 461 million tonnes of carbon dioxide-equivalent, excluding the contribution from black carbon. India biogas market is expected to grow from $1.47 billion in 2022 to $2.25 Billion in 2029 at a CAGR of 6.3% in forecast period (Fig. 14.5).

China has an enormous potential for biomethane production apart from India. China produces around 970 × 106 t/a of crop residues annually of which 495 × 106 t/a are available for bio methanation. Such a huge amount of crop residues has potential to generate 82.25 × 109 m^3/a of biomethane annually. Biomethane thus generated from such an amount of waste have environmentally benign impact can eliminate greenhouse gas emissions to an extent of 197 × 106 t/a (CO_2 equivalent) annually in China.

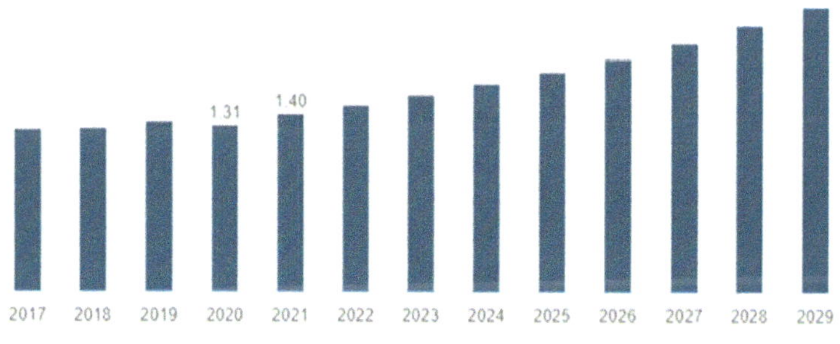

Fig. 14.5 Future potential of Indian biogas market. *Source* Joshi (2023)

14.2.3.2 Palm Oil Waste in Indonesia and Malaysia

Almost 85% of palm oil output is generated in only two nations, with Indonesia producing 46.8 million MT or 58% of world supply, while Malaysia produced 20.9 million MT or 26% (Niall, 2021). As the industry expands rapidly, palm oil production goes hand in hand with its biomass generation. Approximately 70% of fresh fruit bunches become waste in the form of empty fruit bunches, shells, and liquid effluent (Salman, 2021). The statistics of palm oil and biomass production are shown in Fig. 14.6.

One of the main components of cosmetics, such as lipstick, shampoo, deodorant, and food additives, is palm oil. Effluents, grease, empty fruit bunches, palm kernel shells, the stem of the plant, fibre, leaves, and other materials are produced during the extraction of palm oil. Sludge from wastewater treatment and empty fruit bunch shells can be used as mulch, compost, or organic fertiliser if the wastes are managed properly.

Greenhouse gas emissions from palm oil wastes may be reduced by turning them into electricity, which is a useful way to get access to carbon credit revenues for sustainable management of waste. In the palm oil business, the oil palm remnants are unquestionably a cause for worry. For every 10% of oil produced, almost 90% of the residues are overlooked and (Ng et al., 2012) remains as a waste biomass material. Oil palm waste can be used for flexible energy generation as one method of resolving the aforementioned global problems. According to estimates, Oil Palm Kernel Shells (OPKS), Oil Palm Empty Fruit Bunches (OPEFB), and Oil Palm Mesocarp Fiber (OPMF) could produce up to 2400–7460 MW of installed capacity, while Palm Oil Mill Effluent (POME) biogas with 7200 h/year of operational hours might provide up to 410–483 MW of electricity (Salleh et al., 2020). Only two main techniques,

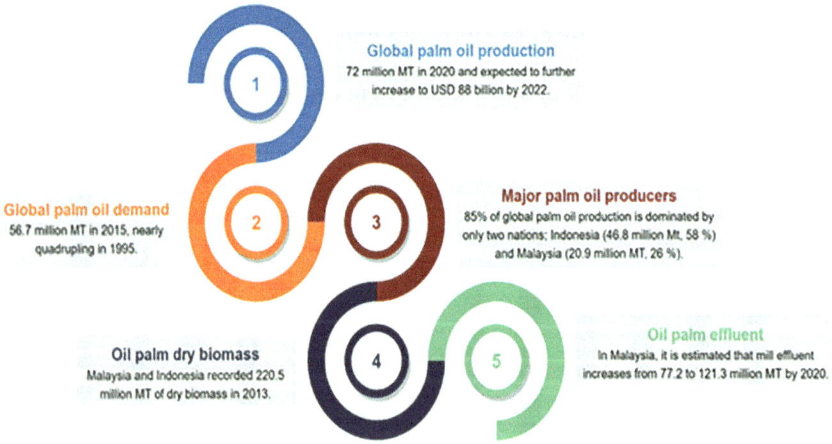

Fig. 14.6 Summary of the current statistics of the palm oil industry and its biomass generation. *Source* Salman (2021)

such as thermochemical processes (gasification, pyrolysis, and direct combustion) and biochemical processes (aerobic digestion, aerobic digestion) to produce solid, liquid, and gaseous biofuels, are well known for their complete usability to transform waste into energy (WtE). Conversion of Palm oil waste into biogas can reduce the rate of 1102 tons of GHG emission released from 1 ton CPO can decrease from 1102.325 to around 192 kg CO_2equiv (Seyed et al., 2015). The consequences for fossil fuels have made the development of biofuel an alternative fuel for reducing the life-cycle emissions of crude oil.

14.2.3.3 Fruit Industry Waste

India is currently the second-largest producer of fruit crops in the world. As per National Horticulture Database during 2021–22, India produced 107.24 million metric tonnes of fruits and 204.84 million metric tonnes of vegetables. Many of these crops are seasonal, perishable, and have a short shelf life. Therefore, fruit crops are lavishly processed into various value-added products in order to keep this crop available throughout the year in addition to being marketed as a fresh commodity. Fruit crop processing does, regrettably, result in a significant quantity of waste being produced as an industrial by-product, even though it reduces post-harvest losses.

In a study published in 2004, Gunaseelan studied biochemical methane potential of 54 fruits and vegetable wastes samples and eight standard biomass samples were determined in order to compare extents and rates of their conversion to methane studied and evaluated biomethane yields potential of various fruit wastes. It was noted that Mango peels of some types, orange wastes, rotting pomegranate seeds, and lemon pressings all produced much more methane than cellulose. The overall temperature had no impact on the methane generation from mango peel waste, even though the conversion kinetics were greater at 35 °C than at 28 °C.

Unusable waste from the cashew nut industry is cashew nutshell oil. The cashew nutshell oil is transformed into cashew nutshell biodiesel through a conventional transesterification process and combined with 98.4% pure pentanol to serve as an oxygenated additive. When 10 and 20% of pentanol is added to cashew nutshell biodiesel, harmful gases like CO, NO_x, HC, and smoke emissions are significantly reduced by 3.1 and 3.4% (Devarajan et al., 2017).

14.2.3.4 Distillery and Brewery Waste

In 2020, the global annual beer production was estimated at 1.82 billion hectorlitres (hL), with North and South America leading in production (615.28 million hL), followed by Asia (550.88 million hL), Europe (500.93 million hL), Africa (131.51 million hL) and Australia/Oceania (20.99 million hL) (Statista, 2021). Based on the global beer production quantities, about 36.4 million tons of BSG would be available worldwide (Zeko-Pivač et al., 2022) Brewer's wasted grain, hot trub, and leftover brewer's yeast are the three primary wastes produced by the brewing sector. Properly handling of these wastes may have positive economic effects and contribute to preventing environmental damage brought on by their excessive build-up. Although it is difficult for manufacturers to dispose of these wastes, the food industry may utilize them. The main waste created during beer production is brewer's spent grain (BSG), which makes up around 85% of all residues produced by the brewing industry (Aliyu & Bala, 2011). BSG is created during the mashing stage of the brewing process and eliminated before the boiling stage. Hot trub is another waste product produced during the brewing of beer. Sediments created during the boiling of the wort during the brewing process are referred known as "hot trub." Hot trub particles are thought to range in size from 30 to 80 μm (Matías et al., 2015). Most studies on BSG pretreatment have been focussed for its use for bioenergy production. One study considers silver nitrate pretreatment of BS G prior to pyrolysis (Ashman et al., 2020), while other studies have focussed upon its conversion to biogas (Dudek et al., 2019; Panjičko et al., 2017), and bioethanol (Caetano et al., 2013), The maximum biogas production without the addition of biochar was 92.3 dm^3/kg of dry organic matter (DOM), while it increased to 122 dm^3/kg DOM with the addition of 5% biochar. The addition of a high percentage of biochar (20%–50%) can inhibit biogas production (Dudek et al., 2019).

In the work of Ortiz et al. (2019) it is shown that by gasification of BSG it is possible to achieve a net economic saving of 22%, if the produced gas would be used for heat production in breweries, with the advantage of minimizing BSG as waste.

14.2.3.5 Paper and Pulp Industry Waste

A global review of manufacturing sectors divulged that 17% of the total global waste comes from paper industries (Karak et al., 2012; Sakai et al., 1996). The wastes from pulp and paper mills are both of organic and inorganic types. Most of the biomass generated from pulp and paper mills is organic residues, with black liquor being the major stream. Black liquor being the major waste other potential wastes include sawdust, sludges of different types, wastepaper, and fly ash. The major inorganic type wastes from pulping and after-treatment stage include the dregs, lime mud, and grits. The different wastes are obtained as extracted from different pulp and paper industries are summarized in Fig. 14.7

Within the pulp and paper mills, many processing units produce different kinds of sludges and wastes. There are two types of wastes: wastes with high levels of carbon and wastes with high levels of calcium oxide compounds. The majority of the waste from the paper industries are flammable. They may be recycled via energy recovery. These wastes have a low calorific value due to their high moisture content (up to 60%). The creation of solid recovered fuel is an alternative to waste incineration. The European pulp and paper sector generates 11 million tonnes of trash each year, of which 70% comes from the deinking of recycled paper (Bajpai, 2015). Given the wide diversity in paper product consumption per person from nation to country, global consumption of paper and paper board is estimated to be roughly 55 kg/person per year on average. Despite the fact that digitization is anticipated to have a variety

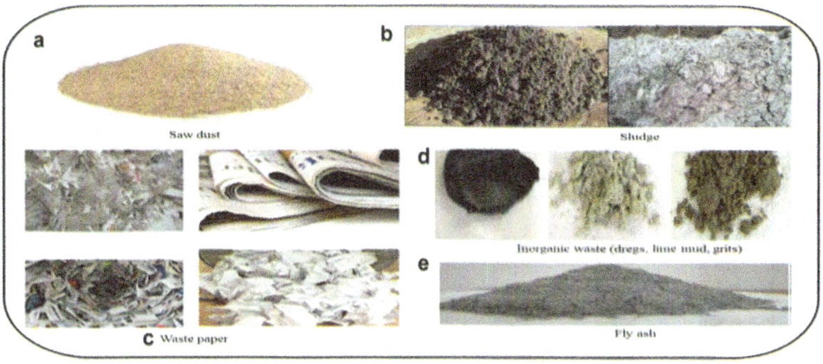

Fig. 14.7 Various waste streams of paper and pulp industry. *Source* Sakai et al. (1996)

of effects on production and consumption, paper and board usage continues to rise. The dominating nations like those in North America, Europe, and Asia as well as low-production and high-consumption nations like those in Africa continue to use a lot of paper.

14.2.3.6 Waste Vegetable Cooking Oil (WCO)

Generally, WCO is generated when various edible vegetable oils are employed to fry food items in households, hotels, restaurants and other catering outlets. Vegetable oils are predominantly triacylglycerols (88–98%) and an important component of a healthy diet that constitutes about 15–20% of total calory intake in most industrialized countries (Orsavova et al., 2015) Global vegetable oil consumption increased from 150 million metric tons (MMT) in 2013/14 to above 200MMT in 2020/21 (Fig. 14.8).

The global WCO market size that was $5.50 billion in 2019 is estimated to reach $8.48 billion by 2027 (Statista, 2021). Another forecast states that the global used cooking oil market is projected to grow from $5.97 billion in 2021 to $10.08 billion in 2028 at a CAGR of 7.76% in forecast period, 2021–2028. Waste vegetable or cooking oils emerged as favourable alternative to pure cooking or vegetable oils for biodiesel production because of their cost effectiveness. Biodiesel feedstocks considerably influences the biodiesel cost of production and as such waste vegetable oils can be obtained at more less than half the retail price of pure vegetable oils (Nomanbhay et al., 2018; Suzihaque et al., 2022). Waste cooking oil are the leftovers from frying food that contains fats from plants or animals. Such vegetable oil-based wastes when exposed to alcohol under the process of transesterification results in monoalkyl esters. It has been estimated that around 5, 0.9, 5.5 and 0.05 to 0.03 million tonnes of plant or animal fat-based vegetable oils are discarded without waste treatment (Azmi, 2019; Kharina et al., 2018; Tsai 2019; Zhang et al., 2020). Apart from transesterification, micro emulsification and pyrolysis are the major chemical processes for biodiesel production from waste vegetable oils.

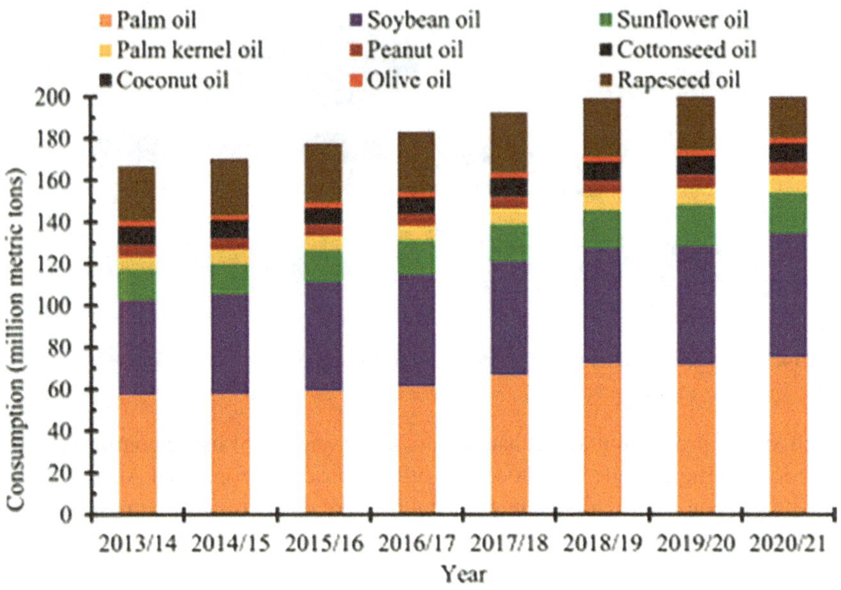

Fig. 14.8 Global vegetable oil consumption data (MMT/Annum). *Source* Orsavova et al. (2015)

Biodiesel production from WCO is environmentally friendly for it recycles waste cooking oil and gives renewable energy with lower pollution. It substitutes some amount of petrochemical oil import and lowers the cost of waste management. Biodiesel production from waste cooking oil has three solutions those are economic, environmental and waste management (Meng et al., 2008; Li et al., 2014). The high cost of biofuels has been one of the factors limiting the full democratization of usage, using WCO as feedstock will substantially make biofuels more affordable, particularly for rural dwellers and local farm owners.

14.2.3.7 Municipal Solid Waste

Ever rising global population and demand for high standard of living has resulted in waste generation at a very rapid rate. Such demand for high standard of living can be reflected in an increasing annual growth rate @3.2–4.5% for municipal solid waste especially in developed countries (Monisha et al., 2022). At such an alarming rate, the global rate of municipal solid waste generation is expected to reach 2.2 billion tons by 2025 (Couto et al., 2016). Municipal solid waste can be a viable energy alternative to fossil fuel-based energy sources. Various fractions of municipal solid waste when exposed to thermochemical processes such as gasification, pyrolysis, torrefaction, and combustion can generate biofuels such as liquid bioethanol, biodiesel, bio-oils and biohydrogen (Chen et al., 2021). Globally, up to 82.9 billion liters of wastepaper-derived cellulosic ethanol can be produced worldwide, replacing 5.36% of gasoline

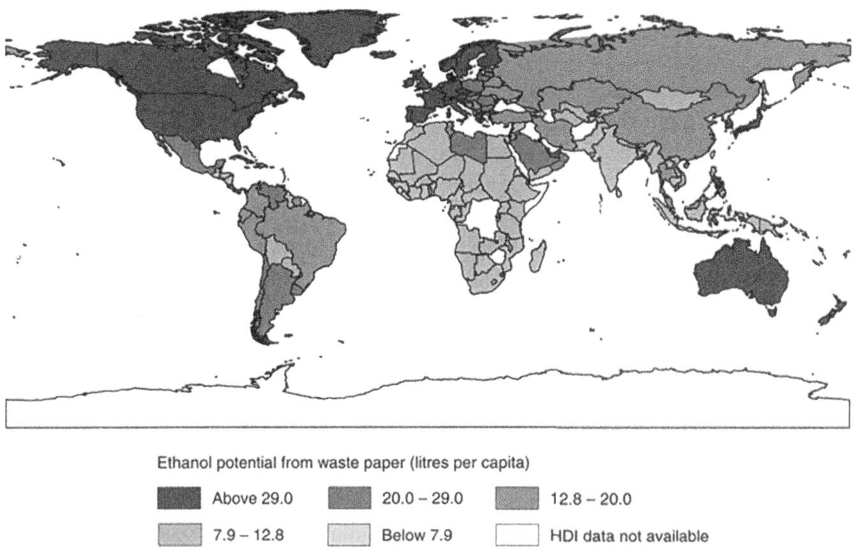

Fig. 14.9 Wastepaper-derived cellulosic ethanol production potential. *Source* Shi et al. (2009)

consumption, with accompanying GHG emissions savings of between 29.2% and 86.1% (Shi et al., 2009) as shown in Fig. 14.9.

Source segregated organic fractions are also a prime source for biogas production. According to International Energy Agency, European Union converts maximum MSW into biogas at 3 million Tons per year, followed by USA at 2 million Tons per year (2022) and China at third place.

14.3 Biochemical or Thermo-Chemical Conversion Routes for Second-Generation Biofuels

Second-generation biofuels are the next step towards net carbon neutrality or even carbon negative in regard to their bearing on CO_2 production. The feedstock used for biofuel production is plant biomass. The biomass, here, refers to non-food raw materials which are rich in lignocellulose and its abundance in the plant makes it a perfect feedstock for biofuel production (Baruah et al., 2018; Gomez et al., 2008; Zabaniotou et al., 2008). So far, planting biomass as a potential source of green energy or biofuel has never been thought of. The basic instinct was to burn the biomass to produce heat and electricity. Since the cell walls are composed of seventy-five% polysaccharides, they have an immense potential to produce liquid biofuels from the pool of sugar (Broda et al., 2022; Pauly & Keegstra, 2008). The composition of ligno-cellulose is a function of varieties of plants, conditions of development, and maturity (Ragauskas et al., 2014). The lignocellulosic-rich biomass may be converted to biofuel either by employing thermochemical processes or biochemical processes (Robak & Balcerek, 2018). The two processes differ in terms of the usage of biological organisms for the conversion of biomass to liquid fuel. The thermochemical process, by its name, uses heat under oxygen-deficient conditions as a catalyst to convert biomass into an intermediate gas inside of a reactor. This intermediate product is transformed into a variety of fuels like methanol, ethanol, hydrogen, and synthetic diesel or aviation fuel. The thermochemical process is based on gasification or pyrolysis. In the gasification process, biomass is reacted with a mixture of air or oxygen in the presence of heat resulting in the production of one of the two gases based on the composition of the component gas used *i.e.*, producer gas and syn gas (Foust et al., 2009; Glushkov et al., 2021). Both these products can be further transformed into various fuels. Pyrolysis is the simple degradation of plant biomass in the absence of oxygen using heat energy ranging from 550–1300 K, which can be either slow, fast, or flash (Demirbas, 2004; Glushkov et al., 2021; Shafizadeh, 1982). Depending on the type of product required, any one of the processes can be utilized to get charcoal or bio-crude oil, respectively. The difference between the three lies in the rate of heating the biomass.

Biochemical processes utilize microbial populations to catalyze the conversion of biomass into biofuel with greater conversion efficiency. This is primarily due to its ability to function under low-temperature conditions in contrast to thermochemical processes and for producing a specific product which is entirely based on the type of organism used to catalyze the conversion (Achinas & Euverink, 2016; Sims et al., 2010). The biomass, hence, can be transformed into a variety of intermediates through the screening of enzymes or microorganisms. The process involves pre-treatment of lignocellulosic material (biomass), enzymatic hydrolysis, fermentation of hydrolyzed products by the activity of specific strains of microorganisms, and distillation of biofuel/bioethanol. The processes ensure that the biomass releases cellulose and hemicellulose for the hydrolytic enzymes to act on them and break them into monomers or simple sugars (Kang et al., 2014; Sarkar et al., 2012). Microorganisms utilize these sugars as their energy source and release bioethanol/biofuel as their byproduct. Distillation is done to separate the biofuel by fermentation from the rest of the products.

14.3.1 Pre-treatment

The purpose of pre-treatment of the lignocellulosic biomass is to separate cellulose and hemicellulose from the lignin which protects them against the hydrolytic action of the enzymes. Lignin is not a desirable component of plant biomass because it is difficult to degrade. It surrounds cellulose and hemicellulose. Chemically, cellulose is a polymer of D-glucose linked by β-1,4-glycosidic linkages whereas hemicellulose, unlike cellulose, comprises a variety of carbohydrate monomers like arabinose, mannose, galactose, glucose, and xylose linked together (Battaglia et al., 2011; Ruel et al., 2012). This makes the plant biomass an ideal feedstock for biofuel production. The pre-treatment also aims to increase the surface area of plant biomass for the enzymes to act more easily (Mosier et al., 2005). It can be achieved either through physical, chemical, or biological treatments. The mechanical treatment is used to reduce the size of the biomass and involves breaking of cell wall crystallinity. This can be achieved through grinding, milling, irradiation of electron beams, etc. (Brodeur et al., 2011). Chemical treatment includes treatment with bases regarded as alkaline pre-treatment, acids (acid pre-treatment), oxygen as an oxidizer (wet oxidation), etc. The physical method is an energy-intensive process while the chemical method produces harmful gases. The cleanest method for pre-treatment biomass would be biological. The process is slow in comparison to its predecessors but is environmentally friendly. The fungi can digest/break lignin using the enzyme called lignin peroxidases belonging to the family of oxidoreductases (Sindhu et al., 2016). With a wide variety of methods available to pretreat, the composition of plant biomass is used as a determinant for the selection of the most suitable method. Plant material with low levels of lignin and high in sugars are not suited to be pretreated with thermal degradation because of the high susceptibility of pentoses to degradation. Steam explosion, one of the physical processes, is recommended for pretreating straws. This method separates fibers through explosive decompression under high pressure and temperatures. It uses less energy than other physical methods of pretreating biomass.

Yu et al. (2018) pretreated corncob using reagents such as H_2O_2, and $FeCl_3$ and were able to recover 94% of the carbohydrates as soluble monosaccharides and 87% of solid lignin was recovered as the filter residue. Upon enzymatic hydrolysis and subsequent fermentation, they were able to yield 79% of the theoretical ethanol yield. Kim (2018) was able to yield 33.8 ± 0.5 g/L ethanol from the alkali-thermal pretreatment method. Sodium hydroxide in the range of 0.5 M to 2.0 M was used in this method and resulted in delignification up to 55.4–56.9%. Singh et al. (2014) experimented with three different pretreatment methods and were able to obtain maximum lignin loss and holocellulose gain with 5% NaOH (alkali), 5% Na–chlorite, and 2% H_2O_2 (oxidative) in rice and wheat straw. Pretreatment using alkali sodium hydroxide (NaOH) at low-temperature Hoşgün et al. (2017) was able to remove 41.18% of lignin from hazelnut shells.

14.3.2 Enzymatic Hydrolysis

The pretreatment is followed by enzymatic hydrolysis which ensures that the polysaccharides, like cellulose and hemicellulose, present in the plant biomass are converted into monosaccharides or fermentable sugars which are further transformed into ethanol during fermentation (Robak & Balcerek, 2018). Based on the pretreatment method employed and the type of material/biomass the concentration of polysaccharides would vary in the solution (Choudhary et al., 2017). Cellulose and hemicellulose undergo hydrolysis by an enzyme called glycosyl-hydrolases. The cellulase enzyme breaks cellulose into cellobiose and then breaks it into glucose. The enzymes are derived from microorganisms such as *Aspergillus niger*, *Aspergillus nidulans*,

Trichoderma reesei, Thermomonospora spp., Cellulomonas spp., Bacillus spp., etc. (Battaglia et al., 2011). In comparison to the more traditional acid hydrolysis, enzymatic hydrolysis needs less energy, is less toxic, has low utility cost (can reuse the organisms), has no inhibitory by-products, and is environment friendly (Sims et al., 2010).

In a study by Pimentel et al. (2021), the effect of enzymatic cocktails, one being 43% *Aspergillus spp.* LMI03 enzyme extract and 57% *T. reesei* QM9414, and the other being 50% *Aspergillus spp.* LMI03 enzyme extract, 25% of *P. citrinum* LMI01, and 25% *T. reesei*, on the lignocellulosic biomass, namely cellulose pulp and cassava peel was assessed. The hydrolytic efficiency with cellulose pulp was 93% with an enzymatic cocktail consisting of 43% *Aspergillus spp.* LMI03 enzyme extract and 57% *T. reesei* QM9414. On the other hand, an efficiency of 78% was observed with a cocktail comprising 50% *Aspergillus spp.* LMI03 enzyme extract, 25% of *P. citrinum* LMI01, and 25% *T. reesei.*

14.3.3 Fermentation

The basic objective of the fermentation process is to efficiently convert sugars released by hydrolysis into ethanol with the help of microorganisms, such as yeast. Depending on the type of substrate, the fermenting organism can vary and so does the efficiency of the conversion. Theoretically, for every gram of hexose sugar consumed a maximum of 0.51 g ethanol can be produced (Gombert & Maris, 2015). The process leads to stressful conditions, such as accumulation of ethanol, decrease in pH, higher microbial population, etc. For fermentation, a pH ranging from 6.5–7.5 is desirable (Hamelinck et al., 2005). For hexose sugars, such as glucose, sucrose, and fructose, microorganisms preferred are *Saccharomyces cerevisiae, Zygomonas spp.,* etc. whereas for pentose sugars microbial organisms involved in the conversion are *Candida shehatae, Pichia stipites,* and *Pachysolen tannophilus* (Mosier et al., 2005). During the fermentation process, it is desirable to have a lower population resulting in higher conversion efficiency. Whenever ATP production increases, the energy is diverted for the growth and development of the microbial population which reduces the conversion efficiency of the biofuel/bioethanol.

14.4 Environmental Performance and Sustainability of Second-Generation Biofuels

The burning of fossil fuels was identified as one of the major reasons for elevated levels of greenhouse gases in the atmosphere during an international summit held in Kyoto in 1997. Biofuels are seen as an alternative from an environmental and energy security point of view (Jeswani et al., 2020; Karmee & Lin, 2014). However, presently commercialized biofuel technology based on first-generation biofuels has given rise to several environmental controversies. Considering this, life cycle assessment (LCA) studies were carried out globally to monitor the environmental impacts of advanced biofuels, with a hypothesis that advanced biofuels could alter the environment-related issues of present biofuel technology (Gnansounou et al., 2009; Mizik & Gyarmati, 2022). Life cycle assessment studies evaluate the environmental impact of biofuels during their life cycle *i.e.*, cradle-to-grave approach. They hold significant value in estimating whether there is a shift in sustainability aspects of biofuels from one stage of the life cycle to another or to the supply chain (Jeswani et al., 2020). Direct greenhouse gas emissions from waste and residue-based biofuels will be discussed here as per LCA studies across different regions of the world to clarify whether second-generation biofuels are sustainable alternatives to current technology or not.

14.4.1 Perennial Grasses

Lignocellulosic biomass is a renewable energy source available in the form of non-edible grasses, and woody and agricultural residues (Sakthivadivel et al., 2019). Lower greenhouse gas emissions during combustion and sequestration of a large amount of atmospheric carbon dioxide during their growth make biofuels based on grasses carbon neutral. Advanced biofuels based on such feedstocks along with novel technologies have the potential to mitigate climate change (Eisentraut et al., 2011; Næss et al., 2023). Significantly lower greenhouse gas emissions are observed with lignocellulosic or perennial grass-based bioethanol in comparison to fossil fuels or first-generation biofuels (Khan et al., 2016). Wang et al., 2012 studied the emissions from ethanol derived from various feedstocks and observed that corn and sugarcane-based ethanol was able to reduce emissions only to an extent of 19–48% and 40–62%, respectively whereas switchgrass, corn stover, and Miscanthus were able to bring down the emissions by 77–97%, 90–103% and 101–115%, respectively. Greenhouse gas emissions can further be reduced with an enhanced yield of such energy crops under the influence of agronomic and genetic interventions (Liu et al., 2020; Schmer et al., 2008). Apart from reduction in emissions, switchgrass can enhance soil organic carbon storage, especially at a soil depth of 30 cm wherein it is less susceptible to mineralization. Miscanthus-based ethanol was further reiterated to reduce greenhouse gas emissions compared to petrol by Dwivedi et al. (2015), Scown et al. (2012). It was also observed that considering soil carbon sequestration and carbon credits added with the supply of electricity to the biorefinery grid, miscanthus can act as feedstock for net greenhouse gas sequestration. Emissions during cultivation can be reduced by reducing or eliminating nitrogen fertilizer applications, especially in high or medium-nitrogen-status soils. Unfertilized miscanthus cultivation can reduce N_2O emissions by one hundred times when compared to intensive pastureland and by five times when compared to arable lands (Mccalmont et al., 2017; Mi et al., 2018).

14.4.2 *Forest and Pulp Residues*

Forest biomass can act as a key feedstock for biofuel production with its abundance, high biomass yield, and lower greenhouse emissions when compared to 1st generation feedstocks or conventional fossil fuels (Wang et al., 2020) Generally, forest biomass can be classified as industrial Roundwood and fuelwood (Raunikar et al., 2010; Yu et al., 2021). Fuelwood has been harvested conventionally for heat, electricity, or biofuel production (Tan et al., 2015). Forest biomass with high content of sugars is attracting considerable attention for energy generation, especially in the European Union wherein there is an increasing trend of consuming forest residues, wood processing waste, and forest crops for energy generation (Urban et al., 2010; Yu et al., 2021). Forest biomass has a vast potential to curtail carbon dioxide emissions in the atmosphere when processed as fuels. For example, forest biomass-based energy sources have the potential to reduce carbon dioxide emissions in Australia by twenty-five million tons annually (Pour et al., 2018). A significant reduction in emissions of CO, NO_x, and HC was reported by Becerra-Ruiz et al. (2019) when forest biomass-based bioethanol was utilized as an alternative to gasoline in a portable engine generator. Jatropha, another non-edible biodiesel feedstock, has wide potential to replace conventional fuels with minimum emissions of greenhouse gases. Jatropha-based biodiesel was reported to have significantly lower emissions compared to conventional fuel diesel and the cut down on emissions was proportionally increasing with an increase in blending ratio (Whitaker & Heath, 2009).

14.5 Source Segregated MSW Organic Wastes and Crop Residues

LCA studies for waste-based biofuel feedstocks are complex considering the production of low-carbon fuels, avoiding emissions and expenditure from waste management, co-products generated, and recycling of valuable materials from such waste (Meng et al., 2019). Although cultivation phase emissions are altogether avoided, transportation-based emissions must be considered (Soam et al., 2016). Food waste can reduce carbon emissions in ethanol production with the generation of compost as a by-product as well as the elimination of methane emissions during unsustainable disposal of such wastes (Ebner et al., 2014). Apart from methane emissions, food waste disposal in landfills is responsible for the release of greenhouse gases of 113 mt CO_2 equivalence annually (Karmee, 2016). A plethora of food waste is generated annually, which was around 931 million tonnes for the year 2019 (UNEP, 2021). Food wastes are generated from various industries such as the vegetable and fruit industry, beverage industry, dairy industry, and cereal grain industry (Baiano, 2014). However, advanced biofuels are required to fulfill the mandate to reduce at least 60% of emissions compared to gasoline. Researchers across the globe have reported their findings on the environmental impact of several types of food waste. Results on the environmental sustainability of a feedstock however can vary as per the evaluation approach considered. For example, regulatory requirements in the European Union consider an allocation approach wherein emissions are allocated between main products and co-products generated whereas the USA follows a system expansion approach (Konti et al., 2020).

Agricultural residues have the vast potential to cut down the global warming potential of bioethanol compared to other feedstocks such as energy crops or fossil fuels. Reduction in greenhouse gas emissions was attributed to the elimination of fertilizer-based emissions with the use of residues as feedstocks as such emissions are allocated to the economic product of the crop. Lignocellulosic biomass is further supplemented with carbon credits based on heat or electricity generated from lignin that can be supplied to the grid or biochar production which can act as a major route of carbon sequestration in soil (Singh et al., 2016). Such an approach can led to the negative global warming potential of waste-based biofuels. Conversion of crop residues to bioethanol can avoid emissions based on unsustainable disposal or removal of residues from crop fields. Indo-Gangetic plains in India face a severe problem with rice straw burning that leads to the release of various greenhouse gases into the atmosphere as well as the malign impact on soil biological activity. Around 50 mt of rice straw is burned in India annually which is responsible for the release of several greenhouse gases such as NO_2, SO_2, CO, CH_4, and especially 150 mt of CO_2 (Kaur et al., 2022). Rice straw acts as a potential feedstock with a high concentration of cellulose and hemicellulose and a theoretical ethanol yield of around 0.42 l/kg of dry straw (Zhu et al., 2006). Carbon–neutral bioethanol is generated by using rice straw as feedstock being responsible for environmental and economic gains of around US$664 ha^{-1} in India (Bhattacharya et al., 2021). Wheat straw-based bioethanol was found to curtail greenhouse gas emissions to an extent of 45% when stem explosion pre-treatment was given to the wheat straw. Even after taking into consideration the environmental burden of wheat straw removal from crop fields, bioethanol with a steam explosion pathway was observed to be environmentally favorable when compared to fossil fuels (Wang et al., 2012). The carbon footprint in the European Union was found to be significantly lower with advanced bioethanol based on wheat and maize residues when compared to fossil fuels (Holmatov et al., 2022). However, emission savings depend on ethanol concentration blends i.e., higher savings for higher blending ratios and vice-versa (Borrion et al., 2012). Wheat straw-based bioethanol blends of 85% were observed to cut down emissions to an extent of 73% whereas a blend of 15% was able to reduce emissions by 13% as per a well-to-wheel study conducted by Borrion et al., 2012. Sugarcane industry by-products such as molasses can be utilized as potential bioethanol feedstock and have the potential to reduce greenhouse gas emissions by 43% when compared to fossil fuels. Further, the environmental sustainability of such fuels can be strengthened with the generation of co-products such as livestock feed and yeast. Similarly, oil palm-based industrial effluents can bring down emissions up to 71% and can support the grid with the generation of heat or electricity (Searle, 2017).

Fruit and vegetable-based industries generate waste annually in the form of solid waste or as effluents. Such wastes are fabulously rich in sugars and low in lignin, therefore, a preferable feedstock for bioethanol production (Ruynal et al., 1998). On the other hand, injudicious disposal of such wastes in landfills or water bodies creates severe environmental issues with the release of greenhouse gases or high biological or chemical oxygen demands in aquatic ecosystems (Das et al., 2016). For example, whey a by-product of the dairy industry can be subjected to two alternatives i.e., can be disposed of in water bodies leading to environmental problems, or can be exposed to lactose metabolizing microbes to produce bioethanol (Mann & Sooch, 2020). Pourbafrani et al. (2013) carried out the LCA studies for citrus-based bioethanol and observed that citrus-based bioethanol has the potential to reduce greenhouse gas emissions to an extent of 134% when compared to gasoline. Similarly, banana rachis-based bioethanol had a lower environmental burden than conventional fossil fuels when used as a transportation fuel (Guerrero & Munoz, 2018).

14.6 Engine Combustion-Based Pollutant Emissions from Second Generation Biofuels

Biofuels release a comparable amount of greenhouse gases into the atmosphere, although the amount of carbon dioxide absorbed, sequestered in soil during its cultivation or avoided due to its mismanagement favors the biofuels (Aljaafari et al., 2022). A reduction to an extent of 45, 47 and 6% can be achieved for particulates, hydrocarbons, and carbon monoxide with burning of biofuels compared to fossil fuel-based energy sources (Santhoshkumar et al., 2019).

However, biofuels such as biodiesel are responsible for release of toxic materials such as sulphur, several aromatic compounds, carcinogens, potential mutagens having malign impact on human and environmental health (Woźniak-Karczewska et al., 2019; Bitire & Jen, 2022). Apart from toxic materials, biofuels may elevate levels of nitrogenous oxides and ozone in the atmosphere (Abed et al., 2019). Solid phase particles released from biofuels have higher toxicity although lower mutagenicity which can be attributed to lower level of polycyclic aromatic compounds and higher levels of carbonyl compounds released into atmosphere after biofuel combustion, respectively (Aljaafari et al., 2022).

14.6.1 Biofuels and Carbon Sequestration

Combining bioenergy production with carbon capture and sequestration can also help to achieve net negative emissions as carbon stored by photosynthesizing biomass growth is sequestered rather than released to the atmosphere (IEA, 2011). The concept was first developed by Obersteiner et al. (2001) as a backstop climate risk measure, and by Keith (2001) as a potential mitigation tool. Since then, biomass energy with carbon capture and sequestration (BECCS) has come to be viewed as a key carbon dioxide removal approach to keep global atmospheric CO_2 concentrations below 500 ppm and avoid catastrophic climate change. BECCS is used by integrated assessment models (IAMs) because its cost is low relative to other low carbon technologies and because the modules to represent other carbon dioxide removal technologies are undeveloped (e.g., direct air capture and soil carbon management). BECCS typically refers to the integration of trees and crops that extract CO_2 from the atmosphere as they grow, the use of this biomass in power plants, and the application of carbon capture and sequestration via CO_2 injection into geological formations.

14.7 Conclusion

Energy insecurity with declining fossil fuel reservoirs and the issue of climate change has led to the commercialization of first-generation biofuels to avoid dependence on non-renewable energy sources. However, better insights into the global warming potential of first-generation biofuels with life cycle assessment studies have exposed the malign impact of such energy sources on climate. A need for alternative fuels with minimum or no land use changes, or environmentally benign production routes has risen to achieve the target of sustainable energy sources and a carbon–neutral economy. Second-generation biofuels derived from organic wastes or crop residues or non-edible lignocellulosic biomass especially from marginal lands ensure simultaneous attainment of circular economy and climate mitigation objectives at global scale. Lignocellulosic biomass and organic wastes are available in abundance to meet partial requirements of energy demands globally. Life cycle assessment studies for such feedstocks have revealed their potential their better environmental sustainability. However, fuel productivity per unit feedstock is still low, therefore, there is a need for mature biochemical or thermochemical conversion routes to achieve higher productivity. There exists a great potential for technology development using multi-feed-based systems, which can prove efficient, economical and sustainable with future growth requirements. Future development may characterize environmental performance; circular aspect of materiality and climate neutrality as the key to unlocking the wider potential of such feedstocks to fulfill energy demands in a sustainable manner.

Funding No funding was availed from any agency for this publication.

Competing Interests There are no competing interests associated with this paper.

References

Abed, K. A., Gad, M. S., El Morsi, A. K., Sayed, M. M., & Elyazeed, S. A. (2019). Effect of biodiesel fuels on diesel engine emissions. *Egyptian Journal of Petroleum, 28*(2), 183–188.

Abrini, J., Naveau, H., & Nyns, E. J. (1994). Clostridium autoethanogenum, sp. nov., an anaerobic bacterium that produces ethanol from carbon monoxide. *Archives of Microbiology, 161*, 345–351.

Achinas, S., & Euverink, G. J. W. (2016). Consolidated briefing of biochemical ethanol production from lignocellulosic biomass. *Electronic Journal of Biotechnology, 23*, 44–53.

Akashdeep, C & Thompson, C.R. (2022) India's biomethane generation potential from wastes and the corresponding greenhouse gas emissions abatement possibilities under three end use scenarios: electricity generation, cooking, and road transport applications. *Sustainable Energy Fuels, 7*, 209–241

Aliyu, S., & Bala, M. (2011). Brewer's spent grain: A review of its potentials and applications. *African Journal of Biotechnology, 10*(3), 324–331.

Aljaafari, A., Fattah, I. M. R., Jahirul, M. I., Gu, Y., Mahlia, T. M. I., Islam, M. A., & Islam, M. S. (2022). Biodiesel Emissions: A State-of-the-Art Review on Health and Environmental Impacts. *Energies, 15*(18), 6854.

Anonymous (2018). https://pib.gov.in

Anonymous (2022). https://www.worldometers.info/world-population

Ashman, C. H., Gao, L., & Goldfarb, J. L. (2020). Silver Nitrate In Situ Upgrades Pyrolysis Biofuels from Brewer's Spent Grain via Biotemplating. *Journal of Analytical and Applied Pyrolysis, 146*, 104729.

Assandri, D., Bagagiolo, G., Cavallo, E., & Pampuro, N. (2022). Replacing Agricultural Diesel Fuel with Biomethane from Agricultural Waste: Assessment of Biomass Availability and Potential Energy Supply in Piedmont (North-West Italy). *Agronomy, 12*(12), 2996.

Ayodele, B. V., Alsaffar, M. A., & Mustapa, S. I. (2020). An overview of integration opportunities for sustainable bioethanol production from first-and second-generation sugar-based feedstocks. *Journal of Cleaner Production, 245*, Article 118857.

Azmi, M. (2019). Malaysians use too much cooking oil, The Moles. https://www.mole.my/malaysians-use-too-much-cooking-oil/

Baiano, A. (2014). Recovery of biomolecules from food wastes—A review. *Molecules, 19*(9), 14821–14842.

Bajpai, P. (2015). *Pulp and paper industry: Chemicals*. Elsevier.

Baruah, J., Nath, B. K., Sharma, R., Kumar, S., Deka, R. C., Baruah, D. C., & Kalita, E. (2018). Contemporary trends in the pretreatment of lignocellulosic biomass for value-added products. *Frontiers in Energy Research, 6*, 141.

Battaglia, E., Hansen, S. F., Leendertse, A., Madrid, S., Mulder, H., Nikolaev, I., & de Vries, R. P. (2011). Regulation of pentose utilisation by AraR, but not XlnR, differs in Aspergillus nidulans and Aspergillus niger. *Applied Microbiology and Biotechnology, 91*, 387–397.

Becerra-Ruiz, J. D., Gonzalez-Huerta, R. G., Gracida, J., Amaro-Reyes, A., & Macias-Bobadilla, G. (2019). Using green-hydrogen and bioethanol fuels in internal combustion engines to reduce emissions. *International Journal of Hydrogen Energy, 44*(24), 12324–12332.

Bhattacharyya, P., Bisen, J., Bhaduri, D., Priyadarsini, S., Munda, S., Chakraborti, M., et al. (2021). Turn the wheel from waste to wealth: Economic and environmental gain of sustainable rice straw management practices over field burning in reference to India. *Science of the Total Environment, 775*, Article 145896.

Bitire, S. O., & Jen, T. C. (2022). Performance and emission analysis of a CI engine fueled with parsley biodiesel–diesel blend. *Materials for Renewable and Sustainable Energy, 11*(2), 143–153.

Borrion, A. L., McManus, M. C., & Hammond, G. P. (2012). Environmental life cycle assessment of bioethanol production from wheat straw. *Biomass and Bioenergy, 47*, 9–19.

Broda, M., Yelle, D. J., & Serwańska, K. (2022). Bioethanol Production from Lignocellulosic Biomass—Challenges and Solutions. *Molecules, 27*(24), 8717.

Brodeur, G., Yau, E., Badal, K., Collier, J., Ramachandran, K. B., & Ramakrishnan, S. (2011). Chemical and physicochemical pretreatment of lignocellulosic biomass: a review. *Enzyme research, 2011.*

Caetano, N. S., Moura, R. F., Meireles, S., Mendes, A. M., & Mata, T. M. (2013). Bioethanol from Brewer's Spent Grains: Acid Pretreatment Optimization. *Chemical Engineering Transactions, 35*, 1021–1026.

Chen, H., Li, T., & Wang, Q. (2019). Ten years of algal biofuel and bioproducts: Gains and pains. *Planta, 249*, 195–219.

Chen, W. H., Lin, B. J., Lin, Y. Y., Chu, Y. S., Ubando, A. T., Show, P. L., et al. (2021). Progress in biomass torrefaction: Principles, applications, and challenges. *Progress in Energy and Combustion Science, 82*, Article 100887.

Choudhary, J., Singh, S., & Nain, L. (2017). Bioprospecting thermotolerant ethanologenic yeasts for simultaneous saccharification and fermentation from diverse environments. *Journal of Bioscience and Bioengineering, 123*(3), 342–346.

Consumption of vegetable oils worldwide from 2013/14 to 2020/2021, by oil type (in million metric tons), Available online, https://www.statista.com/statistics/263937/vegetable-oils-global-consumption/

Couto, N., Silva, V., & Rouboa, A. (2016). Municipal solid waste gasification in semi-industrial conditions using air-CO2 mixtures. *Energy, 104*, 42–52.

Das, M., Raychaudhuri, A., & Ghosh, S. K. (2016). Supply chain of bioethanol production from whey: A review. *Procedia Environmental Sciences, 35*, 833–846.

de Carvalho Freitas, E. S., Xavier, L. H., Oliveira, L. B., & Guarieiro, L. L. N. (2022). System dynamics applied to second generation biofuel in Brazil: A circular economy approach. *Sustainable Energy Technologies and Assessments, 52*, 102288.

Demirbas, A. (2004). Current technologies for the thermo-conversion of biomass into fuels and chemicals. *Energy Sources, 26*(8), 715–730.

Demirbas, M. F. (2009). Biorefineries for biofuel upgrading: A critical review. *Applied Energy, 86*, S151–S161.

Devarajan, Y., Munuswamy, D. B., & Mahalingam, A. (2017). Performance, combustion, and emission analysis on the effect of ferrofluid on neat biodiesel. *Process Safety and Environmental Protection, 111*, 283–291.

Devi, A., Bajar, S., Kour, H., Kothari, R., Pant, D., & Singh, A. (2022). Lignocellulosic biomass valorization for bioethanol production: a circular bioeconomy approach. *Bioenergy Research*, 1–22.

Dudek, M., Swiechowski, K., Manczarski, P., Koziel, J. A., & Białowiec, A. (2019). The Effect of Biochar Addition on the Biogas Production Kinetics from the Anaerobic Digestion of Brewers' Spent Grain. *Energies, 12*, 1518.

Dwivedi, P., Wang, W., Hudiburg, T., Jaiswal, D., Parton, W., Long, S., et al. (2015). Cost of abating greenhouse gas emissions with cellulosic ethanol. *Environmental Science & Technology, 49*(4), 2512–2522.

Ebner, J., Babbitt, C., Winer, M., Hilton, B., & Williamson, A. (2014). Life cycle greenhouse gas (GHG) impacts a novel process for converting food waste to ethanol and co-products. *Applied Energy, 130*, 86–93.

EISA (Energy Independence and Security Act of 2007) (2007). Library of Congress.

Eisentraut, A., Brown, A., & Fulton, L. (2011). *Technology roadmap: Biofuels for transport.* International Energy Agency. https://doi.org/10.1002/bbb.330

Foust, T. D., Aden, A., Dutta, A., & Phillips, S. (2009). An economic and environmental comparison of a biochemical and a thermochemical lignocellulosic ethanol conversion process. *Cellulose, 16*(4), 547–565.

Freitas, E. S. D. C., Guarieiro, L. L. N., da Silva, M. V. I., Amparo, K. K. D. S., Machado, B. A. S., Guerreiro, E. T. D. A., & Torres, E. A. (2022). Emission and performance evaluation of a diesel engine using addition of ethanol to diesel/biodiesel fuel blend. *Energies, 15*(9), 2988.

Glushkov, D., Nyashina, G., Shvets, A., Pereira, A., & Ramanathan, A. (2021). Current status of the pyrolysis and gasification mechanism of biomass. *Energies, 14*(22), 7541.

Gnansounou, E., Dauriat, A., Villegas, J., & Panichelli, L. (2009). Life cycle assessment of biofuels: Energy and greenhouse gas balances. *Bioresource Technology, 100*(21), 4919–4930.

Gombert, A. K., & van Maris, A. J. (2015). Improving conversion yield of fermentable sugars into fuel ethanol in first generation yeast-based production processes. *Current Opinion in Biotechnology, 33*, 81–86.

Gomez, L. D., Steele-King, C. G., & McQueen-Mason, S. J. (2008). Sustainable liquid biofuels from biomass: The writing's on the walls. *New Phytologist, 178*(3), 473–485.

Grady, J. L., & Chen, G. J. (1998). *U.S. Patent No. 5,821,111*. U.S. Patent and Trademark Office.

Guerrero, A. B., & Muñoz, E. (2018). Life cycle assessment of second-generation ethanol derived from banana agricultural waste: Environmental impacts and energy balance. *Journal of Cleaner Production, 174*, 710–717.

Hamelinck, C. N., Van Hooijdonk, G., & Faaij, A. P. (2005). Ethanol from lignocellulosic biomass: Techno-economic performance in short-, middle-and long-term. *Biomass and Bioenergy, 28*(4), 384–410.

Holmatov, B., Hoekstra, A. Y., & Krol, M. S. (2022). EU's bioethanol potential from wheat straw and maize stover and the environmental footprint of residue-based bioethanol. *Mitigation and Adaptation Strategies for Global Change, 27*(1), 1–18.

Hoşgün, E. Z., Berikten, D., Kıvanç, M., & Bozan, B. (2017). Ethanol production from hazelnut shells through enzymatic saccharification and fermentation by low-temperature alkali pretreatment. *Fuel, 196*, 280–287.

Hosseini, S. E., & Wahid, M. A. (2015). Pollutant in palm oil production process. *Journal of the Air & Waste Management Association, 65*(7), 773–781.

IEA (2011). Combining Bioenergy with CCS: Reporting and Accounting for Negative Emissions under UNFCCC and the Kyoto Protocol. OECD/IEA.

Jeswani, H. K., Chilvers, A., & Azapagic, A. (2020). Environmental sustainability of biofuels: A review. *Proceedings of the Royal Society A, 476*(2243), 20200351.

Joshi, A. (2023). Market report India. World Biogas Association. https://www.worldbiogasassociation.org/wp-content/uploads/2023/12/FINAL_COUNTRY_REPORT.with-authors.pdf. Accessed 14 April 2026

Kang, Q., Appels, L., Tan, T., & Dewil, R. (2014). Bioethanol from lignocellulosic biomass: current findings determine research priorities. *The Scientific World Journal, 2014*.

Karak, T., Bhagat, R., & Bhattacharyya, P. (2012). Municipal solid waste generation, composition, and management: The world scenario. *Critical Reviews in Environment Science and Technology, 42*, 1509–1630.

Karmee, S. K. (2016). Liquid biofuels from food waste: Current trends, prospects and limitation. *Renewable and Sustainable Energy Reviews, 53*, 945–953.

Karmee, S. K., & Lin, C. S. K. (2014). Valorisation of food waste to biofuel: Current trends and technological challenges. *Sustainable Chemical Processes, 2*(1), 1–4.

Kaur, M., Malik, D. P., Malhi, G. S., Sardana, V., Bolan, N. S., Lal, R., & Siddique, K. H. (2022). Rice residue management in the Indo-Gangetic Plains for climate and food security: A review. *Agronomy for Sustainable Development, 42*(5), 1–25.

Keith, D. W. (2001). Sinks, energy crops and land use: Coherent climate policy demands an integrated analysis of biomass—An editorial comment. *Climatic Change, 49*(1–2), 1–10. https://doi.org/10.1023/A:1010617015484

Khan, A. A., Usmani, R. A., & Anis Anwar, M. (2016). Cellulosic ethanol (energy security and climate change mitigation): A review. *Journal of Biofuels, 7*(2), 71–78.

Khan, S. U., Al-Khaled, K., Aldabesh, A., Awais, M., & Tlili, I. (2021). Bioconvection flow in accelerated couple stress nanoparticles with activation energy: Bio-fuel applications. *Scientific Reports, 11*(1), 1–15.

Kharina, A., Searle, S., Rachmadini, D., Kurniawan, A. A., & Prionggo, A. (2018). The potential economic, health and greenhouse gas benefits of incorporating used cooking oil into Indonesia's biodiesel. *White Paper, 26*, 2018.

Kim, S. (2018). Enhancing bioethanol productivity using alkali-pretreated empty palm fruit bunch fiber hydrolysate. *BioMed research international, 2018.*

Konti, A., Kekos, D., & Mamma, D. (2020). Life cycle analysis of the bioethanol production from food waste—A review. *Energies, 13*(19), 5206.

Korhonen, J., Honkasalo, A., & Seppälä, J. (2018). Circular economy: The concept and its limitations. *Ecological Economics, 143*, 37–46.

Kowalski, Z., Kulczycka, J., Verhé, R., Desender, L., De Clercq, G., Makara, A., et al. (2022). Second-generation biofuel production from the organic fraction of municipal solid waste.

Li, M., Zheng, Y., Chen, Y., & Zhu, X. (2014). Biodiesel production from waste cooking oil using a heterogeneous catalyst from pyrolyzed rice husk. *Bioresource Technology, 154*, 345–348.

Liu, X., Kwon, H., Northrup, D., & Wang, M. (2020). Shifting agricultural practices to produce sustainable, low carbon intensity feedstocks for biofuel production. *Environmental Research Letters, 15*(8), 084014.

Love, J. (2022). Microbial pathways for advanced biofuel production. *Biochemical Society Transactions, 50*(2), 987–1001.

Łukajtis, R., Hołowacz, I., Kucharska, K., Glinka, M., Rybarczyk, P., Przyjazny, A., & Kamiński, M. (2018). Hydrogen production from biomass using dark fermentation. *Renewable and Sustainable Energy Reviews, 91*, 665–694.

Maness, P. C., & Weaver, P. F. (2002). Hydrogen production from a carbon-monoxide oxidation pathway in Rubrivivax gelatinosus. *International Journal of Hydrogen Energy, 27*(11–12), 1407–1411.

Mann, M. K., & Sooch, B. S. (2020). Emerging trends in food industry waste valorization for bioethanol production. *Biorefineries: A Step Towards Renewable and Clean Energy*, 57–92.

Matías, J., Encinar, J. M., González, J., & González, J. F. (2015). Optimisation of ethanol fermentation of Jerusalem artichoke tuber juice using simple technology for a decentralised and sustainable ethanol production. *Energy for Sustainable Development, 25*, 34–39.

McCalmont, J. P., Hastings, A., McNamara, N. P., Richter, G. M., Robson, P., Donnison, I. S., & Clifton-Brown, J. (2017). Environmental costs and benefits of growing Miscanthus for bioenergy in the UK. *GCB Bioenergy, 9*(3), 489–507.

Mehmood, M.A., Ibrahim, M., Rashid, U., Nawaz, M., Ali, S., Hussain, A., & Gull, M. (2017). Biomass production for bioenergy using marginal lands. *Sustainable Production and Consumption*, 3–21.

Meng, F., Sun, X., Liu, J., McKechnie, J., & Yang, J. (2019). Life cycle energy use and greenhouse gas emission of lightweight vehicle–a body-in-white design. *Journal of Cleaner Production, 220*, 1–8.

Meng, X., Chen, G. & Wang, Y. (2008). Biodiesel production from waste cooking oil via alkali catalyst and its engine test. *Fuel Processing Technology, 9*, 5–11.

Meramo, S., Gonzalez-Quiroga, A., & Gonzalez-Delgado, A. (2022). Technical, Environmental, and Process Safety Assessment of Acetone-Butanol-Ethanol Fermentation of Cassava Residues. *Sustainability, 14*(23), 16185.

Mi, J., Liu, W., Zhao, X., Kang, L., Lin, C., Yan, J., & Sang, T. (2018). N_2O and CH_4 emission from Miscanthus energy crop fields in the infertile Loess Plateau of China. *Biotechnology for Biofuels, 11*(1), 1–14.

Mizik, T., & Gyarmati, G. (2022). Three pillars of advanced biofuels' sustainability. *Fuels, 3*(4), 607–626.

Monisha, R. S., Mani, R. L., Sivaprakash, B., Rajamohan, N., & Vo, D. V. N. (2022). Green remediation of pharmaceutical wastes using biochar: a review. *Environmental Chemistry Letters*, 1–24.

Mosier, N., Wyman, C., Dale, B., Elander, R., Lee, Y. Y., Holtzapple, M., & Ladisch, M. (2005). Features of promising technologies for pretreatment of lignocellulosic biomass. *Bioresource Technology, 96*(6), 673–686.

Næss, J. S., Hu, X., Gvein, M. H., Iordan, C. M., Cavalett, O., Dorber, M., et al. (2023). Climate change mitigation potentials of biofuels produced from perennial crops and natural regrowth on abandoned and degraded cropland in Nordic countries. *Journal of Environmental Management, 325*, 116474.

Naik, S. N., Goud, V. V., Rout, P. K., & Dalai, A. K. (2010). Production of first- and second-generation biofuels: A comprehensive review. *Renewable and Sustainable Energy Reviews, 14*(2), 578–597.

Neupane, D. (2022). Biofuels from Renewable Sources, a Potential Option for Biodiesel Production. *Bioengineering, 10*(1), 29.

Ng, W. P. Q., Lam, H. L., Ng, F. Y., Kamal, M., & Lim, J. H. E. (2012). Waste-to-wealth: Green potential from palm biomass in Malaysia. *Journal of Cleaner Production, 34*, 57–65.

Niall, M. (2021). Which countries produce the most palm oil? https://www.forbes.com/sites/niallmccarthy/2020/10/02/which-countries-produce-the-most-palm-oil-infographic/?sh=64b4de041e42. Accessed 22 December 2021.

Nomanbhay, S., Hussein, R., & Ong, M. Y. (2018). Sustainability of biodiesel production in Malaysia by production of bio-oil from crude glycerol using microwave pyrolysis: A review. *Green Chemistry Letters and Reviews, 11*(2), 135–157.

Obersteiner, M., Azar, C., Kauppi, P., Mollersten, K., Moreira, J., Nilsson, S., Read, P., Riahi, K., Schlamadinger, B., Yamagata, Y., Yan, J., & van Ypersele, J. P. (2001). Managing climate risk. *Science, 294*(5543), 786–787. https://doi.org/10.1126/science.294.5543.786b

Orsavova, J., Misurcova, L., Ambrozova, J. V., Vicha, R., Mlcek, J. (2015). Fatty acids composition of vegetable oils and its contribution to dietary energy intake and dependence of cardiovascular mortality on dietary intake of fatty acids. *International Journal of Molecular Science, 16*, 12871–12890.

Ortiz, I., Torreiro, Y., Molina, G., Marono, M., & Sánchez, J. M. (2019). A feasible application of circular economy: Spent grain energy recovery in the beer industry. *Waste Biomass Valor, 10*, 3809–3819.

Pandey, A., et al. (2021). Health and economic impact of air pollution in the states of India: The global burden of disease study 2019. *Lancet Planet Health, 5*, e25–e38.

Panjičko, M., Zupančič, G. D., Fanedl, L., Logar, R. M., Tišma, M., & Zelic, B. (2017). Biogas production from brewery spent grain as a mono-substrate in a two-stage process composed of solid-state anaerobic digestion and granular biomass reactors. *Journal of Cleaner Production, 166*, 519–529.

Pauly, M., & Keegstra, K. (2008). Cell-wall carbohydrates and their modification as a resource for biofuels. *The Plant Journal, 54*(4), 559–568.

Piccolo, C., & Bezzo, F. (2009). A techno-economic comparison between two technologies for bioethanol production from lignocellulose. *Biomass and Bioenergy, 33*(3), 478–491.

Pimentel, P. S. S. R., de Oliveira, J. B., Astolfi-Filho, S., & Pereira, N., Jr. (2021). Enzymatic hydrolysis of lignocellulosic biomass using an optimized enzymatic cocktail prepared from secretomes of filamentous fungi isolated from Amazonian biodiversity. *Applied Biochemistry and Biotechnology, 193*(12), 3915–3935.

Pour, N., Webley, P. A., & Cook, P. J. (2018). Opportunities for application of BECCS in the Australian power sector. *Applied Energy, 224*, 615–635.

Pourbafrani, M., McKechnie, J., MacLean, H. L., & Saville, B. A. (2013). Life cycle greenhouse gas impacts of ethanol, biomethane and limonene production from citrus waste. *Environmental Research Letters, 8*(1), 015007.

Prasad, S., & Dhanya, M. S. (2011). *Air quality and biofuels, environmental impact of biofuels*. M. A. D. S. Bernardes (ed.). InTech. ISBN: 978-953-307-479-5. https://doi.org/10.5772/17889

Purohit, P., & Dhar, S. (2015). Biofuel roadmap for India—(November 2015)"; UNEP (United Nations Environment Program) DTU (Denmark Technical University) Partnership.

Ragauskas, A., Beckham, G., Biddy, M., Chandra, R., Chen, F., Davis, M., & Keller, M. (2014). GA Tuskan ja CE Wyman. *Science, 344*, 1246843.

Rai, A. K., Al Makishah, N. H., Wen, Z., Gupta, G., Pandit, S., & Prasad, R. (2022). Recent developments in lignocellulosic biofuels, a renewable source of bioenergy. *Fermentation, 8*(4), 161.

Raunikar, R., Buongiorno, J., Turner, J. A., & Zhu, S. (2010). Global outlook for wood and forests with the bioenergy demand implied by scenarios of the Intergovernmental Panel on Climate Change. *Forest Policy and Economics, 12*(1), 48–56.

Resch, M. G., & Emme, B. (2022). Using incremental changes to convert lignocellulosic feedstocks to cellulosic ethanol. *Front. Energy Res., 10*, 1–6. https://doi.org/10.3389/fenrg.2022.835714

Robak, K., & Balcerek, M. (2018). Review of second-generation bioethanol production from residual biomass. *Food Technology and Biotechnology, 56*(2), 174.

Rokicki, T., Koszela, G., Ochnio, L., Perkowska, A., Borawski, P., Beldycka-Borawska, A., et al. (2022). Changes in the production of energy from renewable sources in the countries of Central and Eastern Europe. *Front. Energy Res., 10*, 993547. https://doi.org/10.3389/fenrg.2022.993547

Ruane, J., Sonnino, A., & Agostini, A. (2010). Bioenergy and the potential contribution of agricultural biotechnologies in developing countries. *Biomass and Bioenergy, 34*(10), 1427–1439.

Ruel, K., Nishiyama, Y., & Joseleau, J. P. (2012). Crystalline and amorphous cellulose in the secondary walls of Arabidopsis. *Plant Science, 193*, 48–61.

Ruynal, J., Delgenes, J. P., & Moletta, R. (1998). Two phase anaerobic digestion of solid waste by a multiple liquefaction reactors process. *Bioresource Technology, 65*, 97–103.

Sakai, S., Sawell, S., Chandler, A., Eighmy, T., Kosson, D., Vehlow, J., Van Der Sloot, H., Hartlen, J., & Hjelmar, O. (1996). World trends in municipal solid waste management. *Waste Management, 16*, 45.

Sakthivadivel, D., Kumar, P. G., Vigneswaran, V. S., Meikandan, M., & Iniyan, S. (2019). Performance study of an advanced micro-gasifier stove with coconut shell. *European Journal of Sustainable Development Research, 3*(4), em0101.

Salleh, S. F., Mohd Roslan, M. E., Abd Rahman, A., Shamsuddin, A. H., Tuan Abdullah, T. A. R., & Sovacool, B. K. (2020). Transitioning to a sustainable development framework for bioenergy in Malaysia: Policy suggestions to catalyse the utilisation of palm oil mill residues. *Energy, Sustainability and Society, 10*(1), 1–20.

Salman, Z. (2021). Biomass wastes from palm oil mills. https://www.bioenergyconsult.com/palm-biomass/. Accessed 22 December 2021.

SanthoshKumar, A., Thangarasu, V., & Anand, R. (2019). Performance, combustion, and emission characteristics of DI diesel engine using mahua biodiesel. In *Advanced Biofuels* (pp. 291–327). Woodhead Publishing.

Sarkar, N., Ghosh, S. K., Bannerjee, S., & Aikat, K. (2012). Bioethanol production from agricultural wastes: An overview. *Renewable Energy, 37*(1), 19–27.

Schmer, M. R., Vogel, K. P., Mitchell, R. B., & Perrin, R. K. (2008). Net energy of cellulosic ethanol from switchgrass. *Proceedings of the National Academy of Sciences, 105*(2), 464–469.

Scown, C. D., Nazaroff, W. W. W., Mishra, U., Strogen, B., Lobscheid, A. B., Masanet, E., et al. (2012). Lifecycle greenhouse gas implications of US national scenarios for cellulosic ethanol production. *Environmental Research Letters, 7*(1), 014011.

Searle, J. (2017). Biological naturalism. *The Blackwell companion to consciousness*, 327–336.

Shafizadeh, F. (1982). Introduction to pyrolysis of biomass. *Journal of Analytical and Applied Pyrolysis, 3*(4), 283–305.

Shi, A. Z., Koh, L. P., & Tan, H. T. W. (2009). The biofuel potential of municipal solid waste. *GCB Bioenergy, 1*(5), 317–369.

Sims, R. E., Mabee, W., Saddler, J. N., & Taylor, M. (2010). An overview of second-generation biofuel technologies. *Bioresource Technology, 101*(6), 1570–1580.

Sindhu, R., Binod, P., & Pandey, A. (2016). Biological pretreatment of lignocellulosic biomass—An overview. *Bioresource Technology, 199*, 76–82.

Singh, R., Shukla, A., Tiwari, S., & Srivastava, M. (2014). A review on delignification of lignocellulosic biomass for enhancement of ethanol production potential. *Renewable and Sustainable Energy Reviews, 32*, 713–728.

Singh, R., Srivastava, M., & Shukla, A. (2016). Environmental sustainability of bioethanol production from rice straw in India: A review. *Renewable and Sustainable Energy Reviews, 54*, 202–216.

Soam, S., Kapoor, M., Kumar, R., Borjesson, P., Gupta, R. P., & Tuli, D. K. (2016). Global warming potential and energy analysis of second-generation ethanol production from rice straw in India. *Applied Energy, 184*, 353–364.

Statista (2021). Beer Production Worldwide in 2020, by Region (In Million Hectoliters). https://www.statista.com/statistics/202417/beer-output-volumesof-the-different-continents-in-2010/

Tan, Z., Chen, K., & Liu, P. (2015). Possibilities and challenges of China's forestry biomass resource utilization. *Renewable and Sustainable Energy Reviews, 41*, 368–378.

Tonini, D., Hamelin, L., Alvarado-Morales, M., & Astrup, T. F. (2016). GHG emission factors for bioelectricity, biomethane, and bioethanol quantified for twenty-four biomass substrates with consequential life-cycle assessment. *Bioresource Technology, 208*, 123–133.

Tsai, W. T. (2019). Mandatory recycling of waste cooking oil from residential and commercial sectors in Taiwan. *Resources, 8*(1), 38.

United Nations Environment Programme (2021). Food Waste Index; United Nations, Nairobi, Kenya. ISBN 9789280738513.

Urban, L., Masa, V., Pavlas, M., & Stehlik, P. (2010). Novel type of technology for biomass utilization. *Forestry, 2020*, 2030.

Wang, H., Zhang, S., Bi, X., & Clift, R. (2020). Greenhouse gas emission reduction potential and cost of bioenergy in British Columbia, Canada. *Energy Policy, 138*, 111285.

Wang, M., Han, J., Dunn, J., Cai, H., & Elgowainy, A. (2012). Well-to-wheels energy use and greenhouse gas emissions of ethanol from corn, sugarcane, corn stover, switchgrass, and miscanthus. *Environmental Research Letter, 7*, 045905.

Whitaker, M., & Heath, G. (2009). *Life Cycle Assessment of the use of Jatropha biodiesel in Indian locomotives* (No. NREL-TP-6A2-44428). Golden, CO: National Renewable Energy Laboratory.

Woźniak-Karczewska, M., Lisiecki, P., Białas, W., Owsianiak, M., Piotrowska-Cyplik, A., Wolko, Ł, et al. (2019). Effect of bioaugmentation on long-term biodegradation of diesel/biodiesel blends in soil microcosms. *Science of the Total Environment, 671*, 948–958.

Yu, H. T., Chen, B. Y., Li, B. Y., Tseng, M. C., Han, C. C., & Shyu, S. G. (2018). Efficient pretreatment of lignocellulosic biomass with high recovery of solid lignin and fermentable sugars using Fenton reaction in a mixed solvent. *Biotechnology for Biofuels, 11*(1), 1–11.

Yu, Q., Wang, Y., Van Le, Q., Yang, H., Hosseinzadeh-Bandbafha, H., Yang, Y., et al. (2021). An overview on the conversion of forest biomass into bioenergy. *Frontiers in Energy Research, 9*, 684234.

Zabaniotou, A., Ioannidou, O., & Skoulou, V. (2008). Rapeseed residues are utilized for energy and second-generation biofuels. *Fuel, 87*(8–9), 1492–1502.

Zabed, H., Sahu, J. N., Suely, A., Boyce, A. N., & Faruq, G. (2017). Bioethanol production from renewable sources: Current perspectives and technological progress. *Renewable and Sustainable Energy Reviews, 71*, 475–501.

Zeko-Pivač, A., Tišma, M., Žnidaršič-Plazl, P., Kulisic, B., Sakellaris, G., Hao, J., & Planinić, M. (2022). The potential of brewer's spent grain in the circular bioeconomy: State of the art and future perspectives. *Frontiers in Bioengineering and Biotechnology, 10*, 744–870.

Zhang, W., Ji, H., Song, Y., Ma, S., Xiong, W., Chen, C., et al. (2020). Green preparation of branched biolubricant by chemically modifying waste cooking oil with lipase and ionic liquid. *Journal of Cleaner Production, 274*, 122918.

Zhu, S., Wu, Y., Yu, Z., Wang, C., Yu, F., Jin, S., & Zhang, Y. (2006). Comparison of three microwave/chemical pretreatment processes for enzymatic hydrolysis of rice straw. *Biosystems Engineering, 93*(3), 279–283.

The manufacturer's authorised representative in the EU is Springer Nature Customer Service Centre GmbH, Europaplatz 3, 69115 Heidelberg, Germany. If you have any concerns regarding our products, please contact ProductSafety@springernature.com

Printed and bound by CPI Group (UK) Ltd, Croydon, CR0 4YY

05/10/2026

02228057-0009